LONDON MATHEMATICAL SO͡ ͡ ͡ ͡NT TEXTS

Managing Editor: Ian J. Leary,
Mathematical Sciences, University of Southar

London Mathematical Society Student Texts 95

Representations of Finite Groups of Lie Type

SECOND EDITION

FRANÇOIS DIGNE
Université de Picardie Jules Verne, Amiens

JEAN MICHEL
Centre National de la Recherche Scientifique (CNRS), Paris

CAMBRIDGE
UNIVERSITY PRESS

CAMBRIDGE
UNIVERSITY PRESS

University Printing House, Cambridge CB2 8BS, United Kingdom

One Liberty Plaza, 20th Floor, New York, NY 10006, USA

477 Williamstown Road, Port Melbourne, VIC 3207, Australia

314–321, 3rd Floor, Plot 3, Splendor Forum, Jasola District Centre,
New Delhi – 110025, India

79 Anson Road, #06-04/06, Singapore 079906

Cambridge University Press is part of the University of Cambridge.

It furthers the University's mission by disseminating knowledge in the pursuit of
education, learning, and research at the highest international levels of excellence.

www.cambridge.org
Information on this title: www.cambridge.org/9781108481489
DOI: 10.1017/9781108673655

First edition © Cambridge University Press 1991
Second edition © François Digne and Jean Michel 2020

First published 1991
Second edition 2020

Printed in the United Kingdom by TJ International Ltd. Padstow Cornwall

A catalogue record for this publication is available from the British Library.

Library of Congress Cataloging-in-Publication Data
Names: Digne, François, author. | Michel, Jean, author.
Title: Representations of finite groups of Lie type / François Digne,
Université de Picardie Jules Verne, Amiens, Jean Michel, Centre
National de la Recherche Scientifique (CNRS), Paris.
Description: Second edition. | Cambridge, United Kingdom ; New York, NY :
Cambridge University Press, 2020. | Series: London mathematical society
student texts ; 95 | Includes bibliographical references and index.
Identifiers: LCCN 2019035714 | ISBN 9781108481489 (hardback) | ISBN
9781108673655 (ebook)
Subjects: LCSH: Lie groups. | Representations of groups.
Classification: LCC QA387 .D54 2020 | DDC 512/.482--dc23
LC record available at https://lccn.loc.gov/2019035714

ISBN 978-1-108-48148-9 Hardback
ISBN 978-1-108-72262-9 Paperback

Contents

Introduction to the Second Edition

We had two main aims in writing this edition:

- Be more self-contained where possible. For instance, we have added brief overviews of Coxeter groups and root systems, and given some more details about the theory of algebraic groups.
- While retaining the same level of exposition as in the first edition, we have given a more complete account of the representation theory of finite groups of Lie type.

In view of the second aim, we have added the following topics to our exposition:

- We cover Ree and Suzuki groups extending our exposition of Frobenius morphisms to the more general case of Frobenius roots.
- We have added to Harish-Chandra theory the topic of Hecke algebras and given as many results as we could easily do for fields of arbitrary characteristic prime to q, in view of applications to modular representations.
- We have added a chapter on the computation of Green functions, with a brief review of invariant theory of reflection groups, and a chapter on the decomposition of unipotent Deligne–Lusztig characters.

Acknowledgements

In addition to the people we thank in the introduction to the first edition, we have benefitted from the input of younger colleagues such as Cédric Bonnafé, Olivier Dudas, Meinolf Geck, Daniel Juteau, and Raphaël Rouquier, and we give special thanks to Gunter Malle for a thorough proofreading of this edition.

We also thank the many people who pointed out to us misprints and other errors in the first edition.

From the Introduction to the First Edition

These notes follow a course given at the Paris VII university during the spring semester of academic year 1987–88. Their purpose is to expound basic results in the representation theory of finite groups of Lie type (a precise definition of this concept will be given in the chapter "Rationality, Frobenius").

Let us start with some notations. We denote by $\mathbb{F}_q$ a finite field of characteristic p with q elements (q is a power of p). The typical groups we will look at are the linear, unitary, symplectic, orthogonal, ..., groups over $\mathbb{F}_q$. We will consider these groups as the subgroups of points with coefficients in $\mathbb{F}_q$ of the corresponding groups over an algebraic closure $\overline{\mathbb{F}}_q$ (which are algebraic reductive groups). More precisely, the group over $\mathbb{F}_q$ is the set of fixed points of the group over $\overline{\mathbb{F}}_q$ under an endomorphism F called the Frobenius endomorphism; this will be explained in the chapter "Rationality, Frobenius". In the following paragraphs of this introduction we will try to describe, by some examples, a sample of the methods used to study the complex representations of these groups.

Induction from Subgroups

Let us start with the example where $G = \mathbf{GL}_n(\mathbb{F}_q)$ is the general linear group over $\mathbb{F}_q$. Let T be the subgroup of diagonal matrices; it is a subgroup of the group B of upper triangular matrices, and there is a semidirect product decomposition $B = U \rtimes T$, where U is the subgroup of the upper triangular matrices which have all their diagonal entries equal to 1. The representation theory of T is easy, since it is a commutative group (actually isomorphic to a product of n copies of the multiplicative group $\mathbb{F}_q^\times$). Composition with the natural homomorphism from B to T (quotient by U) lifts representations of T to representations of B. Inducing these representations from B to the whole of the general linear group gives representations of G (whose irreducible constituents are called "principal series representations"). More generally, we can replace T with a group L of block-diagonal matrices, B with the group of corresponding upper block-triangular matrices P, and we have a semi-direct product decomposition (called a Levi decomposition) $P = V \rtimes L$, where V is the subgroup of P whose diagonal blocks are identity matrices; we may as before induce from P to G representations of L lifted to P. The point of this method is that L is isomorphic to a direct product of general linear groups of smaller degrees than n. We thus have an inductive process to get representations of G if we know how to decompose induced representations from P to G. This approach has been developed in the works of Harish-Chandra, Howlett and Lehrer, and is introduced in the chapter "Harish-Chandra Theory".

Cohomological Methods

Let us now consider the example of $G = U_n$, the unitary group. It can be defined as the subgroup of matrices $A \in \mathbf{GL}_n(\mathbb{F}_{q^2})$ such that ${}^t A^{[q]} = A^{-1}$, where $A^{[q]}$ denotes the matrix whose entries are those of A raised to the qth power. It is thus the subgroup of $\mathbf{GL}_n(\overline{\mathbb{F}}_q)$ consisting of the fixed points of the endomorphism $F : A \mapsto ({}^t A^{[q]})^{-1}$.

A subgroup L of block-diagonal matrices in U_n is again a product of unitary groups of smaller degree. But this time we cannot construct a bigger group P having L as a quotient. More precisely, the group $\mathbf{V}$ of upper block-triangular matrices with entries in $\overline{\mathbb{F}}_q$ and whose diagonal blocks are the identity matrix has no fixed points other than the identity under F.

To get a suitable theory, Harish-Chandra's construction must be generalised; instead of inducing from $V \rtimes L$ to G, we construct a variety attached to $\mathbf{V}$ on which both L and G act with commuting actions, and the cohomology of that variety with ℓ-adic coefficients gives a (virtual) bi-module which defines a "generalised induction" from L to G. This approach, due to Deligne and Lusztig, will be developed in the chapters from "ℓ-adic Cohomology" to "Geometric Conjugacy and Lusztig Series".

Gelfand–Graev Representations

Using the above methods, a lot of information can be obtained about the characters of the groups $\mathbf{G}(\mathbb{F}_q)$, when $\mathbf{G}$ has a connected centre. The situation is not so clear when the centre of $\mathbf{G}$ is not connected. In this case one can use the Gelfand–Graev representations, which are obtained by inducing a linear character "in general position" of a maximal unipotent subgroup (in $\mathbf{GL}_n$ the subgroup of upper triangular matrices with ones on the diagonal is such a subgroup). These representations are closely tied to the theory of regular unipotent elements. They are multiplicity-free and contain rather large cross-sections of the set of irreducible characters, and so give useful additional information in the nonconnected centre case. (In the connected centre case, they are linear combinations of Deligne–Lusztig characters.)

For instance, in $\mathbf{SL}_2(\mathbb{F}_q)$ they are obtained by inducing a nontrivial linear character of the group of matrices of the form $\begin{pmatrix} 1 & u \\ 0 & 1 \end{pmatrix}$: such a character corresponds to a nontrivial additive character of $\mathbb{F}_q$; there are two classes of such characters under $\mathbf{SL}_2(\mathbb{F}_q)$, which corresponds to the fact that the centre of $\mathbf{SL}_2$ has two connected components (its two elements).

The theory of regular elements and Gelfand–Graev representations is expounded in chapter "Regular Elements; Gelfand–Graev Representations", with, as an application, the computation of the values of all irreducible characters on regular unipotent elements.

Acknowledgements

We would like to thank the "équipe des groupes finis" and the mathematics department of the École Normale Supérieure, who provided us with a stimulating working environment, and adequate facilities for composing this book. We thank also the Paris VII university, which gave us the opportunity to give the course which started this book. We thank all those who carefully read the earlier drafts and suggested improvements, particularly Michel Enguehard, Guy Rousseau, Jean-Yves Hée and the editor. We thank Michel Broué and Jacques Tits, who provided us with various ideas and information, and above all George Lusztig, who invented most of the theory.

1

Basic Results on Algebraic Groups

A finite group of Lie type – sometimes also called a finite reductive group – is (in a first approximation) the group of points over a finite field $\mathbb{F}_q$ of a (usually connected) reductive algebraic group over an algebraic closure $\overline{\mathbb{F}}_q$. We begin by recalling the definition of these terms. See 4.2.6 for a precise definition of finite groups of Lie type.

Let us first establish some notations and conventions we use throughout. If g is an automorphism of a set (resp. variety, group, ...) X, we will denote by X^g the set of fixed points of g, and by ${}^g x$ or $g(x)$ the image of the element $x \in X$ by g. A group G acts naturally on itself by conjugation, and we will write ${}^g h$ for ghg^{-1}, where g and h are elements of G and will write $\mathrm{ad}\, g$ for the morphism $h \mapsto {}^g h$. We will write $Z(G)$ for the centre of G; if X is a subset of a set on which G acts, we put $N_G(X) = \{g \in G \mid {}^g X = X\} = \{g \in G \mid \forall x \in X, {}^g x \in X\}$ and $C_G(X) = \{g \in G \mid \forall x \in X, {}^g x = x\}$. If A is a ring and X a set (resp. a group), we denote by AX the free A-module with basis X (resp. the group algebra of X with coefficients in A). We write $H \rtimes G$ for a semi-direct product of groups where H is normal in the product.

We will generally use bold letters for algebraic groups and varieties.

1.1 Basic Results on Algebraic Groups

An algebraic group is an algebraic variety $\mathbf{G}$ endowed with a group structure such that the multiplication and inverse maps are algebraic. In this book, we will need only affine algebraic groups over an algebraically closed field k (which will be taken to be $\overline{\mathbb{F}}_p$ from Chapter 4 onwards): that is, affine algebraic varieties $\mathbf{G} = \mathrm{Spec}\, A$, where A is a reduced k-algebra; that is, without non-zero nilpotent elements. The group structure gives a coalgebra structure on A, or even a Hopf algebra structure. For such a group $\mathbf{G}$, we will call elements of $\mathbf{G}$

5

the elements of the group $\mathbf{G}(k)$ of k-valued points of $\mathbf{G}$; that is, the morphisms $A \to k$. Since an algebraic variety over k is determined by the set of its k-points, we will sometimes identify $\mathbf{G}$ with the set of its elements.

Examples 1.1.1 (of affine algebraic groups)

(i) The **additive group** $\mathbb{G}_a$, defined by the algebra $k[t]$, with comultiplication $t \mapsto t \otimes 1 + 1 \otimes t$. We have $\mathbb{G}_a(k) \simeq k^+$.

(ii) The **multiplicative group** $\mathbb{G}_m$, defined by the algebra $k[t, t^{-1}]$, with co-multiplication $t \mapsto t \otimes t$. We have $\mathbb{G}_m(k) \simeq k^\times$.

(iii) The **general linear group** $\mathbf{GL}_n$, defined by the algebra $k[\{t_{i,j}\}_{1 \le i,j \le n}, \det(t_{i,j})^{-1}]$. The comultiplication is given by $t_{i,j} \mapsto \sum_k t_{i,k} \otimes t_{k,j}$. The group $\mathbf{GL}_n(k)$ identifies with the group of invertible $n \times n$-matrices with entries in k. As an open subvariety of an affine space, $\mathbf{GL}_n$ is connected.

(iv) A finite group Γ is algebraic; its algebra is the algebra of k-valued functions on Γ, which can be identified to $\mathrm{Hom}_k(k\Gamma, k)$. A basis of orthogonal idempotents is formed of the Dirac functions $\delta_g(g') = \delta_{g,g'}$. The comultiplication is $\delta_g \mapsto \sum_{\{g',g'' \in \Gamma \mid g'g''=g\}} \delta_{g'} \otimes \delta_{g''}$.

Actually every connected one-dimensional affine algebraic group is isomorphic to $\mathbb{G}_m$ or $\mathbb{G}_a$ – this is surprisingly difficult to prove; see for instance Springer (1998, 3.4.9).

A morphism of algebraic groups is a morphism of varieties which is also a group homomorphism. A closed subvariety which is a subgroup is naturally an algebraic subgroup; that is, the inclusion map is a morphism of algebraic groups.

Remark 1.1.2 For some problems about algebraic groups, working with varieties instead of schemes loses information. For instance, in our setting, when $k = \overline{\mathbb{F}}_p$, an algebraic closure of the prime field $\mathbb{F}_p$, the kernel of the morphism $x \mapsto x^p : \mathbb{G}_a \to \mathbb{G}_a$ is trivial. However, in the setting of schemes, it can be computed as $\mathrm{Spec}(k[t]/t^p)$, which is not allowed in the setting of varieties since it is a non-reduced algebra.

Proposition 1.1.3 *Let $\{\mathbf{V}_i\}_{i \in I}$ be a family of irreducible subvarieties all containing the identity element of an algebraic group $\mathbf{G}$; then the smallest closed subgroup $\mathbf{H}$ of $\mathbf{G}$ containing the $\mathbf{V}_i$ is equal to the product $\mathbf{W}_{i_1} \ldots \mathbf{W}_{i_k}$ for some finite sequence $(i_1, \ldots, i_k)$ of elements of I, where either $\mathbf{W}_i = \mathbf{V}_i$ or $\mathbf{W}_i = \{x^{-1} \mid x \in \mathbf{V}_i\}$.*

Reference See Borel (1991, I, 2.2). □

H is called the subgroup **generated** by the $\mathbf{V}_i$. It is clear from 1.1.3 that the subgroup generated by a family of connected subvarieties is connected.

An algebraic group is called **linear** if it is isomorphic to a closed subgroup of $\mathbf{GL}_n$. It is clear from the definition that a linear algebraic group is affine; the converse is also true; see Springer (1998, 2.3.7(i)). All algebraic groups considered in the sequel will be linear; an example of non-linear algebraic groups are the elliptic curves.

Proposition 1.1.4 *Let* **G** *be a linear algebraic group and* **H** *a closed subgroup; then:*

(i) *The quotient* **G**/**H** *exists and is a quasi-projective variety (that is, an open subvariety of projective variety).*

(ii) *If* **H** *is a normal subgroup, then* **G**/**H** *is an affine variety and a linear algebraic group for the induced group structure.*

References   See Springer (1998, 5.5.5 and 5.5.10).   □

The connected components of an algebraic group **G** are finite in number and coincide with its irreducible components; the component containing the identity element of **G** is called the **identity component** and denoted by $\mathbf{G}^0$. It is a characteristic subgroup of **G**, and $\mathbf{G}/\mathbf{G}^0$ identifies with the set of connected components of **G**. Conversely, every normal closed subgroup of finite index contains $\mathbf{G}^0$ – these properties are elementary; see, for example, Springer (1998, 2.2.1).

An element of a linear algebraic group **G** is called **semi-simple** (resp. **unipotent**) if its image in some embedding of **G** as a closed subgroup of some $\mathbf{GL}_n$ is semi-simple; that is, conjugate to a diagonal matrix (resp. unipotent, that is, conjugate to an upper unitriangular matrix). This property does not depend on the embedding. Every element has a unique decomposition, its **Jordan decomposition**, as the product of two commuting semi-simple and unipotent elements; see Springer (1998, 2.4.8).

Proposition 1.1.5 *Let* **G** *be a linear algebraic group over* $\overline{\mathbb{F}}_p$, *where* p *is a prime. Then every element of* **G** *has finite order: the semi-simple elements are the* p'-*elements, and the unipotent elements are the* p-*elements.*

Proof   This results from the fact that the above result holds for $\mathbf{GL}_n(\overline{\mathbb{F}}_p)$. Indeed, the diagonal entries of a diagonal matrix are in $\overline{\mathbb{F}}_p^{\times}$, thus of order prime to p, and a matrix X is unipotent if and only if $N = X - \mathrm{I}_n$ is nilpotent, equivalently $X^{p^a} = N^{p^a} + \mathrm{I}_n = \mathrm{I}_n$ for a large enough.   □

1.2 Diagonalisable Groups, Tori, $X(\mathbf{T}), Y(\mathbf{T})$

Definition 1.2.1

(i) A **torus** is an algebraic group $\mathbf{T}$ isomorphic to $\mathbb{G}_m^r$ for some r called the **rank** of $\mathbf{T}$.

(ii) A **diagonalisable group** is an algebraic group isomorphic to a closed subgroup of a torus.

(iii) A **rational character** of an algebraic group $\mathbf{G}$ is a morphism of algebraic groups $\mathbf{G} \to \mathbb{G}_m$.

(iv) The **character group** $X(\mathbf{G})$ of an algebraic group $\mathbf{G}$ is the group of rational characters of $\mathbf{G}$.

Proposition 1.2.2 *A diagonalisable group* $\mathbf{D}$ *is equal to* $\mathrm{Spec}(kX(\mathbf{D}))$.

Proof We first prove the result for a torus $\mathbf{T}$ of rank r; its algebra is $A = k[t_1, \ldots, t_r, t_1^{-1}, \ldots, t_r^{-1}] \simeq k[t, t^{-1}]^{\otimes r}$. An element of $X(\mathbf{T})$ is given by a morphism of both algebras and coalgebras $k[t, t^{-1}] \to A$, which is defined by giving the image of t which must be invertible, thus must be a monomial; this monomial must be unitary for the morphism to be a coalgebra morphism. The group law on $X(\mathbf{T})$, given by pointwise multiplication, corresponds to the multiplication of monomials; thus $X(\mathbf{T})$ is isomorphic to the group of unitary monomials, and $A \simeq kX(\mathbf{T})$, the group algebra of the group $X(\mathbf{T})$.

Let now $\mathbf{D}$ be a closed subgroup of a torus $\mathbf{T}$; its algebra is a quotient of the algebra $kX(\mathbf{T})$ of $\mathbf{T}$, hence is spanned by the images of the rational characters of $\mathbf{T}$; that is, by the restrictions to $\mathbf{D}$ of these characters. By the linear independence of the characters of a group as functions to k – see, for example, Lang (2002, Chapter VI, Theorem 4.1) – these restrictions form a basis of the algebra of $\mathbf{D}$, whence the result.   □

Proposition 1.2.3

(i) *A diagonalisable group is the direct product of a torus by a finite abelian* p'*-group, where* p *is the characteristic of* k.

(ii) *The quotient of a torus by a closed subgroup is a torus.*

References See Springer (1998, 3.2.7(i)) and Borel (1991, Corollary of 8.4).

□

Proposition 1.2.4 *Seen as a functor from diagonalisable groups to* $\mathbb{Z}$*-modules,* X *is an exact functor.*

Proof Left exactness is a general fact for Hom functors. Now, a similar argument to that in the proof of 1.2.2 shows that if $\mathbf{D} \subset \mathbf{D}'$ are diagonalisable groups, then the restriction $X(\mathbf{D}') \to X(\mathbf{D})$ is surjective.   □

Exercise 1.2.5   Show that X is an antiequivalence of abelian categories from diagonalisable groups over k to $\mathbb{Z}$-modules of finite type without p-torsion, where p is the characteristic of k, and that X restricts to an antiequivalence of additive categories from tori to finite rank lattices (here "lattice" means free $\mathbb{Z}$-module). Use the fact that the inverse antiequivalence maps a $\mathbb{Z}$-module M to Spec $k[M]$.

Proposition 1.2.6 *Any exact sequence of tori is split; in particular, a subtorus of a torus is a direct factor.*

Proof   By the antiequivalence of categories X (see 1.2.5) an exact sequence of tori corresponds to an exact sequence of lattices. Such a sequence is always split.   □

Proposition 1.2.7 *Given an algebraic action of a torus* **T** *on an affine variety* **X***, there exists an element* $t \in \mathbf{T}$ *such that* $\mathbf{X}^t = \mathbf{X}^{\mathbf{T}}$.

Proof   It is known that an algebraic group action can be linearised; that is, there exists an embedding of **X** into a finite-dimensional k-vector space V and an embedding of **T** into $\mathbf{GL}(V)$ such that the action of **T** factors through this embedding; see Slodowy (1980, I, 1.3). The space V has a decomposition into irreducible **T**-submodules, each defining some $\chi \in X(\mathbf{T})$. The kernel of a non-trivial such χ has codimension 1 in **T**, since its algebra is the quotient of the algebra of **T** by the ideal generated by χ (identifying the algebra of **T** with the group algebra of $X(\mathbf{T})$). It follows that there is some element $t \in \mathbf{T}$ which lies outside the finitely many kernels of the non-trivial characters of **T** occurring in V; the fixed points of t on V (and so on **X**) are the same as those of **T**, whence the result.   □

Proposition 1.2.8

(i) *A torus consists of semi-simple elements.*
(ii) *A connected linear algebraic group containing only semi-simple elements is a torus.*

Proof   Proposition 1.2.8 (i) comes from the fact that $\mathbb{G}_m^r$ can be embedded as the group of diagonal matrices in $\mathbf{GL}_r$. For (ii) see Springer (1998, 6.3.6).   □

Definition 1.2.9

(i) A **one-parameter subgroup** of an algebraic group **G** is a morphism of algebraic groups $\mathbb{G}_m \to \mathbf{G}$.
(ii) The abelian group of one-parameter subgroups of a torus **T** is denoted by $Y(\mathbf{T})$.

An element of $Y(\mathbf{T})$ is given by a morphism $k[t_1,\ldots,t_r,t_1^{-1},\ldots,t_r^{-1}] \rightarrow k[t,t^{-1}]$, determined by the images of the t_i which must be invertible, thus monomials. These monomials must be unitary, thus powers of t, for the morphism to be a coalgebra morphism. The group law on one-parameter subgroups corresponds to multiplying these powers, so $Y(\mathbf{T})$ is isomorphic to a product of r copies of the group of powers of t.

There is an exact pairing between $X(\mathbf{T})$ and $Y(\mathbf{T})$ (that is, a bilinear map $X(\mathbf{T}) \times Y(\mathbf{T}) \rightarrow \mathbb{Z}$ making each one the $\mathbb{Z}$-dual of the other) obtained as follows: given $\chi \in X(\mathbf{T})$ and $\psi \in Y(\mathbf{T})$, the composite map $\chi \circ \psi$ is a homomorphism from $\mathbb{G}_m$ to itself, so is the map $t \mapsto t^n$ for some $n \in \mathbb{Z}$; the pairing is defined by $(\chi,\psi) \mapsto n$.

Proposition 1.2.10   *The map $y \otimes x \mapsto y(x)$ is a group isomorphism:*

$$Y(\mathbf{T}) \otimes_{\mathbb{Z}} k^{\times} \xrightarrow{\sim} \mathbf{T}(k).$$

Proof Let $(x_i)_{i=1,\ldots,n}$ and $(y_i)_{i=1,\ldots,n}$ be two dual bases of $X(\mathbf{T})$ and $Y(\mathbf{T})$, respectively. It is easy to check that $t \mapsto \sum_{i=1}^{i=n} y_i \otimes x_i(t) : \mathbf{T}(k) \rightarrow Y(\mathbf{T}) \otimes_{\mathbb{Z}} k^{\times}$ is the inverse of the map of the statement, using the fact that $\bigcap_{x \in X(\mathbf{T})} \operatorname{Ker} x = \{1\}$. (This follows from the isomorphism between the algebra of the variety $\mathbf{T}$ and $kX(\mathbf{T})$; see 1.2.2.) $\square$

We now study the relationship between closed subgroups of $\mathbf{T}$ and subgroups of $X(\mathbf{T})$.

Lemma 1.2.11   *Given a torus $\mathbf{T}$, let $x_1,\ldots,x_n$ be linearly independent elements of $X(\mathbf{T})$ and $\lambda_1,\ldots,\lambda_n$ arbitrary elements of $k^{\times}$; then there exists $s \in \mathbf{T}$ such that $x_i(s) = \lambda_i$ for $i = 1,\ldots,n$.*

Reference   See, for example, Humphreys (1975, 16.2, Lemma C). $\square$

Definition 1.2.12   Given a torus $\mathbf{T}$ and a closed subgroup $\mathbf{S}$ of $\mathbf{T}$, we define $\mathbf{S}_{X(\mathbf{T})}^{\perp} = \{x \in X(\mathbf{T}) \mid \forall s \in \mathbf{S}, x(s) = 1\}$; conversely, given a subgroup A of $X(\mathbf{T})$, we define a subgroup of $\mathbf{T}$ by $A_{\mathbf{T}}^{\perp} = \{s \in \mathbf{T} \mid \forall x \in A, x(s) = 1\}$ (which is closed, since A is finitely generated).

Proposition 1.2.13   *If k has characteristic p, given a torus $\mathbf{T}$ and a subgroup A of $X(\mathbf{T})$, the group $(A_{\mathbf{T}}^{\perp})_{X(\mathbf{T})}^{\perp}/A$ is the p-torsion subgroup of $X(\mathbf{T})/A$.*

Proof   First notice that, for any closed subgroup $\mathbf{S}$ of $\mathbf{T}$, the group $X(\mathbf{T})/\mathbf{S}_{X(\mathbf{T})}^{\perp}$ has no p-torsion. Indeed,

$$p^n x \in \mathbf{S}_{X(\mathbf{T})}^{\perp} \Leftrightarrow \forall s \in \mathbf{S}, x(s)^{p^n} = 1 \Leftrightarrow \forall s \in \mathbf{S}, x(s) = 1 \Leftrightarrow x \in \mathbf{S}_{X(\mathbf{T})}^{\perp},$$

where the middle equivalence holds, since $x \mapsto x^{p^n}$ is an automorphism of k.

Thus it is enough to see that $(A_{\mathbf{T}}^{\perp})_{X(\mathbf{T})}^{\perp}/A$ is a p-group. Let $x \in (A_{\mathbf{T}}^{\perp})_{X(\mathbf{T})}^{\perp} - A$. It is a standard result on submodules of free $\mathbb{Z}$-modules that there is a basis $(x, x_1, \ldots, x_r)$ of $\langle A, x \rangle$ such that $(mx, x_1, \ldots, x_r)$ is a basis of A, with $m \in \mathbb{Z}$, possibly $m = 0$ (which means omit mx). The result follows if we can prove that m is a power of p. Let us assume otherwise: then there exists $\lambda \in k^{\times}$ such that $\lambda \neq 1$ and $\lambda^m = 1$ (even if $m = 0$). By 1.2.11, there exists some $s \in \mathbf{T}$ such that $x(s) = \lambda, x_1(s) = 1, \ldots, x_r(s) = 1$. Thus $mx(s) = 1$, so $s \in A_{\mathbf{T}}^{\perp}$, but $x(s) \neq 1$, which contradicts $x \in (A_{\mathbf{T}}^{\perp})_{X(\mathbf{T})}^{\perp}$. $\qquad \square$

1.3 Solvable Groups, Borel Subgroups

We will denote by $\mathbf{G}_u$ the set of unipotent elements of an algebraic group $\mathbf{G}$.

Proposition 1.3.1 *Let $\mathbf{G}$ be a connected solvable algebraic group; then:*

(i) *$\mathbf{G}_u$ is a normal connected subgroup of $\mathbf{G}$.*
(ii) *For every maximal torus $\mathbf{T}$ of $\mathbf{G}$, there is a semi-direct product decomposition $\mathbf{G} = \mathbf{G}_u \rtimes \mathbf{T}$.*
(iii) *Let $\mathbf{S}$ be a subtorus of $\mathbf{G}$ – then $N_{\mathbf{G}}(\mathbf{S}) = C_{\mathbf{G}}(\mathbf{S})$.*

Proof For (i) and (ii) see Springer (1998, 6.3.3(ii) and 6.3.5(iv)); these assertions follow from the theorem of Lie–Kolchin, which states that every closed solvable subgroup of $\mathbf{GL}_n(k)$ is conjugate to a subgroup of the group of upper triangular matrices.

For (iii), if $n \in N_{\mathbf{G}}(\mathbf{S})$, $s \in \mathbf{S}$, then $[n, s] \in [\mathbf{G}, \mathbf{G}] \cap \mathbf{S} \subset \mathbf{G}_u \cap \mathbf{S} = 1$, where the first inclusion comes from (ii). $\qquad \square$

Let us note that this proposition implies that every connected solvable algebraic group containing no unipotent elements is a torus.

Definition 1.3.2 Maximal closed connected solvable subgroups of an algebraic group are called **Borel subgroups**.

These groups have paramount importance in the theory. The next theorem states their basic properties.

Theorem 1.3.3 *Let $\mathbf{G}$ be a connected algebraic group; then:*

(i) *All Borel subgroups of $\mathbf{G}$ are conjugate.*
(ii) *Every element of $\mathbf{G}$ is in some Borel subgroup.*
(iii) *The centraliser in $\mathbf{G}$ of any torus is connected.*
(iv) *A Borel subgroup is equal to its normaliser in $\mathbf{G}$.*
(v) *Every maximal torus of $\mathbf{G}$ is in some Borel subgroup.*

Outline of proof   To show (i) – see Springer (1998, 6.2.7(iii)) – one first shows that $\mathbf{G}/\mathbf{B}$ is a complete variety; a complete variety has the property that a connected solvable group acting on it always has a fixed point. Thus another Borel subgroup $\mathbf{B}'$ acting on $\mathbf{G}/\mathbf{B}$ by left translation has a fixed point; that is, there exists $g \in \mathbf{G}$ such that $\mathbf{B}'g\mathbf{B} = g\mathbf{B}$, so $g^{-1}\mathbf{B}'g \subset \mathbf{B}$, whence the result. Property (ii) similarly results from properties of complete varieties; see Springer (1998, 6.4.5(i)). For (iii) see Springer (1998, 6.4.7(i)). For (iv) see Springer (1998, 6.4.9). Property (v) results from the definition of a Borel subgroup.   □

It follows from (iv) above and the remark that the closure of a solvable group is solvable – see Borel (1991, I, 2.4) – that the words "closed connected" can be omitted from the definition of a Borel subgroup when $\mathbf{G}$ is connected.

Theorem 1.3.4   *Let* $\mathbf{T}$ *be a torus of a connected algebraic group* $\mathbf{G}$, *then:*

(i) $N_{\mathbf{G}}(\mathbf{T})^0 = C_{\mathbf{G}}(\mathbf{T}) = C_{\mathbf{G}}(\mathbf{T})^0$.
(ii) *The Borel subgroups of* $C_{\mathbf{G}}(\mathbf{T})$ *are the groups* $C_{\mathbf{G}}(\mathbf{T}) \cap \mathbf{B}$, *where* $\mathbf{B}$ *runs over the Borel subgroups of* $\mathbf{G}$ *containing* $\mathbf{T}$.

References   See, for example Springer (1998, 3.2.9 and 6.4.7).   □

It follows from (i) that the quotient $N_{\mathbf{G}}(\mathbf{T})/C_{\mathbf{G}}(\mathbf{T})$ is finite.

Definition 1.3.5   A closed subgroup of a connected algebraic group $\mathbf{G}$ which contains a Borel subgroup is called a **parabolic subgroup**.

Corollary 1.3.6   *Let* $\mathbf{G}$ *be a connected algebraic group:*

(i)   *Any parabolic subgroup is equal to its normaliser in* $\mathbf{G}$ *and is connected.*
(ii) *Two parabolic subgroups containing the same Borel subgroup and* $\mathbf{G}$-*conjugate are equal.*
(iii) *All maximal tori of* $\mathbf{G}$ *are conjugate; every semi-simple element of* $\mathbf{G}$ *is in some maximal torus.*
(iv) *Two elements of a maximal torus* $\mathbf{T}$ *are* $\mathbf{G}$-*conjugate if and only if they are* $N_{\mathbf{G}}(\mathbf{T})$-*conjugate.*

Proof   Let us prove (i). As the Borel subgroups are connected, $\mathbf{P}^0$ contains a Borel subgroup $\mathbf{B}$. As another Borel subgroup of $\mathbf{G}$ in $\mathbf{P}^0$ is $\mathbf{P}^0$-conjugate to $\mathbf{B}$, we have $N_{\mathbf{G}}(\mathbf{P}^0) = \mathbf{P}^0 N_{\mathbf{G}}(\mathbf{B}) = \mathbf{P}^0\mathbf{B} = \mathbf{P}^0$. As $\mathbf{P} \subset N_{\mathbf{G}}(\mathbf{P}^0)$ we have $\mathbf{P} = \mathbf{P}^0$.

Let us prove (ii). Using again that Borel subgroups of $\mathbf{P}$ are $\mathbf{P}$-conjugate, we get that two conjugate parabolic subgroups containing the same Borel subgroup are $N_{\mathbf{G}}(\mathbf{B})$-conjugate, thus are equal, since $N_{\mathbf{G}}(\mathbf{B}) = \mathbf{B}$.

Since any maximal torus is in some Borel subgroup, the first assertion in (iii) results from 1.3.3 (i) and the same property for connected solvable groups;

see Springer (1998, 6.3.5 (iii)). The second assertion of (iii) comes from 1.3.3 (ii) and the same property for connected solvable groups; see Springer (1998, 6.3.5 (i)).

Let us prove (iv). If s and $^g s$ both lie in $\mathbf{T}$, then $\mathbf{T}$ and $^{g^{-1}}\mathbf{T}$ are two maximal tori containing s, thus are two maximal tori of the group $C_{\mathbf{G}}(s)$, and since they are connected they lie in the identity component $C_{\mathbf{G}}(s)^0$. By (iii) they are conjugate by some element $x \in C_{\mathbf{G}}(s)^0$; that is, $^x\mathbf{T} = {}^{g^{-1}}\mathbf{T}$ so $gx \in N_{\mathbf{G}}(\mathbf{T})$. As $^{gx}s = {}^g s$, we get the result. □

Statement 1.3.6(iii) allows to give the following definition.

Definition 1.3.7   The **rank** of an algebraic group is the rank of its maximal tori.

Exercise 1.3.8   If a closed subgroup $\mathbf{H}$ of the connected algebraic group $\mathbf{G}$ contains a maximal torus $\mathbf{T}$ of $\mathbf{G}$, then $N_{\mathbf{G}}(\mathbf{H}) \subset \mathbf{H}^0 \cdot N_{\mathbf{G}}(\mathbf{T})$.

1.4 Unipotent Groups, Radical, Reductive and Semi-Simple Groups

A **unipotent** algebraic group is a group containing only unipotent elements.

Proposition 1.4.1   *Every unipotent subgroup of an affine algebraic group* $\mathbf{G}$ *is nilpotent.*

Proof   By embedding $\mathbf{G}$ into $\mathbf{GL}_n$, we can reduce to the case $\mathbf{G} = \mathbf{GL}_n$; any unipotent subgroup of $\mathbf{GL}_n(k)$ is conjugate to a subgroup of the group of upper triangular matrices which have all their diagonal entries equal to 1 – see Springer (1998, 2.4.12) for a proof – whence the result. □

Proposition 1.4.2   *As an algebraic variety, a connected unipotent group is isomorphic to an affine space; in characteristic* 0 *a unipotent group is necessarily connected.*

Proof   For the first part, see Springer (1998, 14.2.7 and 14.3.10). If the characteristic is 0 and $\mathbf{U}$ is a unipotent algebraic group, then $\mathbf{U}/\mathbf{U}^0$ is a finite unipotent group. But a unipotent element in characteristic 0 has infinite order, since this holds in $\mathbf{GL}_n$, whence the second part. □

Corollary 1.4.3   *The maximal connected unipotent subgroups of an algebraic group* $\mathbf{G}$ *are the groups* $\mathbf{B}_u$ *where* $\mathbf{B}$ *is a Borel subgroup of* $\mathbf{G}$.

Proof   By 1.4.1 such a subgroup is nilpotent, thus in a Borel subgroup, whence the result by 1.3.1(i). □

Proposition 1.4.4   *Let* **G** *be an algebraic group.*

(i) *The product of the closed connected normal solvable subgroups of* **G** *is also a closed connected normal solvable subgroup of* **G**, *called the* **radical** *of* **G** *and denoted by* $R(\mathbf{G})$.

(ii) *Similarly, the set of all closed connected normal unipotent subgroups of* **G** *has a unique maximal element called the* **unipotent radical** *of* **G** *and denoted by* $R_u(\mathbf{G})$.

(iii) $R_u(\mathbf{G}) = R(\mathbf{G})_u$ (*where* $R(\mathbf{G})_u$ *is defined as in 1.3.1(i)*).

Proof   Using 1.1.3 it follows that the product in (i) is actually finite. This implies (i), since the product of two solvable groups normalising each other is still solvable. To see (ii) and (iii), we first remark that a closed connected normal unipotent subgroup is in $R(\mathbf{G})$, thus in $R(\mathbf{G})_u$. We then observe that $R(\mathbf{G})_u$ is normal in **G**, being characteristic in $R(\mathbf{G})$, and is connected by 1.3.1(i).   □

Corollary 1.4.5 $R(\mathbf{G})$ *is the identity component of the intersection of all Borel subgroups.*

Proof   Indeed, $R(\mathbf{G})$ is contained in at least one Borel subgroup. Since it is normal and all Borel subgroups are conjugate, it is contained in their intersection. Since it is connected, it is contained in the identity component of this intersection. Conversely, this component is solvable and normal.   □

Definition 1.4.6   An algebraic group is called **reductive** if its unipotent radical is trivial, and **semi-simple** if its radical is trivial.

Exercise 1.4.7   For an algebraic group **G**, the group $\mathbf{G}/R(\mathbf{G})$ is semi-simple and the group $\mathbf{G}/R_u(\mathbf{G})$ is reductive.

Proposition 1.4.8   *If* **G** *is connected and reductive, then* $R(\mathbf{G}) = Z(\mathbf{G})^0$, *the identity component of the centre of* **G**.

Proof   By 1.3.1(iii), since $R_u(\mathbf{G})$ is trivial, $R(\mathbf{G})$ is a torus, normal in **G**. Since **G** is connected and for any torus **T** we have $N_{\mathbf{G}}(\mathbf{T})^0 = C_{\mathbf{G}}(\mathbf{T})^0$ (see 1.3.4), $R(\mathbf{G})$ is central in **G**, and, being connected, is in $Z(\mathbf{G})^0$. Conversely, $Z(\mathbf{G})^0$, being normal solvable and connected, is contained in $R(\mathbf{G})$.   □

Proposition 1.4.9   *If* **G** *is reductive and connected, its derived group* Der(**G**) *is semi-simple and has a finite intersection with* $Z(\mathbf{G})$.

Proof   The first assertion results from the second one, since the radical $R(\text{Der}(\mathbf{G}))$, being characteristic by 1.4.8, is in $R(\mathbf{G})$, thus in $Z(\mathbf{G}) \cap \text{Der}(\mathbf{G})$; it is trivial, since it is connected and this last group is finite.

To see the second assertion, we may embed $\mathbf{G}$ in some $\mathbf{GL}(V)$; the space V is a direct sum of isotypic spaces V_χ for some $\chi \in X(Z(\mathbf{G}))$. The action of $\mathbf{G}$ preserves this decomposition, so the image of $\mathbf{G}$ is in $\prod_\chi \mathbf{GL}(V_\chi)$ and that of Der($\mathbf{G}$) in $\prod \mathbf{SL}(V_\chi)$, while that of $Z(\mathbf{G})$ consists of products of scalar matrices in each V_χ, whence the result. □

1.5 Examples of Reductive Groups

Example 1.5.1   The general linear group $\mathbf{GL}_n$; see 1.1.1(iii). The theorem of Lie–Kolchin (see the proof of 1.3.1) shows that the upper triangular matrices in $\mathbf{GL}_n$, defined by the algebra $k[\{t_{i,j}\}_{1 \le i \le j \le n}, \{t_{i,i}^{-1}\}_{1 \le i \le n}]$, form a Borel subgroup $\mathbf{B}$ – this group is connected as open in an affine space. The group of diagonal matrices is a torus $\mathbf{T}$. It is maximal in $\mathbf{B}$ since it is equal to the quotient of $\mathbf{B}$ by the upper unitriangular matrices, which are unipotent, and a torus contains only semi-simple elements; see 1.2.8(i). It is thus maximal in $\mathbf{G}$ by 1.3.3(v) and (i).

The lower triangular matrices – conjugate to the upper triangular by the matrix of the permutation $(1,n)(2,n-1)\dots$ – form another Borel subgroup, whose intersection with $\mathbf{B}$ is $\mathbf{T}$. Thus $\mathbf{GL}_n$ is reductive by 1.4.5 and 1.4.4(iii).

However, $\mathbf{GL}_n$ is not semi-simple: the centre is the group of scalar matrices, which is connected, thus equal to $R(\mathbf{G})$ by 1.4.8.

The Borel subgroups are the conjugates of the upper triangular matrices, hence identify as the stabilisers of the **complete flags**: that is, the increasing sequences of vector subspaces $0 = F_0 \subsetneq F_1 \subsetneq \dots \subsetneq F_n = k^n$. This allows to identify the variety $\mathbf{G}/\mathbf{B}$ with the **flag variety**.

Example 1.5.2   The **special linear group** $\mathbf{SL}_n = \operatorname{Spec} k[t_{i,j}]/(\det(t_{i,j}) - 1)$. Since a Borel subgroup of $\mathbf{SL}_n$ is a subgroup of a Borel subgroup of $\mathbf{GL}_n$ by maximality, and since the upper triangular matrices of $\mathbf{SL}_n$ form a connected group, by the same argument as in $\mathbf{GL}_n$ they form a Borel subgroup. A similar argument shows that the diagonal matrices of $\mathbf{SL}_n$ form a maximal torus. The group $\mathbf{SL}_n$ is reductive by the same argument as $\mathbf{GL}_n$, thus is semi-simple, as it has a finite centre (see 1.4.8).

Example 1.5.3 $\mathbf{PGL}_n$ is the quotient of $\mathbf{GL}_n$ by $\mathbb{G}_m$ embedded as the centre of $\mathbf{GL}_n$. To see it is an affine variety, we can either refer to Proposition 1.1.4 or identify it with the subgroup of $g \in \mathbf{GL}(M_n(k))$, which are algebra automorphisms; that is, such that $g(E_{i,j})g(E_{k,l}) = \delta_{j,k}g(E_{i,l})$ where $E_{i,j}$ is the **elementary matrix** defined by $\{E_{i,j}\}_{k,l} = \delta_{i,j}\delta_{k,l}$. The image of a maximal torus (resp. a Borel subgroup) of $\mathbf{GL}_n$ is a maximal torus (resp. a Borel subgroup) of $\mathbf{PGL}_n$, since the centre of $\mathbf{GL}_n$ is contained in all Borel subgroups and all maximal tori.

If char $k = p$, the centre $Z(\mathbf{SL}_p)$ is $\mathrm{Spec}\, k[t]/(t^p - 1) = \mathrm{Spec}\, k[t]/(t-1)^p$, which as a variety has a single point and thus is the trivial group, but it is not trivial as a scheme! In characteristic p the natural morphism $\mathbf{SL}_p \to \mathbf{PGL}_p$ is not separable; it is a bijection on the points over k but is not an isomorphism of group schemes, an instance of the problem we mentioned in 1.1.2.

Example 1.5.4 The **symplectic group** $\mathbf{Sp}_{2n}$. On $V = k^{2n}$ with ordered basis $(e_1, \ldots, e_n, e'_n, \ldots, e'_1)$, we define the **symplectic** bilinear form $\langle e_i, e_j \rangle = \langle e'_i, e'_j \rangle = 0$, $\langle e_i, e'_j \rangle = -\langle e'_j, e_i \rangle = \delta_{i,j}$. The group $\mathbf{Sp}_{2n}$ is the (clearly closed) subgroup of $\mathbf{GL}_{2n}$ which preserves this form. If J' is the $n \times n$ matrix $J' = \begin{pmatrix} & & 1 \\ & \cdot^{\cdot^{\cdot}} & \\ 1 & & \end{pmatrix}$ and $J = \begin{pmatrix} & J' \\ -J' & \end{pmatrix}$, we have $\langle v, v' \rangle = {}^t v J v'$; thus g is symplectic if and only if ${}^t g J g = J$. The torus $\mathbf{T} = \{\mathrm{diag}(t_1, \ldots, t_n, t_n^{-1}, \ldots, t_1^{-1})\}$ is a maximal torus: it is in a unique maximal torus $\mathbf{T}_1$ of $\mathbf{GL}_{2n}$, the torus of diagonal matrices, since $\mathbf{T}_1 = C_{\mathbf{GL}_{2n}}(\mathbf{T})$ and $\mathbf{T}$ consists of all symplectic diagonal matrices. The symplectic upper triangular matrices form a subgroup $\mathbf{B}$ consisting of the matrices $\begin{pmatrix} B & BJ'S \\ 0 & J'{}^t B^{-1} J' \end{pmatrix}$, where B is upper triangular and S is symmetric. The group $\mathbf{B}$ is connected since it is the product of the connected varieties of the upper triangular and the symmetric matrices. It is solvable since it is a subgroup of the upper triangular matrices of $\mathbf{GL}_{2n}$. Thus it is contained in some Borel subgroup of $\mathbf{GL}_{2n}$; that is, in the stabiliser of a complete flag. But the only line stable by $\mathbf{B}$ is $\langle e_1 \rangle$, and by induction, considering $e_1^\perp / \langle e_1 \rangle$, the only flag stabilised by $\mathbf{B}$ is the flag $\langle e_1 \rangle \subset \langle e_1, e_2 \rangle \subset \cdots \subset \langle e_1, \ldots, e_n \rangle \subset \langle e_1, \ldots, e_n, e'_n \rangle \subset \cdots \subset \langle e_1, \ldots, e_n, e'_n, \ldots, e'_1 \rangle$. Thus the only Borel subgroup of $\mathbf{GL}_{2n}$ containing $\mathbf{B}$ is the group $\mathbf{B}_1$ of (invertible) upper triangular matrices; thus $\mathbf{B}$ is a Borel subgroup of $\mathbf{Sp}_{2n}$, since any larger solvable connected subgroup would be in $\mathbf{B}_1$, and $\mathbf{B} = \mathbf{B}_1 \cap \mathbf{Sp}_{2n}$.

The group $\mathbf{B}$ is the stabiliser of a complete (that is, maximal) flag of isotropic subspaces $\langle e_1 \rangle \subset \langle e_1, e_2 \rangle \subset \cdots \subset \langle e_1, \ldots, e_n \rangle$. By conjugation in $\mathbf{Sp}_{2n}$, we see that any stabiliser of a complete isotropic flag is a Borel subgroup.

We deduce that $\mathbf{Sp}_{2n}$ is connected: it is enough to see that every element is in a Borel subgroup. Any $g \in \mathbf{Sp}_{2n}$ has at least one eigenvector $x \in V$; as g is symplectic, it induces a symplectic automorphism h of $V_1 = \langle x \rangle^\perp / \langle x \rangle$ (which is naturally endowed with a symplectic form induced by the initial form on V). By induction on $\dim V$, we may assume that $\mathbf{Sp}(V_1)$ is connected, and thus h is in some Borel subgroup of $\mathbf{Sp}(V_1)$; that is, it stabilises a complete isotropic flag of V_1. The inverse image in V of this flag, completed by $\langle x \rangle$, is a complete isotropic flag of V stabilised by g, whence the result. The induction starts with the group $\mathbf{Sp}_2$, which is equal to $\mathbf{SL}_2$, hence connected.

The permutation $(1, 2n)(2, 2n - 1) \ldots$ is symplectic and conjugates $\mathbf{B}$ to the symplectic lower triangular matrices; thus by the same argument as for $\mathbf{GL}_n$, we find that $\mathbf{Sp}_{2n}$ is reductive. The centre of $\mathbf{Sp}_{2n}$ is formed of the scalar symplectic matrices, which are only I_{2n} and $-I_{2n}$; thus $\mathbf{Sp}_{2n}$ is semi-simple (see also the end of Example 2.3.14 for a computation with roots).

Example 1.5.5 The **orthogonal** and **special orthogonal** groups. We will assume the characteristic of k different from 2. On $V = k^n$ $(n \geq 1)$ with ordered basis $(e_1, \ldots, e_n)$, we define the symmetric bilinear form

$$\langle e_i, e_j \rangle = \begin{cases} 1 & \text{if } i + j = n + 1, \\ 0 & \text{otherwise.} \end{cases}$$

The group $\mathbf{O}_n$ is the (closed) subgroup of $\mathbf{GL}_n$ which preserves this form. Similarly as for the symplectic group, we consider the $n \times n$ matrix

$$J = \begin{pmatrix} & & 1 \\ & \cdot^{\cdot^{\cdot}} & \\ 1 & & \end{pmatrix};$$ we have $\langle v, v' \rangle = {}^t v J v'$, and g is orthogonal if ${}^t g J g = J$.
A diagonal matrix $\mathrm{diag}(t_1, \ldots, t_n)$ is orthogonal if and only if $t_i t_{n+1-i} = 1$ for all i. These matrices form a torus if n is even. If n is odd, they verify $t_{(n+1)/2} = \pm 1$ and form a non-connected group whose identity component is defined by $t_{(n+1)/2} = 1$. Using the same argument as for the symplectic group, we get that this identity component is a maximal torus of $\mathbf{O}_n$.

The orthogonal upper triangular matrices form a subgroup $\mathbf{B}$; if $n = 2m$ is even, it consists of the matrices $\begin{pmatrix} B & BJ'A \\ 0 & J'^t B^{-1} J' \end{pmatrix}$, where A and B are matrices of size $m \times m$ with B upper triangular and A antisymmetric, and J' is the matrix defined in Example 1.5.4. If $n = 2m + 1$ is odd, $\mathbf{B}$ consists of the matrices $\begin{pmatrix} B & -BJ'^t w & BJ'A \\ 0 & 1 & w \\ 0 & 0 & J'^t B^{-1} J' \end{pmatrix}$, where A, B, and J' are the same as in the even case and w is a row vector of size m.

The group $\mathbf{B}$ is connected since it is the product of the connected varieties of the upper triangular matrices and of antisymmetric matrices, and in addition, in the odd case, the product by $\mathbb{G}_a^m$.

By the same argument as for the symplectic group, we get that $\mathbf{B}$ is a Borel subgroup of $\mathbf{O}_n$.

In the even case, the group $\mathbf{B}$ is also the stabiliser of the maximal flag of isotropic subspaces $\langle e_1 \rangle \subset \langle e_1, e_2 \rangle \subset \cdots \subset \langle e_1, \ldots, e_{n/2} \rangle$. By conjugating in $\mathbf{O}_n$, we see that any stabiliser of a complete isotropic flag is a Borel subgroup. A similar characterisation can be made in the odd case, since $\mathbf{B}$ is then the identity component of the stabiliser of the maximal flag of isotropic subspaces $\langle e_1 \rangle \subset \langle e_1, e_2 \rangle \subset \cdots \subset \langle e_1, \ldots, e_{(n-1)/2} \rangle$.

The matrix equation defining $\mathbf{O}_n$ shows that the determinant of an orthogonal matrix has to be ± 1. There are orthogonal elements of determinant -1, for example, if $n \geq 2$, $\begin{pmatrix} & & 1 \\ & I_{n-2} & \\ 1 & & \end{pmatrix}$. Hence the group $\mathbf{O}_n$ is not connected. We define the **special orthogonal group** $\mathbf{SO}_n$ as $\mathbf{O}_n \cap \mathbf{SL}_n$. The proof that $\mathbf{SO}_n$ is connected goes along the same lines as for $\mathbf{Sp}_{2n}$; see Exercise 1.5.6 below. In the even case, the induction starts with $\mathbf{SO}_2$, which is the one dimensional torus consisting of the matrices $\begin{pmatrix} a & 0 \\ 0 & a^{-1} \end{pmatrix}$. In the odd case, the induction starts with $\mathbf{SO}_1 = \{1\}$.

The permutation $(1,n)(2, n-1)\ldots$ is orthogonal and conjugates $\mathbf{B}$ to orthogonal lower triangular matrices; thus, as in the above examples, we deduce that $\mathbf{SO}_n$ is reductive. The centre of $\mathbf{SO}_n$ is formed of the scalar orthogonal matrices, which are $\{\pm I_n\}$ if n is even and I_n if n is odd; thus $\mathbf{SO}_n$ is semi-simple (see the end of Examples 2.3.15 and 2.3.16 for a computation with roots).

Exercise 1.5.6   Show that any $g \in \mathbf{SO}_n$ has an isotropic eigenvector x and that the restriction of g to $V_1 = \langle x \rangle^{\perp}/\langle x \rangle$ is in $\mathbf{SO}(V_1)$. (Hint: decompose $V = \langle x \rangle \oplus V_1' \oplus \langle y \rangle$ with $y \in V - \langle x \rangle^{\perp}$ and $\langle x \rangle^{\perp} = \langle x \rangle \oplus V_1'$.) Deduce the connectedness of $\mathbf{SO}_n$ by induction on $\dim V$.

Notes

There are a number of books we recommend that provide a good exposition of the basic theory of algebraic groups; for instance, the "classical" books of Borel (1991), Springer (1998), or Milne (2017), or the less complete but more pedagogical approach of Geck (2003).

2

Structure Theorems for Reductive Groups

We give in this chapter the main structure theorems for reductive groups. In order to do that, we first recall the definition and some properties of Coxeter groups and root systems (refer to 2.3.1(i) and (iii) to see why root systems appear in this context).

2.1 Coxeter Groups

Let W be a group generated by a set S of elements left stable by taking inverses. Let S^* be the free monoid on S; that is, the set of words on S (finite sequences of elements of S).

Let $w \in W$ be the image of $s_1 \ldots s_k \in S^*$. The word $s_1 \ldots s_k \in S^*$ is called a **reduced expression** for w if it has minimal length among the words representing w; we then write $l(w) = k$. We call l the **length** on W with respect to S.

We assume now the set S which generates W consists of involutions; that is, each element of S is its own inverse (and is different from 1). Notice that reversing words is then equivalent to taking inverses in W. For $s, s' \in S$ we will denote by $\Delta_{s,s'}^{(m)}$ the word $\underbrace{ss'ss' \ldots}_{m \text{ terms}}$. If the product ss' has finite order m, we will just write $\Delta_{s,s'}$ for $\Delta_{s,s'}^{(m)}$; then the relation $\Delta_{s,s'} = \Delta_{s',s}$ holds in W. Writing the relation $(ss')^m = 1$ this way has the advantage that transforming a word by the use of this relation does not change the length – this will be useful later. Relations of this kind are called **braid relations** because they define the braid groups, which are groups related to the Coxeter groups but also have a topological definition.

Definition 2.1.1 A pair (W, S) where S is a set of involutions generating the group W is a **Coxeter system** if

$$\langle s \in S \mid s^2 = 1, \Delta_{s,s'} = \Delta_{s',s} \text{ for } s, s' \in S \text{ with } ss' \text{ of finite order} \rangle$$

is a presentation of W.

⚡ We may ask if a presentation of the above kind always defines a Coxeter system. That is, given a presentation with relations $\Delta_{s,s'}^{(m)} = \Delta_{s',s}^{(m)}$, is m the order of ss' in the defined group and is the order of s equal to 2? This is always the case – see Bourbaki (1968, V, §4.3); but it is not obvious.

If (W,S) is a Coxeter system, we say that W is a **Coxeter group** and that S is a **Coxeter generating set** of W.

According to, for example, Bourbaki (1968, V, §4.4 corollaire 2) a Coxeter group has a faithful representation in which the elements of S act as reflections (see 2.2.1), so we also call the elements of S the **generating reflections** of W, and the W-conjugates of elements of S the **reflections** of W, their set being denoted by $\mathrm{Ref}(W)$.

Characterisations of Coxeter Groups

Theorem 2.1.2 *Let W be a group generated by a set S of involutions. The following are equivalent:*

(i) *(W,S) is a Coxeter system.*

(ii) *There exists a (unique) map N from W to the set of subsets of $\mathrm{Ref}(W)$, such that $N(s) = \{s\}$ for $s \in S$ and for $x,y \in W$ we have $N(xy) = N(y)\dotplus y^{-1}N(x)y$, where $\dotplus$ denotes the symmetric difference of two sets (the sum modulo 2 of the characteristic functions).*

(iii) *(Exchange condition) If $s_1 \ldots s_{l(w)}$ is a reduced expression for $w \in W$ and $s \in S$ is such that $l(sw) \leq l(w)$, then there exists i such that $sw = s_1 \ldots \hat{s}_i \ldots s_{l(w)}$, where $\hat{s}_i$ denotes a missing term.*

(iv) *W satisfies $l(sw) \neq l(w)$ for $s \in S$, $w \in W$, and (Matsumoto's lemma) two reduced expressions of the same word can be transformed into each other by using just the braid relations. Formally, given any monoid M and any morphism $f : S^* \to M$ such that $f(\Delta_{s,s'}) = f(\Delta_{s',s})$ when ss' has finite order, then f is constant on the reduced expressions of any $w \in W$.*

We will see along the way that $|N(w)| = l(w)$. Note that (iii) could be called the "left exchange condition". By symmetry there is a right exchange condition where sw is replaced by ws.

Proof   We first show that (i)⇒(ii). The definition of N may look technical and mysterious, but the intuition is – see 2.2.11(i) below – that W has a reflection representation permuting a set of "root vectors" (there are two opposite roots attached to each reflection), that these roots are divided into positive and negative by a linear form which does not vanish on any root, and that $N(w)$ records the reflections whose roots change sign by the action of w.

Lemma 2.1.3 *If N is as in (ii), then*

$$N(s_1 \ldots s_k) = \{s_k\}\dotplus\{{}^{s_k}s_{k-1}\}\dotplus \cdots \dotplus\{{}^{s_ks_{k-1}\cdots s_2}s_1\}.$$

Proof This is just an inductive application of the formulae of (ii). □

We get (ii) if we show that the function defined by 2.1.3 on S^* factors through W. To do that we need N to be compatible with the relations defining W, that is $N(ss) = \emptyset$ and $N(\Delta_{s,s'}) = N(\Delta_{s',s})$. This is straightforward.

We now show (ii)⇒(iii). We will actually check the right exchange condition; by symmetry if (i) implies this condition it also implies the left condition. We first show that if $s_1 \dots s_k$ is a reduced expression for w, then $|N(w)| = k$; that is, all the elements of $\mathrm{Ref}(W)$ which appear on the right-hand side of 2.1.3 are distinct. Otherwise, there would exist $i < j$ such that $s_k \dots s_i \dots s_k = s_k \dots s_j \dots s_k$; then $s_i s_{i+1} \dots s_j = s_{i+1} s_{i+2} \dots s_{j-1}$ which contradicts that the expression is reduced.

We next observe that $l(ws) \le l(w)$ implies $l(ws) < l(w)$. Indeed $N(ws) = \{s\} \dotplus s^{-1}N(w)s$, thus by the properties of $\dotplus$ we have $l(ws) = l(w) \pm 1$. Also, if $l(ws) < l(w)$, we must have $s \in s^{-1}N(w)s$ or equivalently $s \in N(w)$. It follows that there exists i such that $s = s_k \dots s_i \dots s_k$, which multiplying on the left by w gives $ws = s_1 \dots \hat{s}_i \dots s_k$, whence (iii).

We now show (iii)⇒(iv). The exchange condition implies $l(sw) \ne l(w)$ because if $l(sw) \le l(w)$ it gives $l(sw) < l(w)$. Given $f : S^* \to M$ as in (iv) we use induction on $l(w)$ to show that f is constant on reduced expressions. Otherwise, let $s_1 \dots s_k$ and $s'_1 \dots s'_k$ be two reduced expressions for the same element w whose images by f differ. By the exchange condition there exists i such that $s'_1 s_1 \dots s_k = s_1 \dots \hat{s}_i \dots s_k$ in W, thus $s'_1 s_1 \dots \hat{s}_i \dots s_k$ is another reduced expression for w. If $i \ne k$ we may apply induction to deduce that $f(s_1 \dots s_k) = f(s'_1 s_1 \dots \hat{s}_i \dots s_k)$ and similarly apply induction to deduce that $f(s'_1 \dots s'_k) = f(s'_1 s_1 \dots \hat{s}_i \dots s_k)$, a contradiction. Thus $i = k$ and $s'_1 s_1 \dots s_{k-1}$ is a reduced expression for w such that $f(s'_1 s_1 \dots s_{k-1}) \ne f(s_1 \dots s_k)$.

Arguing the same way, starting this time from the pair of expressions $s_1 \dots s_k$ and $s'_1 s_1 \dots s_{k-1}$, we get that $s_1 s'_1 s_1 \dots s_{k-2}$ is a reduced expression for w such that

$$f(s_1 s'_1 s_1 \dots s_{k-2}) \ne f(s'_1 s_1 \dots s_{k-1}).$$

Going on, this process will stop when we get two reduced expressions of the form $\Delta^{(m)}_{s_1,s'_1}, \Delta^{(m)}_{s'_1,s_1}$, such that $f(\Delta^{(m)}_{s_1,s'_1}) \ne f(\Delta^{(m)}_{s'_1,s_1})$. We cannot have m greater than the order of $s_1 s'_1$ since the expressions are reduced, nor less than that order, because the order would be smaller. And we cannot have m equal to the order of $s_1 s'_1$ because this contradicts the assumption.

We finally show (iv)⇒(i). (i) can be stated as: given any group G and a morphism of monoids $f : S^* \to G$ such that $f(s)^2 = 1$ and $f(\Delta_{s,s'}) = f(\Delta_{s',s})$

then f factors through a morphism $g : W \to G$. Let us define g by $g(w) = f(s_1 \ldots s_k)$ when $s_1 \ldots s_k$ is a reduced expression for w. By (iv) the map g is well defined. To see that g factors f we need to show that for any expression $w = s_1 \ldots s_k$ we have $g(w) = f(s_1 \ldots s_k)$. This will follow by induction on the length of the expression if we show that $f(s)g(w) = g(sw)$ for $s \in S, w \in W$. If $l(sw) > l(w)$ this equality is immediate from the definition of g. If $l(sw) < l(w)$ we use $f(s)^2 = 1$ to rewrite the equality $g(w) = f(s)g(sw)$ and we apply the reasoning of the first case. Finally $l(sw) = l(w)$ is excluded by assumption. $\square$

Exercise 2.1.4   Show, using the exchange property, that if (W,S) is a Coxeter system and for some $s,t \in S$ and $w \in W$ we have $l(swt) = l(w)$ and $l(sw) = l(wt)$, then $sw = wt$.

The Longest Element

Proposition 2.1.5 *Let (W,S) be a Coxeter system. Then the following properties are equivalent for an element $w_0 \in W$:*

(i) $l(w_0 s) < l(w_0)$ *for all $s \in S$.*
(ii) $l(w_0 w) = l(w_0) - l(w)$ *for all $w \in W$.*
(iii) w_0 *has maximal length amongst elements of W.*

If such a w_0 exists, it is unique, $w_0^2 = 1$, W is finite, and conjugating by w_0 is an automorphism of (W,S).

Proof   It is clear that (ii) implies (iii) and that (iii) implies (i).

To see that (i) implies (ii), we will show by induction on $l(w)$ that w_0 as in (i) has a reduced expression ending by a reduced expression for w^{-1}. Write $w^{-1} = vs$ where $l(v) + l(s) = l(w)$. By induction we may write $w_0 = yv$ where $l(w_0) = l(y) + l(v)$. The (right) exchange condition, using that $l(w_0 s) < l(w_0)$ but $l(vs) > l(v)$, shows that $w_0 s = \hat{y}v$ where $\hat{y}$ represents y with one letter omitted. It follows that $\hat{y}vs$ is a reduced expression for w_0.

For an element satisfying (ii) we have $w_0^2 = 1$ since $l(w_0^2) = l(w_0) - l(w_0) = 0$ and w_0 is unique since another w_1 satisfying (ii) has the same length by (iii) thus $l(w_0 w_1) = l(w_0) - l(w_1) = 0$ thus $w_1 = w_0^{-1} = w_0$.

If w_0 as in (i) exists, then S is finite since $S \subset N(w_0)$ and W is then finite by (iii).

Finally, if $s \in S$ and $w_0 s w_0^{-1} = t$, then from $w_0 s = tw_0$ and (ii) and its symmetric we get $l(w_0) - l(s) = l(w_0) - l(t)$ thus $l(t) = l(s) = 1$ and $t \in S$. $\square$

Lemma 2.1.6

 (i) *For $r \in \mathrm{Ref}(W)$, we have $r \in N(r)$.*

 (ii) *For $w \in W$, we have $N(w) = \{r \in \mathrm{Ref}(W) \mid l(wr) < l(w)\}$.*

(iii) *(Generalised exchange condition) If $s_1 \ldots s_k$ is a reduced expression for $w \in W$ and $r \in \mathrm{Ref}(W)$ is such that $l(wr) < l(w)$, then there exists i such that $wr = s_1 \ldots \hat{s}_i \ldots s_k$.*

(iv) *If $s_1 \ldots s_{2k+1}$ is a reduced expression for $r \in \mathrm{Ref}(W)$, then $s_1 \ldots s_k s_{k+1}$ $s_k \ldots s_1$ is another one.*

 (v) *If W is finite, $N(w_0) = \mathrm{Ref}(W)$.*

Proof Let $r \in \mathrm{Ref}(W)$ and choose an expression $r = vsv^{-1}$ with $s \in S$ and $l(v)$ minimal; that is, considering a reduced expression for v, choose an expression $r = s_k s_{k-1} \ldots s_1 s_2 \ldots s_k$ with $s_i \in S$ and k minimal. Let $r_1 = s_k, r_2 = {}^{s_k} s_{k-1}, \ldots$ be the terms in the right-hand side of 2.1.3, so that $N(r) = \{r_1\} \dotplus \cdots \dotplus \{r_{2k-1}\}$. Then it is easy to check that $r = r_i$ for $i < k$ would imply that k is not minimal, and since $r_{2k-i} = r r_i r$, we also cannot have $r = r_{2k-i}$ for $i < k$ since it would imply $r = r_i$. Since $r = r_k$ it follows that $r \in N(r)$.

Now for (ii), if $r \in N(w)$ then $r = {}^{s_k s_{k-1} \cdots s_{i+1}} s_i$ for some $i \leq k$, where $w = s_1 \ldots s_k$ is a reduced expression for w. Then $wr = s_1 \ldots s_{i-1} s_{i+1} \ldots s_k$ has a length strictly smaller than w. Conversely, if $r \notin N(w)$, then $r \notin r^{-1} N(w) r$ and since $r \in N(r)$ by (i), we have $r \in N(r) \dotplus r^{-1} N(w) r = N(wr)$, hence $l(w) = l((wr)r) < l(wr)$ by the direct part of the proof.

(iii) is a consequence of (ii), since if $l(wr) < l(w)$ by (ii) r is equal to one of the terms on the right-hand side of 2.1.3, and $r = {}^{s_k \cdots s_{i+1}} s_i$ implies $s_i s_{i+1} \ldots s_k r = s_{i+1} \ldots s_k$.

For (iv), let $x = s_k s_{k-1} \ldots s_1$ and $y = s_{k+2} \ldots s_{2k+1}$. It is sufficient to show $x = y$. Since $s_{k+1} y r = x$, by (iii) we have $x = s_{k+1} \ldots \hat{s}_i \ldots s_{2k+1}$ for some i, and since $l(x) = k$ this is a reduce expression. Since $s_{k+1} x$ is part of another reduced expression (the reverse of the one given for r), we have $l(s_{k+1}x) > l(x)$ which forces $i = k + 1$ thus $x = y$.

Finally (v) is an immediate consequence of (ii). □

Coxeter Diagrams

We will now describe finite Coxeter groups, thus we assume $S = \{s_1, \ldots, s_n\}$ is finite. A Coxeter system (W, S) is encoded by a graph with vertices $1, \ldots, n$ and edges encoding the order of ss' when it is greater than 2. This order is encoded by a single edge when equal to 3, a double edge when equal to 4, a triple edge when equal to 6, and an edge decorated by the order when equal to 5 or greater than 6.

Classification 2.1.7   The finite **irreducible** (meaning that the diagram is connected) Coxeter groups are classified as follows (see Bourbaki (1968, VI, §4.1 Théorème 1)):

A finite Coxeter group is called a **Weyl group** if it is a reflection group in a $\mathbb{Q}$-vector space. This selects in the above list exactly the diagrams where the order of ss' is always in $\{2,3,4,6\}$. The group $I_2(6)$ is also denoted by G_2.

2.2 Finite Root Systems

In this section V is a finite dimensional vector space on a subfield $K \subset \mathbb{R}$ and V^* is its dual.

Definition 2.2.1   A **reflection** $s \in \mathbf{GL}(V)$ is an element of order 2 such that $\mathrm{Ker}(s - \mathrm{I}_V)$ is a hyperplane.

It follows that s has an eigenvalue -1 with multiplicity 1, and that if $\alpha \in V$ is an eigenvector for -1 and $\alpha^{\vee} \in V^*$ is a linear form of kernel $\mathrm{Ker}(s - \mathrm{I}_V)$, chosen such that $\alpha^{\vee}(\alpha) = 2$, then $s(x) = x - \alpha^{\vee}(x)\alpha$.

We call α a **root** attached to the reflection s and $\alpha^\vee$ the corresponding **coroot**. They are unique up to scalings inverse to each other. Conversely any pair of vectors $\alpha \in V, \alpha^\vee \in V^*$ such that $\alpha^\vee(\alpha) = 2$ defines a reflection.

Definition 2.2.2

- A (finite) **root system** is a finite set $\Phi \subset V$ with a bijection $\alpha \mapsto \alpha^\vee : \Phi \to \Phi^\vee \subset V^*$ such that for any $\alpha \in \Phi$ we have $\alpha^\vee(\alpha) = 2$ and Φ is stabilised by the reflection s_α of root α and coroot $\alpha^\vee$.
- The system is **crystallographic** if $\alpha^\vee(\beta) \in \mathbb{Z}$ for all $\alpha, \beta \in \Phi$.
- The system is **reduced** if for any $\alpha \in \Phi$ we have $\Phi \cap K\alpha = \{\alpha, -\alpha\}$.
- The system is **irreducible** if there is no partition into two non-empty subsets $\Phi = \Phi_1 \coprod \Phi_2$ such that $\alpha_1^\vee(\alpha_2) = 0$ for all $\alpha_1 \in \Phi_1$ and $\alpha_2 \in \Phi_2$.

If the system is crystallographic, Φ and $\Phi^\vee$ generate dual lattices.

We omitted two conditions compared to the definition given in Bourbaki (1968, VI, §1), where it is added that Φ spans V and is crystallographic; we will see in 2.2.10 that a root system is associated with a Coxeter group; any finite Coxeter group has a root system as we defined it, but only the Weyl groups have crystallographic ones.

In the following we fix a root system Φ and denote by W the group generated by $\{s_\alpha\}_{\alpha \in \Phi}$. It is finite since its elements are determined by the permutation of Φ they induce. Thus there exists a W-invariant scalar product $(.,.)$ on V.

Lemma 2.2.3 *Using $(.,.)$ to identify V^* with V, we have $\alpha^\vee = \frac{2\alpha}{(\alpha,\alpha)}$.*

Proof Using the invariance of $(.,.)$ we get for all $v \in V$ that $(\alpha, v) = (s_\alpha \alpha, s_\alpha v) = (-\alpha, v - \alpha^\vee(v)\alpha)$ which gives $\alpha^\vee(v) = \frac{2(\alpha,v)}{(\alpha,\alpha)}$. $\qquad \square$

Using the identification of Lemma 2.2.3 allows us to work in a Euclidean space and forget $\Phi^\vee$; but keeping V^* allows the extension of the theory to infinite root systems.

In the following we assume Φ is reduced, in order to simplify somewhat the statements and proofs – a non-reduced system BC_n occurs in certain parts of reductive groups theory that we will not cover.

We will need an order on Φ induced by a total order on V. It is sufficient for our purpose to determine which roots are positive or negative for this order, and for that it is sufficient to have a linear form λ on V which does not vanish on Φ: the positive roots are those on which λ takes positive values.

Theorem 2.2.4 *Given a linear form which does not vanish on Φ, denote by Φ^+ the corresponding set of positive roots. Then there exists a unique minimal subset $\Pi \subset \Phi^+$ such that $\Phi^+ = \Phi \cap K_{\geq 0}\Pi$. The set Π thus defined is a basis of the subspace $V_0 \subset V$ generated by Φ.*

Proof Note first that there exists a minimal subset $\Pi \subset \Phi^+$ such that $\Phi^+ = \Phi \cap K_{\geq 0}\Pi$: to obtain such a subset, starting from Φ^+, just iteratively remove elements that are a positive linear combination of others in the considered subset.

Lemma 2.2.5 *For a minimal Π as above $(\alpha, \beta) \leq 0$ for $\alpha, \beta \in \Pi, \alpha \neq \beta$.*

Proof Assume by contradiction that $(\alpha, \beta) > 0$. Then $s_\alpha(\beta) = \beta - c\alpha$ where $c = \frac{2(\alpha, \beta)}{(\alpha, \alpha)} > 0$. Either $s_\alpha(\beta) \in \Phi^+$ or $-s_\alpha(\beta) \in \Phi^+$.

In the first case by assumption $s_\alpha(\beta) = \sum_{\gamma \in \Pi} c_\gamma \gamma$ with $c_\gamma \geq 0$; we rewrite this as $\sum_{\gamma \in \Pi - \{\beta\}} c_\gamma \gamma + c\alpha + (c_\beta - 1)\beta = 0$. We cannot have $c_\beta - 1 \geq 0$ since a non-empty sum of positive vectors cannot be zero. Thus we expressed β as an element of $K_{\geq 0}(\Pi - \{\beta\})$ which contradicts the minimality of Π.

In the second case we similarly rewrite $-s_\alpha(\beta) = \sum_{\gamma \in \Pi} c_\gamma \gamma$ with $c_\gamma \geq 0$ as $\sum_{\gamma \in \Pi - \{\alpha\}} c_\gamma \gamma + \beta + (c_\alpha - c)\alpha = 0$, and similarly we must have $c_\alpha - c < 0$ giving an expression of α as an element of $K_{\geq 0}(\Pi - \{\alpha\})$ which again contradicts the minimality of Π. □

Let us see now that Π is a basis. We know it spans V_0 since Φ does. We have to exclude a linear dependence amongst its elements. Such a relation can be written $v = \sum_{\alpha \in \Pi_1} c_\alpha \alpha = \sum_{\beta \in \Pi_2} c_\beta \beta$ where v is a non-zero vector, where $c_\alpha, c_\beta \geq 0$ and where $\Pi = \Pi_1 \sqcup \Pi_2$. But then we have $0 < (v, v) = (\sum_{\alpha \in \Pi_1} c_\alpha \alpha, \sum_{\beta \in \Pi_2} c_\beta \beta)$ which contradicts Lemma 2.2.5.

We finally show that Π is unique: if there are two such bases $\Pi \neq \Pi'$ let us consider $\alpha \in \Pi - \Pi'$; express it on Π' as $\alpha = \sum_{\beta \in \Pi'} c_\beta \beta$ then express each involved β on Π: since $\beta \neq \alpha$ these expressions will involve a root in $\Pi - \alpha$ (we use here that the system is reduced) and this root will remain when doing the sum, since the coefficients are positive; this is a contradiction to the minimality of Π. □

A Φ^+ as above is called a **positive subsystem** and a Π as above is called a **basis** of Φ. The cardinality of Π is called the **rank** of Φ.

Note that in the basis Π the entries of the matrix s_α are 1 or $-\alpha^\vee(\beta)$, thus in this basis we have $W \subset \mathbf{GL}_n(\mathbb{Z})$ if the root system is crystallographic, where $n = |\Pi|$.

Proposition 2.2.6 *Two positive subsystems (resp. bases) are W-conjugate.*

Proof It is enough to consider positive subsystems since they determine bases.

Lemma 2.2.7 *For $\alpha \in \Pi$ and any $\beta \in \Phi^+ - \{\alpha\}$ we have $s_\alpha(\beta) \in \Phi^+$.*

Proof If $\beta \in \Phi^+ - \{\alpha\}$, then $\beta = \sum_{\gamma \in \Pi} c_\gamma \gamma$ where at least one $c_\gamma > 0$ with $\gamma \neq \alpha$, otherwise $\beta \in \Phi^+ \cap K_{\geq 0}\alpha = \{\alpha\}$. But then $s_\alpha(\beta) = \beta - \alpha^\vee(\beta)\alpha$ has the same coefficient on γ, and, since any root has all non-zero coefficients on Π of the same sign, the root $s_\alpha(\beta)$ is positive. □

We use the lemma to conjugate another positive subsystem Φ' on Φ^+, using induction on $|\Phi^+ \cap -\Phi'|$. If this number is positive, then $\Pi \cap -\Phi' \neq \emptyset$, otherwise $\Pi \subset \Phi'$ which implies $\Phi^+ \subset \Phi'$ which implies $\Phi^+ = \Phi'$ since all positive subsystems have same cardinality $|\Phi|/2$. Choose thus $\alpha \in \Pi \cap -\Phi'$; since $s_\alpha(\Phi^+) = (\Phi^+ - \{\alpha\}) \coprod \{-\alpha\}$, the set $s_\alpha(\Phi^+)$ is a positive subsystem such that $|s_\alpha(\Phi^+) \cap -\Phi'| = |\Phi^+ \cap -\Phi'| - 1$. $\qquad\square$

Corollary 2.2.8 *Every root α is in the W-orbit of Π, and either α or $-\alpha$ is a sum of elements of Π.*

Proof It is enough to show it for every positive root since $s_\alpha(\alpha) = -\alpha$. Take $\alpha = \sum_{\gamma \in \Pi} c_\gamma \gamma \in \Phi^+ - \Pi$; as $0 < (\alpha, \alpha) = \sum_{\gamma \in \Pi} c_\gamma(\alpha, \gamma)$ there exists $\gamma \in \Pi$ such that $(\alpha, \gamma) > 0$. Then $\alpha' = s_\gamma(\alpha)$ is still positive by 2.2.7 and is obtained by removing a positive multiple of γ from α. Thus if we set $h(\alpha) = \sum_\gamma c_\gamma$ we have $h(\alpha') < h(\alpha)$. We can repeat this process as long as $\alpha' \notin \Pi$. As Φ^+ is finite this process must eventually stop, at a root in Π. $\qquad\square$

The proof of the corollary shows more, that every root is conjugate to an element of Π by a sequence of $s_\gamma, \gamma \in \Pi$. In particular every s_α is in the group generated by $\{s_\gamma\}_{\gamma \in \Pi}$, thus W itself is generated by $\{s_\gamma\}_{\gamma \in \Pi}$.

We show now that W is a Coxeter group using yet another characterisation of Coxeter groups.

Lemma 2.2.9 *Let W be group generated by a set S of involutions and let $\{D_s\}_{s \in S}$ be a set of subsets of W such that:*

- $D_s \ni 1$.
- $D_s \cap sD_s = \emptyset$.
- *If for $s, s' \in S$ we have $w \in D_s, ws' \notin D_s$ then $ws' = sw$.*

Then (W, S) is a Coxeter system, and $D_s = \{w \in W \mid l(sw) > l(w)\}$.

Proof We will show the exchange condition. Let $s_1 \ldots s_k$ be a reduced expression for $w \notin D_s$ and let i be minimal such that $s_1 \ldots s_i \notin D_s$; we have $i > 0$ since $1 \in D_s$. From $s_1 \ldots s_{i-1} \in D_s$ and $s_1 \ldots s_i \notin D_s$ we get $ss_1 \ldots s_{i-1} = s_1 \ldots s_i$, whence $sw = s_1 \ldots \hat{s}_i \ldots s_k$ thus $l(sw) < l(w)$ and we have checked the exchange condition in this case. If $w \in D_s$ then $sw \notin D_s$ and by the first part $l(w) < l(sw)$ so we have nothing to check. $\qquad\square$

Proposition 2.2.10

(i) (W, S) where $S = \{s_\alpha \mid \alpha \in \Pi\}$ *is a Coxeter system; we have* $\mathrm{Ref}(W) = \{s_\alpha \mid \alpha \in \Phi^+\}$.

(ii) *A root system is irreducible if and only if the Coxeter group is irreducible.*

Proof We apply 2.2.9 with $D_{s_\alpha} = \{w \in W \mid w^{-1}(\alpha) > 0\}$ for $\alpha \in \Pi$. That $D_{s_\alpha} \cap s_\alpha D_{s_\alpha} = \emptyset$ is clear. Now take $w \in D_{s_\alpha}$ and $\alpha' \in \Pi$ such that $ws_{\alpha'} \notin D_{s_\alpha}$, that is $w^{-1}(\alpha) > 0$ and $s_{\alpha'}w^{-1}(\alpha) < 0$. As by 2.2.7 $s_{\alpha'}$ changes the sign of only α', we must have $w^{-1}(\alpha) = \alpha'$. As w preserves the scalar product, it conjugates $s_{\alpha'}$ to s_α, whence (i). The last sentence of (i) reflects 2.2.8.

(ii) comes from the fact that in the Coxeter diagram two nodes are linked if and only if the corresponding reflections do not commute which is equivalent to the corresponding roots not being orthogonal. □

Lemma 2.2.11 *For W as in 2.2.10:*

(i) *The set $N(w)$ of 2.1.2(ii) is $\{s_\alpha \mid \alpha \in \Phi^+, w(\alpha) < 0\}$.*

(ii) *The element w_0 of 2.1.5 is such that $w_0(\Phi^+) = \Phi^-$ where we write Φ^- for $-\Phi^+$.*

Proof We set $N'(w) = \{s_\alpha \mid \alpha \in \Phi^+, w(\alpha) < 0\}$ and show by induction on $l(w)$ that $N(w) = N'(w)$: let $w = vs$ with $s \in S$, $l(w) > l(v)$; by 2.1.2(ii) and its proof we have $N(w) = s \coprod sN(v)s$; by the right counterparts of 2.2.9 and of the proof of 2.2.10, we have $N'(w) = s \coprod sN'(v)s$. This proves (i).

(ii) is a consequence of 2.1.6(v). □

The classification of finite crystallographic root systems is the same as the classification of finite Weyl groups, except that there are two root systems B_n and C_n corresponding to the same Weyl group. The list of irreducible crystallographic root systems is thus, A_n, B_n, C_n, D_n, E_6, E_7, E_8, F_4 and G_2; see Bourbaki (1968, VI, §4). We describe now the first four types.

Example 2.2.12 Root system of type A_n.

Let $\{e_1, \ldots, e_{n+1}\}$ be an orthonormal basis of $\mathbb{R}^{n+1}$. Then $\Phi = \{e_i - e_j\}_{i,j \in [1,\ldots,n+1], i \neq j}$ is a root system of cardinality $n(n+1)$. The set $\{e_i - e_j \mid i < j\}$ is a positive subsystem relative to the linear form $x \mapsto (x, (n+1)e_1 + ne_2 + \cdots + e_{n+1})$. We have $\Pi = \{e_i - e_{i+1}\}_{i=1,\ldots,n}$. If we set $\alpha_i = e_i - e_{i+1}$, we have $e_i - e_j = \alpha_i + \alpha_{i+1} + \cdots + \alpha_j$ for $i < j$. The group W is the symmetric group $\mathfrak{S}_{n+1}$, permuting the e_k: the reflection s_i of root α_i interchanges e_i and e_{i+1} and fixes the others e_k; these reflections form a Coxeter system defining the Coxeter graph A_n of 2.1.7.

Example 2.2.13 Root system of type C_n.

It consists of the $2n^2$ roots in $\mathbb{R}^n$ given by $\pm 2e_i$ and $\pm e_i \pm e_j$ ($i \neq j$) with (e_i) as in Example 2.2.12. For the same linear form as above we have $\Phi^+ = \{2e_i\}_i \cup \{e_i \pm e_j\}_{i<j}$ and $\Pi = \{e_1 - e_2, \ldots, e_{n-1} - e_n, 2e_n\}$. The reflection s_i of root $e_i - e_{i+1}$ exchanges e_i and e_{i+1} and fixes the other e_k; the reflection s_{2e_n} maps

e_n to $-e_n$ and fixes the other e_k; we get for W the **hyperoctahedral** group: the group of linear maps which permutes the $\pm e_i$; the Coxeter graph defined by these reflections is the graph C_n of 2.1.7.

Exercise 2.2.14   Replacing $2e_i$ with e_i in C_n still gives a root system, called B_n, which has the same Coxeter graph.

Example 2.2.15   Root system of type D_n.

D_n can be seen as a subsystem of a system of type B_n or C_n; it is for example the root subsystem of the system of type B_n consisting of the $n(n-1)$ vectors $\pm e_i \pm e_j$ with $i \neq j$. The vectors $e_i + e_j$ and $e_i - e_j$ with $i < j$ form a positive subsystem, the corresponding basis being $\{e_1 - e_2, e_2 - e_3, \ldots, e_{n-1} - e_n, e_{n-1} + e_n\}$. The reflection s_i ($i < n$) of root $e_i - e_{i+1}$ exchanges e_i and e_{i+1} and fixes the other e_k. The reflection s_n of root $e_{n-1} + e_n$ maps e_n to $-e_{n-1}$ and e_{n-1} to $-e_n$ and fixes the other e_k. The Weyl group is the subgroup of the hyperoctahedral group consisting of the elements such that the permutation of the $\pm e_i$ has an even number of sign changes. The Coxeter graph is the graph D_n of 2.1.7.

2.3 Structure of Reductive Groups

Theorem 2.3.1   *Let* **G** *be a connected reductive group over k, and let* **T** *be a maximal torus of* **G**. *Then*

(i) *Any non-trivial minimal closed unipotent subgroup of* **G** *normalised by* **T** *is isomorphic to* $\mathbb{G}_a$. *Choosing for such a group* **U** *an isomorphism* $x \mapsto \mathbf{u}(x) : \mathbb{G}_a \xrightarrow{\sim} \mathbf{U}$, *we define* $\alpha \in X(\mathbf{T})$ *by* $t\mathbf{u}(x)t^{-1} = \mathbf{u}(\alpha(t)x)$ *for* $t \in \mathbf{T}$. *The collection* Φ *of* α — *thus obtained when* **U** *varies* — *has no repetition; its elements are called the* **roots of G relative to T**. *We denote by* $\mathbf{U}_\alpha$ *the group* **U** *inducing a given* $\alpha \in \Phi$.

(ii) $\Phi = -\Phi$, *and for any* $\alpha \in \Phi$, *there exists a homomorphism* $\phi_\alpha : \mathbf{SL}_2 \to \mathbf{G}$ *whose image is* $\langle \mathbf{U}_\alpha, \mathbf{U}_{-\alpha} \rangle$, *and which is injective or has kernel* $\{\pm I_2\} = Z(\mathbf{SL}_2)$, *and is such that*

$$\phi_\alpha \begin{pmatrix} 1 & * \\ 0 & 1 \end{pmatrix} = \mathbf{U}_\alpha, \quad \phi_\alpha \begin{pmatrix} 1 & 0 \\ * & 1 \end{pmatrix} = \mathbf{U}_{-\alpha}.$$

(iii) Φ *is a reduced crystallographic root system in* $X(\mathbf{T}) \otimes \mathbb{R}$; *for* $\alpha \in \Phi$ *the one-parameter subgroup* $\check{\alpha} \in Y(\mathbf{T})$ *defined by* $x \mapsto \phi_\alpha \begin{pmatrix} x & 0 \\ 0 & x^{-1} \end{pmatrix}$ *is the corresponding coroot. We have* $C_\mathbf{G}(\mathbf{T}) = \mathbf{T}$ *and the natural map* $W(\mathbf{T}) := N_\mathbf{G}(\mathbf{T})/\mathbf{T} \to \mathbf{GL}(X(\mathbf{T}) \otimes \mathbb{R})$ *identifies* $W(\mathbf{T})$, *called the* **Weyl group** *of* **G** *with respect to* **T**, *with the reflection group defined by* Φ; s_α *is the image* *of $\dot{s}_\alpha := \phi_\alpha \begin{pmatrix} 0 & 1 \\ -1 & 0 \end{pmatrix}$. We will write $W_{\mathbf{G}}(\mathbf{T})$ for $W(\mathbf{T})$ if we want to specify the ambient group.*

(iv) *Any closed connected subgroup $\mathbf{H}$ of $\mathbf{G}$ normalised by $\mathbf{T}$ is generated by $(\mathbf{T} \cap \mathbf{H})^0$ and the $\mathbf{U}_\alpha$ it contains; in particular, $\mathbf{G}$ is generated by $\mathbf{T}$ and the $\mathbf{U}_\alpha$.*

(v) *A unipotent subgroup $\mathbf{H}$ of $\mathbf{G}$ normalised by $\mathbf{T}$ is closed, connected and equal to $\prod_{\mathbf{U}_\alpha \subset \mathbf{H}} \mathbf{U}_\alpha$ in any order, where this product (made by the group law) is a product of varieties.*

(vi) *Borel subgroups containing $\mathbf{T}$ are in bijection with positive subsystems of Φ: if $\mathbf{B}$ corresponds to Φ^+, then $R_u(\mathbf{B}) = \prod_{\alpha \in \Phi^+} \mathbf{U}_\alpha$.*

(vii) *If $\alpha, \beta \in \Phi$, $\alpha \neq -\beta$, then $[\mathbf{U}_\alpha, \mathbf{U}_\beta] \subset \prod_{\{\lambda, \mu \in \mathbb{N}^* \,|\, \lambda\alpha + \mu\beta \in \Phi\}} \mathbf{U}_{\lambda\alpha + \mu\beta}$.*

Sketch of proof and references   Most items are proved in Springer (1998, 7.1 to 8.2.10).

Lemma 2.3.2

(i) *The unipotent radical of an algebraic group $\mathbf{G}$ is equal to the unipotent radical of the intersection of the Borel subgroups containing a given maximal torus.*

(ii) *In a connected reductive group, the centraliser of any torus is reductive (and connected by 1.3.3(iii)).*

(iii) *In a connected reductive group, every maximal torus is equal to its centraliser.*

Proof   (i) follows from 1.4.5. See Springer (1998, 7.6.4) for the rest.   □

For $\alpha \in X(\mathbf{T})$ let $\mathbf{G}_\alpha = C_{\mathbf{G}}(\mathrm{Ker}(\alpha)^0)$; it is a reductive group by 2.3.2(ii). Using the adjoint action of $\mathbf{T}$ on the Lie algebra $\mathfrak{G}$ of $\mathbf{G}$, one shows that:

- There are finitely many α such that $\mathbf{G}_\alpha \neq C_{\mathbf{G}}(\mathbf{T})$, since such an α must be a weight of $\mathbf{T}$ on $\mathfrak{G}$.
- The $\mathbf{G}_\alpha$ generate $\mathbf{G}$, since the group they generate has same Lie algebra as $\mathbf{G}$.

Then one shows that if $\mathbf{G}_\alpha$ is not solvable, then its Weyl group is equal to $\mathbb{Z}/2\mathbb{Z}$. It follows that $\mathbf{G}_\alpha/R(\mathbf{G}_\alpha)$ is a semi-simple group of rank 1, thus is $\mathbf{SL}_2$ or $\mathbf{PGL}_2$; see Springer (1998, 7.2.4).

The next step is to show that if $\Phi = \{\alpha \mid \mathbf{G}_\alpha \text{ is not solvable}\}$, then $\mathbf{G}_\alpha/R(\mathbf{G}_\alpha)$ lifts to a subgroup of $\mathbf{G}_\alpha$, which defines ϕ_α; see Springer (1998, 7.3). Each $\mathbf{G}_\alpha$ for $\alpha \in \Phi$ contributes a reflection $s_\alpha \in W$ – see (iii) of the theorem – and a copy $\mathbf{U}_\alpha$ of $\mathbb{G}_a$ normalised by $\mathbf{T}$, and on which $\mathbf{T}$ acts via α.

It is then shown that the s_α ($\alpha \in \Phi$) generate W – see Springer (1998, 7.1.9) – and that Φ is a reduced crystallographic root system; see Springer

(1998, 7.4.4). The stability by s_α results from the definition and the system is reduced since $\mathbf{G}_{n\alpha} = \mathbf{G}_\alpha$ – the identity component of the kernel is the same for $n\alpha$ and α. The fact that $\alpha^\vee$ is the coroot corresponding to α results from checking 2.2.3 in $\mathbf{G}_\alpha$.

By 1.3.4(ii) the solvable $\mathbf{G}_\alpha$ are in all Borel subgroups containing $\mathrm{Ker}(\alpha)^0$ hence in all Borel subgroups containing $\mathbf{T}$. If C is the intersection of such Borel subgroups, then $R_u(C) = R_u(\mathbf{G})$ by 2.3.2(i). Thus if $\mathbf{G}$ is reductive, no $\mathbf{G}_\alpha$ is solvable; the radical of $\mathbf{G}_\alpha$ is in a maximal torus of $\mathbf{G}_\alpha$ so is in $\mathbf{T}$, thus $\mathbf{G}_\alpha$ is generated by $\mathbf{U}_\alpha$, $\mathbf{U}_{-\alpha}$ and $\mathbf{T}$.

Property (iv) when $\mathbf{H} \supset \mathbf{T}$ results from the fact that such an $\mathbf{H}$ is generated by the $\mathbf{G}_\alpha$ it contains. For (iv) in general see Borel and Tits (1965, 3.4).

For property (v) see Chevalley (2005, 13.2, Théorème 1(d)).

The essential point of (vi) is that a Borel subgroup containing $\mathbf{T}$ defines a linear form on $X(\mathbf{T})$; see Springer (1998, 7.4.6).

For (vii), as $\alpha \neq -\beta$ we may assume both positive for some order, thus in some $R_u(\mathbf{B})$. If we denote by $\mathbf{u}_\alpha$ the isomorphism of (i) corresponding to α, in the isomorphism (v) the algebraic group law for $R_u(\mathbf{B})$ gives that there exists polynomials p_γ such that $[\mathbf{u}_\alpha(x), \mathbf{u}_\beta(y)] = \prod_\gamma \mathbf{u}_\gamma(p_\gamma(x,y))$. The action of $t \in \mathbf{T}$ gives $p_\gamma(\alpha(t)x, \beta(t)y) = \gamma(t)p_\gamma(x,y)$. The linear independence of characters implies that γ is of the form $\lambda\alpha + \mu\beta$ where $p_\gamma = c_\gamma x^\lambda y^\mu$. And λ, μ cannot vanish if $p_\gamma \neq 0$; if, for instance $\mu = 0$, then p_γ does not depend on y and thus does not vanish for $\mathbf{u}_\beta(y) = 1$, a contradiction. $\qquad\square$

We will sometimes write W for $W(\mathbf{T})$ when there is no ambiguity about $\mathbf{T}$. Let us note that by 1.3.6(iii) the group $W(\mathbf{T})$ and Φ do not depend on $\mathbf{T}$ up to isomorphism. Note also that it results from 2.3.1(i) that for $w \in W$ and $\alpha \in \Phi$, we have ${}^w\mathbf{U}_\alpha = \mathbf{U}_{w(\alpha)}$.

Corollary 2.3.3 *Two Borel subgroups containing $\mathbf{T}$ are conjugate under $W(\mathbf{T})$.*

Proof This is a consequence of Theorem 2.3.1(vi) and (iii) and of Proposition 2.2.6. $\qquad\square$

Until the end of this chapter, $\mathbf{G}$ is a connected reductive group and the notation is as in 2.3.1.

Proposition 2.3.4

(i) *The centre $Z(\mathbf{G})$ is the intersection of the kernels in $\mathbf{T}$ of the roots of $\mathbf{G}$ relative to $\mathbf{T}$.*

(ii) *If $\mathbf{H}$ is a closed connected subgroup of $\mathbf{G}$, the centre of $\mathbf{H}/\mathbf{H} \cap Z(\mathbf{G})$ is $Z(\mathbf{H})/(Z(\mathbf{H}) \cap Z(\mathbf{G}))$. In particular the centre of $\mathbf{G}/Z(\mathbf{G})$ is trivial.*

Proof We have $Z(\mathbf{G}) \subset C_{\mathbf{G}}(\mathbf{T}) = \mathbf{T}$, see 2.3.2(iii). An element of $\mathbf{T}$ central in $\mathbf{G}$ must act trivially on all $\mathbf{U}_\alpha$, and so be in the kernel of all the roots. Conversely, since by 2.3.1(iv) the $\mathbf{U}_\alpha$ and $\mathbf{T}$ generate $\mathbf{G}$, such an element centralises $\mathbf{G}$, whence (i). Let us prove (ii). If $\mathbf{H}$ is a torus, then the property is true. Otherwise, we have to prove that if $h \in \mathbf{H}$ commutes with all elements up to $Z(\mathbf{G})$, then $h \in Z(\mathbf{H})$. By 1.2.8(ii), there exists a unipotent element $u \in \mathbf{H}$. If $huh^{-1} = uz$ with $z \in Z(\mathbf{G})$, since the left-hand side is a unipotent element, its semi-simple part z has to be 1.   □

Proposition 2.3.5 *Let $\mathbf{G}$ be a connected reductive group. Then each of the following properties is equivalent to $\mathbf{G}$ being semi-simple:*

(i) $\mathbf{G}$ *is generated by the* $\mathbf{U}_\alpha$.
(ii) $\mathbf{G} = \mathrm{Der}\,\mathbf{G}$.
(iii) *For any maximal torus $\mathbf{T}$, the roots of $\mathbf{G}$ with respect to $\mathbf{T}$ span $X(\mathbf{T}) \otimes \mathbb{Q}$.*
(iv) *For any maximal torus $\mathbf{T}$, the coroots of $\mathbf{G}$ with respect to $\mathbf{T}$ span $Y(\mathbf{T}) \otimes \mathbb{Q}$.*

Proof Let Φ be the set of roots of $\mathbf{G}$ with respect to the maximal torus $\mathbf{T}$. Then by 2.3.4(i), Φ spans the kernel of the restriction map $X(\mathbf{T}) \to X(Z(\mathbf{G}))$; if $\mathbf{G}$ is semi-simple, by 1.4.8 the group $X(Z(\mathbf{G}))$ is finite thus this kernel spans $X(\mathbf{T})$ over $\mathbb{Q}$, thus (iii) holds.

(iii) is equivalent to (iv) by duality.

Assume now $\mathbf{G}$ satisfies (iv). Then $\cup_{\alpha \in \Phi} \mathrm{Im}\,\alpha^\vee$ spans $\mathbf{T}$ by 1.2.10. But by the relation 2.4.11(iii) (a computation in $\mathbf{SL}_2$) the group generated by the $\mathbf{U}_\alpha$ contains $\mathrm{Im}\,\alpha^\vee$, thus contains $\mathbf{T}$ and by 2.3.1(iv) we have (i).

For any connected reductive group $\mathbf{G}$ the $\mathbf{U}_\alpha$ are in $\mathrm{Der}\,\mathbf{G}$ since for $t \in \mathbf{T}$ we have $t\mathbf{u}_\alpha(x)t^{-1}\mathbf{u}_\alpha(x)^{-1} = \mathbf{u}_\alpha((\alpha(t) - 1)x)$, thus (i) implies (ii).

Finally a group satisfying (ii) is semi-simple by 1.4.9.   □

Proposition 2.3.6 *A connected semi-simple group $\mathbf{G}$ has finitely many minimal non-trivial normal connected subgroups. Any two of them commute and each of them has a finite intersection with the product of the others; the product of all of them is equal to the whole of $\mathbf{G}$.*

Reference See Springer (1998, 8.1.5).   □

Definition 2.3.7   A connected semi-simple group which has no proper nontrivial normal connected subgroup is called **quasi-simple**.

Proposition 2.3.8 *A connected reductive group is the product of its derived group and its radical.*

Proof It is enough to show that the composite morphism $\mathrm{Der}\,\mathbf{G} \hookrightarrow \mathbf{G} \rightarrow \mathbf{G}/R\mathbf{G}$ is surjective. This results from $\mathrm{Der}\,\mathbf{G}/(R\mathbf{G} \cap \mathrm{Der}\,\mathbf{G}) = \mathrm{Der}(\mathbf{G}/R\mathbf{G}) = \mathbf{G}/R\mathbf{G}$, the last equality by 2.3.5(ii) since $\mathbf{G}/R\mathbf{G}$ is semi-simple. $\square$

By 1.4.9 the product in 2.3.8 is almost direct, that is the intersection is finite.

Corollary 2.3.9 *Every connected reductive group is the almost direct product of its radical and of a finite number of quasi-simple groups.*

Proof This is just 2.3.6 and 2.3.8 put together. $\square$

Proposition 2.3.10 *Let* $\mathbf{T}$ *be a maximal torus of* $\mathbf{G}$*; we have*

(i) $X(\mathbf{T}/Z(\mathbf{G})^0) = (\Phi^\perp)^\perp$.
(ii) $Y(\mathbf{T} \cap \mathrm{Der}\,\mathbf{G}) = (\Phi^{\vee\perp})^\perp$.

Proof We have $Y(Z(\mathbf{G})^0) \subset Y(\mathbf{T})$. The image of $y \in Y(\mathbf{T})$ is a subtorus of rank 1 which is in $Z(\mathbf{G})^0$ if and only if it is in $Z(\mathbf{G})$, hence if and only if it lies in the kernel of all roots by 2.3.4(i). Thus $Y(Z(\mathbf{G})^0) = \Phi^\perp$. This gives (i) since $x \in X(\mathbf{T})$ is in $X(\mathbf{T}/Z(\mathbf{G})^0)$ if and only if it is zero on the images of all $y \in Y(Z(\mathbf{G})^0)$.

We prove (ii). Note that $\mathbf{T} \cap \mathrm{Der}\,\mathbf{G}$ is a maximal torus of $\mathrm{Der}\,\mathbf{G}$ since, being the centraliser in $\mathrm{Der}\,\mathbf{G}$ of a torus, it is connected.

By 2.3.8 we have $\mathbf{T} = (\mathbf{T} \cap \mathrm{Der}\,\mathbf{G})Z(\mathbf{G})^0$. As seen in the proof of 2.3.5 the groups $\mathbf{G}$ and $\mathrm{Der}(\mathbf{G})$ have same roots and coroots. Since by 1.4.9 $\mathrm{Der}\,\mathbf{G}$ is semi-simple, the coroots span $Y(\mathbf{T} \cap \mathrm{Der}\,\mathbf{G}) \otimes \mathbb{Q}$ by 2.3.5(iv), that is $Y(\mathbf{T} \cap \mathrm{Der}(\mathbf{G})) = (\langle \Phi^\vee \rangle \otimes \mathbb{Q}) \cap Y(\mathbf{T})$. This gives the result since $(\langle \Phi^\vee \rangle \otimes \mathbb{Q}) \cap Y(\mathbf{T}) = (\Phi^{\vee\perp})^\perp$. $\square$

We will use the following notation.

Notation 2.3.11 *Suppose we are given two algebraic varieties* $\mathbf{X}$ *and* $\mathbf{Y}$*, and a finite group G acting on the right on* $\mathbf{X}$ *and on the left on* $\mathbf{Y}$*; we will denote by* $\mathbf{X} \times_G \mathbf{Y}$ *the quotient of* $\mathbf{X} \times \mathbf{Y}$ *by the diagonal (left) action of G where $g \in G$ acts by (g^{-1}, g).*

Proposition 2.3.12 *A connected semi-simple reductive group* $\mathbf{G}$ *is quasi-simple if and only if its root system is irreducible – which is equivalent to the Weyl group being irreducible by Proposition 2.2.10(ii).*

Proof Using 2.3.1(iv) it is easy to see that if the root system Φ of $\mathbf{G}$ with respect to $\mathbf{T}$ has a non-trivial orthogonal decomposition $\Phi_1 \coprod \Phi_2$, then each Φ_i gives rise to a non-central proper normal connected subgroup.

Conversely, assume that $\mathbf{G} = \mathbf{G}_1 \times_Z \mathbf{G}_2$, an almost direct product with $Z = \mathbf{G}_1 \cap \mathbf{G}_2$ a finite central subgroup. Using the isomorphism $\mathbf{G}/Z \simeq \mathbf{G}_1/Z \times \mathbf{G}_2/Z$, it is easy to see that maximal tori of $\mathbf{G}$ are of the form $\mathbf{T}_1 \times_Z \mathbf{T}_2$ where

$\mathbf{T}_1$ (resp. $\mathbf{T}_2$) is a maximal torus of $\mathbf{G}_1$ (resp. $\mathbf{G}_2$), and that the set of roots of $\mathbf{G}$ relative to $\mathbf{T}$ is the orthogonal union of the set of roots of $\mathbf{G}_1$ relative to $\mathbf{T}_1$ and of the set of roots of $\mathbf{G}_2$ relative to $\mathbf{T}_2$, so if $\mathbf{G}$ is not quasi-simple its root system is not irreducible.   □

Example 2.3.13   We determine the root system of the general linear group $\mathbf{GL}_n$ (see 1.5.1) relative to the maximal torus $\mathbf{T}$ of diagonal matrices.

Let $\mathbf{U}_{i,j}$ be the group isomorphic to $\mathbb{G}_a$ given by $\{I_n + \lambda E_{i,j} \mid \lambda \in k^+\}$ where $E_{i,j}$ is an elementary matrix as in 1.5.3. This group is normalised by $\mathbf{T}$, thus is a root subgroup $\mathbf{U}_\alpha$. Let $e_i \in X(\mathbf{T})$ be the character mapping $t = \mathrm{diag}(t_1, \ldots, t_n)$ to t_i; we have $X(\mathbf{T}) \simeq \mathbb{Z}^n$ with basis (e_i), in which $\Phi = \{e_i - e_j\}_{1 \le i,j \le n, i \ne j}$ form a root system of type A_{n-1}. The action of t on $\mathbf{U}_{i,j}$ is by t_i/t_j, thus $\mathbf{U}_{i,j} = \mathbf{U}_\alpha$ where $\alpha = e_i - e_j$. Let $\mathbf{U}$ be the nilpotent group of unipotent upper triangular matrices; that is, upper triangular matrices whose diagonal entries are all 1. This group is isomorphic as a variety to an affine space of dimension $n(n-1)/2$. The $n(n-1)/2$ positive roots, for the order defined by the Borel subgroup $\mathbf{B} = \mathbf{U} \rtimes \mathbf{T}$, are all the positive roots, as they are in number equal to $\dim(\mathbf{U})$. The simple roots for this order is the set Π of 2.2.12. The normaliser $N_{\mathbf{G}}(\mathbf{T})$ is the subgroup of monomial matrices, and the Weyl group is isomorphic to the symmetric group $\mathfrak{S}_n$ acting on $\mathbf{T}$ by permuting the diagonal entries.

Example 2.3.14   We determine the root system of $\mathbf{Sp}_{2n}$ (see 1.5.4) relative to the maximal torus $\mathbf{T}$ of diagonal symplectic matrices.

Let $\mathbf{B}$ be the group of upper triangular symplectic matrices; its unipotent radical $\mathbf{U}$ is the subgroup of unipotent matrices – that is, the group of all matrices

$$\begin{pmatrix} B & BJS \\ 0 & J^t B^{-1} J \end{pmatrix}$$

where S is symmetric and B upper triangular and unipotent.

We denote by i,j elements in $1, \ldots, n$ and by i' and j' the corresponding elements in the second set of indices $1', \ldots, n'$. We denote by $\mathbf{U}_{i,j}$ (resp. $\mathbf{U}_{i,j'}$, resp. $\mathbf{U}_{i,i'}$) the group of unipotent matrices whose only non-zero off-diagonal entries are in positions (i,j) and (j',i') and have opposed values (resp. whose only non-zero off-diagonal entries are in positions (i,j') and (j,i') and have equal values, resp. whose only non-zero off-diagonal entry is in position (i,i')). These are all subgroups of $\mathbf{U}$ isomorphic to $\mathbb{G}_a$, normalised by $\mathbf{T}$; the element $t = \mathrm{diag}(t_1, \ldots, t_n, t_n^{-1}, \ldots, t_1^{-1}) \in \mathbf{T}$ acts by multiplication by t_i/t_j (resp. $t_i t_j$, resp. t_i^2). We thus get n^2 distinct root subgroups. They represent all the positive roots since by the above description $\mathbf{U}$ is also of dimension n^2. The simple roots are the α_i (where $1 \le i \le n-1$) which map t to t_i/t_{i+1}, and α_n, which maps t to t_n^2. The root system thus obtained has type C_n.

Example 2.3.15   We now look at $\mathbf{SO}_{2n}$ in characteristic different from 2. We use the notation of Example 1.5.5 except that we number the basis of k^{2n} by $1, 2, \ldots, n, n', \ldots, 2', 1'$. We denote by **T** the maximal torus of orthogonal diagonal matrices. The unipotent radical **U** of **B** consists of matrices $\begin{pmatrix} B & BJ'A \\ 0 & J'^tB^{-1}J' \end{pmatrix} \in$ **B** where B is upper triangular unipotent and A is antisymmetric. Its dimension is thus $n(n-1)$. The root subgroups are the groups $\mathbf{U}_{i,j}$ with $i < j$ (resp. $\mathbf{U}_{i,j'}$ with $i \neq j$) of matrices whose only non-zero off-diagonal entries are in positions (i, j) and (j', i') (resp. (i, j') and (j, i')) and have opposed values. The element $t = \mathrm{diag}(t_1, \ldots, t_n, t_n^{-1}, \ldots, t_1^{-1}) \in \mathbf{T}$ acts on $\mathbf{U}_{i,j}$ (resp. $\mathbf{U}_{i,j'}$) through the root t_i/t_j (resp. $t_i t_j$). We thus get a set of $n^2 - n$ distinct positive roots of $\mathbf{SO}_{2n}$ with respect to **T**; it is the whole set of positive roots since the dimension of **U** is $n^2 - n$. The simple roots are $\alpha_i(t) = t_i/t_{i+1}$ for $1 \leq i \leq n-1$ and $\alpha_n(t) = t_{n-1}t_n$. If $n \geq 2$ the root system thus obtained has type D_n; the centre of the group is the intersection of the kernels of the roots, which is $\pm I_{2n}$; so $\mathbf{SO}_{2n}$ is semi-simple.

Example 2.3.16   We now look at $\mathbf{SO}_{2n+1}$. We use also the notation of Example 1.5.5 but we number the basis by $1, 2, \ldots, n, n+1, n', \ldots, 2', 1'$. We denote by **T** the maximal torus of diagonal special orthogonal matrices. The unipotent radical **U** of **B** consists of matrices $\begin{pmatrix} B & -BJ'^t w & BJ'A \\ 0 & 1 & w \\ 0 & 0 & J'^tB^{-1}J' \end{pmatrix} \in$ **B** where B is upper triangular unipotent and A is antisymmetric. Its dimension is thus n^2. The root subgroups are the groups $\mathbf{U}_{i,j}$ with $1 \leq i < j \leq n$, (resp. $\mathbf{U}_{i,j'}$ with $1 \leq i \neq j \leq n$, resp. $\mathbf{U}_{i,n+1}$ with $1 \leq i \leq n$) of matrices whose only non-zero off-diagonal entries are in positions (i, j) and (j', i') (resp. (i, j') and (j, i'), resp. $(i, n+1)$ and $(n+1, i')$) and have opposed values. The element $t = \mathrm{diag}(t_1, \ldots, t_n, 1, t_n^{-1}, \ldots, t_1^{-1}) \in \mathbf{T}$ acts on $\mathbf{U}_{i,j}$ (resp. $\mathbf{U}_{i,j'}$, resp. $\mathbf{U}_{i,n+1}$) through the root t_i/t_j (resp. $t_i t_j$, resp. t_i). We thus get a set of n^2 distinct positive roots of $\mathbf{SO}_{2n+1}$ with respect to **T**; it is the whole set of positive roots since the dimension of **U** is n^2. The simple roots are $\alpha_i(t) = t_i/t_{i+1}$ for $1 \leq i \leq n-1$ and $\alpha_n(t) = t_n$. If $n \geq 2$ the root system thus obtained has type B_n; the centre of the group is the intersection of the kernels of the roots, which is trivial so $\mathbf{SO}_{2n+1}$ is semi-simple.

2.4 Root Data, Isogenies, Presentation of G

If **G** is a connected reductive group and **T** is a maximal torus, we call **root datum** of **G** attached to **T** the quadruple $(X, Y, \Phi, \Phi^\vee)$ where $X = X(\mathbf{T})$, $Y = Y(\mathbf{T})$ and Φ (resp. $\Phi^\vee$) are the roots (resp. coroots) of **G** relative to **T**. Abstractly we define the following:

Definition 2.4.1   A **root datum** is a quadruple $(X, Y, \Phi, \Phi^\vee)$ where X and Y are dual lattices, and $\Phi \subset X$ is given with a bijection $\alpha \mapsto \alpha^\vee : \Phi \to \Phi^\vee \subset Y$ which defines a root system in the vector space $X \otimes \mathbb{Q}$. The root datum is **crystallographic**, resp. **reduced**, if the root system Φ is crystallographic.

Theorem 2.4.2   *Let k be an algebraically closed field. Any reduced crystallographic root datum is the root datum of a connected reductive group over k; this group is unique up to isomorphism.*

Reference For a proof of this theorem of Chevalley, see Springer (1998, 10.1.1).
                                                                                          □

    This theorem gives a classification of connected reductive groups over algebraically closed fields.

Definition 2.4.3   A root datum $(X, Y, \Phi, \Phi^\vee)$ is:

  (i) **Semi-simple** if Φ spans $X \otimes \mathbb{Q}$, or equivalently $\Phi^\vee$ spans $Y \otimes \mathbb{Q}$.
 (ii) **Adjoint** if Φ spans X over $\mathbb{Z}$.
(iii) **Simply connected** if $\Phi^\vee$ spans Y over $\mathbb{Z}$.

    The above definition of semi-simple agrees with 1.4.6; see 2.3.5(iii). We say that **G** is **adjoint** if its root datum is adjoint.

Proposition 2.4.4   *If **G** is adjoint then $Z(\mathbf{G}) = 1$. If $Z(\mathbf{G}) = 1$ and the characteristic is p, then the quotient $X(\mathbf{T})/\langle \Phi \rangle$ is a p-group.*

Proof By 1.2.4 the restriction maps surjectively $X(\mathbf{T})$ onto the characters of $Z(\mathbf{G})$. Hence if Φ spans $X(\mathbf{T})$ over $\mathbb{Z}$, by 2.3.4(i) any character of $Z(\mathbf{G})$ is trivial so that $Z(\mathbf{G}) = \{1\}$. Conversely if there exists $x \in X(\mathbf{T})$ whose image is a p'-element in $X(\mathbf{T})/\langle \Phi \rangle$, then by a similar argument to 1.2.13, there exists a basis $\{x, x_1, \ldots, x_r\}$ of $X(\mathbf{T})$ such that $\{nx, x_1, \ldots, x_r\}$ is a basis of $\langle \Phi \rangle$ with n prime to p. Hence we can find an element $t \in \mathbf{T}$ such that $x(t) \neq 1$, $x_i(t) = 1$ for all i and $x(t^n) = 1$ so that $\alpha(t) = 1$ for all $\alpha \in \Phi$, that is t is a non-trivial element of $Z(\mathbf{G})$.   □

    A semi-simple algebraic group **G** is said to be **simply connected** if its root datum is simply connected.

Example 2.4.5   Let $\mathbf{G} = \mathbf{SL}_p$ over $\overline{\mathbb{F}}_p$. Then $Z(\mathbf{G}) = \{1\}$ and $X(\mathbf{T})/\langle \Phi \rangle \simeq \mathbb{Z}/p\mathbb{Z}$.

Isogenies

Definition 2.4.6   An **isogeny** is a surjective morphism of algebraic groups with finite kernel.

The kernel of an isogeny is central since any finite normal subgroup of **G** is central; indeed, conjugacy being continuous is trivial on a finite, thus discrete, group. A central subgroup of a reductive group is contained in every maximal torus.

Definition 2.4.7   Let p be a prime number or 0. A p-**morphism** of root data, $(X, Y, \Phi, \Phi^\vee) \xrightarrow{f} (X_1, Y_1, \Phi_1, \Phi_1^\vee)$ is an injective morphism $X_1 \xrightarrow{f} X$ with finite cokernel inducing a bijection $\Phi \xrightarrow{\sigma} \Phi_1$ such that $f(\sigma(\alpha)) = q_\alpha \alpha$ and $f^\vee(\alpha^\vee) = q_\alpha \sigma(\alpha)^\vee$ where $f^\vee : Y \to Y_1$ is the transposed of f and where q_α is a power of p ($q_\alpha = 1$ if $p = 0$).

Theorem 2.4.8   *Let* **G** $\xrightarrow{\phi}$ **G**$_1$ *be an isogeny between reductive groups over* k, *let* p *be the characteristic of* k *and set* **T**$_1 := \phi(\mathbf{T})$. *Then* ϕ *induces a p-morphism* $(X(\mathbf{T}), Y(\mathbf{T}), \Phi, \Phi^\vee) \to (X_1(\mathbf{T}_1), Y_1(\mathbf{T}_1), \Phi_1, \Phi_1^\vee)$ *where* σ *and the* q_α *are determined by the formula* $\phi(\mathbf{u}_\alpha(x)) = \mathbf{u}_{\sigma(\alpha)}(\lambda_\alpha x^{q_\alpha})$ *for some non-zero scalars* λ_α. *Conversely, every p-morphism is induced by an isogeny, unique up to conjugacy by an element of* **T**.

Proof of the first part   The isogeny ϕ induces $X(\mathbf{T}_1) \xrightarrow{f} X(\mathbf{T})$ given by $\alpha \mapsto \alpha \circ \phi$ and $Y(\mathbf{T}) \xrightarrow{f^\vee} Y(\mathbf{T}_1)$ given by $\alpha^\vee \mapsto \phi \circ \alpha^\vee$. If $\mathbf{U}_\alpha$ is a root subgroup of **G**, then $\phi(\mathbf{U}_\alpha)$ is a root subgroup $\mathbf{U}_{\sigma(\alpha)}$ of **G**$_1$; this defines a bijection σ. We define a polynomial P by $\phi(\mathbf{u}_\alpha(x)) = \mathbf{u}_{\sigma(\alpha)}(P(x))$; the compatibility with the action of $t \in \mathbf{T}$ gives $\phi({}^t\mathbf{u}_\alpha(x)) = \phi(\mathbf{u}_\alpha(\alpha(t)x)) = \mathbf{u}_{\sigma(\alpha)}(P(\alpha(t)x))$ and $\phi({}^t\mathbf{u}_\alpha(x)) = {}^{\phi(t)}\mathbf{u}_{\sigma(\alpha)}(P(x)) = \mathbf{u}_{\sigma(\alpha)}(\sigma(\alpha)(\phi(t))P(x))$ whence $P(\alpha(t)x) = \sigma(\alpha)(\phi(t))P(x)$ which implies that P is a monomial; the compatibility with the group law of $\mathbb{G}_a$ gives $P(x + y) = P(x) + P(y)$. This forces $P = \lambda x^{q_\alpha}$ where q_α is a power of $p = \operatorname{char} k$ and λ a constant ($q_\alpha = 1$ if $\operatorname{char} k = 0$). The constants λ are not 0 since the kernel of ϕ is finite; they can be changed by composing ϕ with the conjugation by an element of **T**.

For a proof of the second part, see Springer (1998, 9.6.5). $\square$

We give now an example of isogeny, defined by the corresponding p-morphism. We will see more examples in Chapter 4.

Example 2.4.9   The **opposition automorphism**: $q_\alpha = 1$ and $\sigma(\alpha) = -\alpha$ for all α. It is the conjugation ad $\dot{w}_0$ by a representative of w_0 if w_0 is central in W, and transpose $\circ$ inverse $\circ$ ad $\dot{w}_0$ in **GL**$_n$, **SL**$_n$ or **PGL**$_n$.

Example 2.4.10　An **automorphism of a root system** – that is a simultaneous permutation of the roots and coroots preserving the correspondence $\alpha \mapsto \alpha^\vee$ – is, for any prime p, a p-morphism with σ being the automorphism and $q_\alpha = 1$.

Presentation of G(k)

We sketch here a presentation of **G**(k), due to Steinberg – see Steinberg (2016) – where **G** is a connected reductive group **G** over k with root datum $(X, Y, \Phi, \Phi^\vee)$ – which is used to prove its existence. We first define a torus **T** by $\mathbf{T}(k) = \mathrm{Hom}(X, k^\times)$. Here for $\alpha \in X, t \in \mathbf{T}(k)$, the element $\alpha(t) \in k$ is the image of α by t. An element $\alpha^\vee \in Y$ defines $\alpha^\vee : k^\times \to \mathbf{T}$ by $\alpha^\vee(x)(\beta) = x^{\alpha^\vee(\beta)}$.

For each $\alpha \in \Phi$ we construct a copy $\mathbf{U}_\alpha$ of $\mathbb{G}_a$ with a given isomorphism $k^+ \to \mathbf{U}_\alpha(k) : x \mapsto \mathbf{u}_\alpha(x)$.

Theorem 2.4.11　*The group* **G**(k) *is the group with generators* $\{t \in \mathbf{T}\}$, $\{\mathbf{u}_\alpha(x)\}_{\alpha \in \Phi, x \in k}$ *presented, if we let* $s_\alpha = \mathbf{u}_\alpha(1)\mathbf{u}_{-\alpha}(1)^{-1}\mathbf{u}_\alpha(1)$, *by the relations:*

(i) $\mathbf{u}_\alpha(x)\mathbf{u}_\alpha(y) = \mathbf{u}_\alpha(x + y)$.
(ii) $s_\alpha\mathbf{u}_\alpha(x)s_\alpha^{-1} = \mathbf{u}_{-\alpha}(x)^{-1}$.
(iii) $\mathbf{u}_\alpha(x)\mathbf{u}_{-\alpha}(x^{-1})^{-1}\mathbf{u}_\alpha(x) = \alpha^\vee(x)s_\alpha$.
(iv) $t\mathbf{u}_\alpha(x)t^{-1} = \mathbf{u}_\alpha(\alpha(t)x)$.
(v) $s_\alpha^2 \in \mathbf{T}, \beta(s_\alpha^2) = (-1)^{\alpha^\vee(\beta)}$.
(vi) $s_\alpha s_\beta \ldots = s_\beta s_\alpha \ldots$ *(the braid relations of* $W(\mathbf{T})$*)*.
(vii) $[\mathbf{u}_\alpha(x), \mathbf{u}_\beta(y)] = \prod_{i\alpha+j\beta \in \Phi, i,j>0} \mathbf{u}_{i\alpha+j\beta}(c_{\alpha,\beta,i,j}x^i y^j)$, *where the* $c_{\alpha,\beta,i,j}$ *are integral constants.*

The $c_{\alpha,\beta,i,j}$ depend on the choice of a subset of $\Phi^+ \times \Phi^+$; when changing this subset their sign may vary, but their absolute value depends only on Φ.

Notes

A classic reference for properties of Coxeter groups and root systems is Bourbaki (1968).

The construction and classification of reductive groups was first done around 1955, see Chevalley (2005).

3

(B,N)-Pairs; Parabolic, Levi, and Reductive Subgroups; Centralisers of Semi-Simple Elements

3.1 (B,N)-Pairs

We review properties of reductive groups related to existence of a (B,N)-pair. For an abstract group, having a (B,N)-pair is a very strong condition; many of the theorems we will give for reductive groups follow from this single property.

Definition 3.1.1   We say that two subgroups B and N of a group G form a **(B,N)-pair** (also called a **Tits system**) for G if:

 (i) B and N generate G and $T := B \cap N$ is normal in N.
(ii) The group $W := N/T$ is generated by a set S of involutions such that:

    (a) For $s \in S$, $w \in W$ we have $BsB.BwB \subset BwB \cup BswB$.
    (b) For $s \in S$, we have $sBs \nsubseteq B$.

The group W is called the **Weyl group** of the (B,N)-pair. Note that we write elements of W – instead of representatives of them in N – in expressions representing subsets of G when these expressions do not depend upon the chosen representative.

We will see in 3.1.3(v) that under the assumptions of 3.1.1 we have $S = \{w \in W - \{1\} \mid B \cup BwB$ is a group$\}$, thus S is determined by (B,N).

Proposition 3.1.2   *If* $\mathbf{G}$ *is a connected reductive group and* $\mathbf{T} \subset \mathbf{B}$ *is a pair of a maximal torus and a Borel subgroup, then* $(\mathbf{B}, N_{\mathbf{G}}(\mathbf{T}))$ *is a (B,N)-pair for* $\mathbf{G}$.

Proof   We show first that $\mathbf{B} \cap N_{\mathbf{G}}(\mathbf{T}) = \mathbf{T}$. By 1.3.1(iii) we have $N_{\mathbf{B}}(\mathbf{T}) = C_{\mathbf{B}}(\mathbf{T}) \subset C_{\mathbf{G}}(\mathbf{T}) = \mathbf{T}$ (see 2.3.1(iii)). By definition $\mathbf{T}$ is normal in $N_{\mathbf{G}}(\mathbf{T})$. To prove (i) it remains to show that $\mathbf{B}$ and $N_{\mathbf{G}}(\mathbf{T})$ generate $\mathbf{G}$. Let Φ^+ be the positive subsystem defined by $\mathbf{B}$. By 2.3.1(vi), $\mathbf{B}$ contains all the $\mathbf{U}_\alpha$ ($\alpha \in \Phi^+$). Since s_α conjugates $\mathbf{U}_\alpha$ to $\mathbf{U}_{s_\alpha(\alpha)} = \mathbf{U}_{-\alpha}$, the group generated by $\mathbf{B}$ and $N_{\mathbf{G}}(\mathbf{T})$ contains $\mathbf{T}$ and all the $\mathbf{U}_\alpha$ ($\alpha \in \Phi$), thus by 2.3.1(v) this group is equal to $\mathbf{G}$.

If Π is the basis defined by the ordering Φ^+, (ii) is obtained by taking for S the $\{s_\alpha \mid \alpha \in \Pi\}$.

(ii)(b) reflects that ${}^{s_\alpha}\mathbf{U}_\alpha = \mathbf{U}_{-\alpha}$ is not in $\mathbf{B}$.

It remains to show (ii)(a). Let $s = s_\alpha$, and write $\mathbf{B} = \mathbf{T}\prod_{\beta\in\Phi^+}\mathbf{U}_\beta$. As s normalises $\mathbf{T}$, as ${}^s\mathbf{U}_\beta = \mathbf{U}_{s_\alpha(\beta)}$ and as $s_\alpha(\beta) \in \Phi^+$ if $\beta \in \Phi^+ - \{\alpha\}$, we get $\mathbf{B}s\mathbf{B}w\mathbf{B} = \mathbf{B}s\mathbf{U}_\alpha w\mathbf{B}$. If $w^{-1}(\alpha) \in \Phi^+$ the right hand side is equal to $\mathbf{B}sw\mathbf{B}$. Otherwise we write it as $\mathbf{B}s\mathbf{U}_\alpha ssw\mathbf{B}$ where this time $(sw)^{-1}(\alpha) \in \Phi^+$. Let $\mathbf{B}_\alpha$ be the image by ϕ_α (see 2.3.1(ii)) of the Borel subgroup of $\mathbf{SL}_2$ of upper triangular matrices. If $c \neq 0$ we have in $\mathbf{SL}_2$:

$$\begin{pmatrix} a & b \\ c & d \end{pmatrix} = \begin{pmatrix} -1/c & -a \\ 0 & -c \end{pmatrix}\begin{pmatrix} 0 & 1 \\ -1 & 0 \end{pmatrix}\begin{pmatrix} 1 & d/c \\ 0 & 1 \end{pmatrix}$$

which taking images shows that $s\mathbf{U}_\alpha s \subset \operatorname{Im}\phi_\alpha = \mathbf{B}_\alpha \cup \mathbf{B}_\alpha s\mathbf{U}_\alpha$, whence $\mathbf{B}s\mathbf{U}_\alpha ssw\mathbf{B} \subset \mathbf{B}s\mathbf{U}_\alpha sw\mathbf{B} \cup \mathbf{B}sw\mathbf{B}$ where the first term in the right-hand side is $\mathbf{B}w\mathbf{B}$ since $(sw)^{-1}(\alpha) \in \Phi^+$. □

Theorem 3.1.3 *If G has a (B,N)-pair, then*

(i) $G = \coprod_{w\in W} BwB$ (**Bruhat decomposition**).

(ii) (W,S) *is a Coxeter group.*

(iii) *Condition (ii)(a) of 3.1.1 can be refined to*

$$BsB.BwB = \begin{cases} BswB & \text{if } l(sw) = l(w) + 1, \\ BswB \cup BwB & \text{otherwise.} \end{cases}$$

(iv) *For any $t \in N(w)$ (see 2.1.2(ii)), we have $BtB \subset Bw^{-1}BwB$.*

(v) $S = \{w \in W - \{1\} \mid B \cup BwB \text{ is a group}\}$.

(vi) *We have $N_G(B) = B$.*

Proof Let us show (i). As B and N generate G, we have $G = \cup_i (BNB)^i$. Since $BNB = BWB$ we will get $G = BWB$ if we show that $BWBWB = BWB$. For this it is enough to show that $BwBWB \subset BWB$ for $w \in W$; writing $w = s_1 \ldots s_n$ with $s_i \in S$, since $BwB \subset Bs_1B\ldots Bs_nB$ it is enough to show $BsBWB \subset BWB$ for $s \in S$; but this results from 3.1.1(ii)(a). It remains to show that $BwB \neq Bw'B$ if $w \neq w'$. We show this by induction on $\inf(l(w), l(w'))$, where l is the length with respect to S; assume for instance that $l(w) \leq l(w')$. The start of the induction is $l(w) = 0$ and the result comes from $w' \notin B$. Otherwise, taking $s \in S$ such that $l(sw) < l(w)$, by induction $BswB$ is equal neither to $Bw'B$ nor to $Bsw'B$ thus $BswB \cap BsB.Bw'B = \emptyset$; as $BswB \subset BsB.BwB$ it follows that $BwB \neq Bw'B$.

For (ii), we use 2.2.9 with $D_s = \{w \in W \mid BsBwB = BswB\}$ (note that if $w \notin D_s$ then $BsBwB = BswB \coprod BwB$). Clearly $D_s \ni 1$.

If w, $sw \in D_s$, then from $BsBwB = BswB$ and $BsBswB = BwB$ we get $BsBsBwB = BwB$ which is a contradiction since multiplying on the right by BwB the equality $BsBsB = BsB \coprod B$ (since $sBs \not\subset B$ by 3.1.1(ii)(b)), we get $BsBsBwB = BswB \coprod BwB$.

It remains to show for (ii) that $w \in D_s, ws' \notin D_s$ implies $ws' = sw$. The assumption $ws' \notin D_s$ implies $BsBws'B = Bsws'B \coprod Bws'B$; in particular $BsBws'$ meets $Bws'B$; multiplying on the right by $s'B$ it follows that $BsBwB$ meets $Bws'Bs'B \subset (BwB \coprod Bws'B)$ (this last inclusion follows from 3.1.1(ii)(a) reversed, which is obtained by taking inverses). Thus $BswB = BsBwB$ (since $w \in D_s$) is equal to $Bws'B$, or to BwB. The latter cannot happen since $w \neq sw$, thus $sw = ws'$ as was to be shown. We have also shown (iii) by the property of D_s given in the last sentence of 2.2.9.

Let us show (iv). If $w = s_1 \dots s_k$ is a reduced expression, for all i we can write by (iii) $BwB = Bs_1 \dots s_{i-1}Bs_iBs_{i+1} \dots s_kB$ and similarly for $Bw^{-1}B$ whence

$$Bw^{-1}BwB = Bs_k \dots s_{i+1}Bs_iBs_{i-1} \dots s_1Bs_1 \dots s_{i-1}Bs_iBs_{i+1} \dots s_kB$$
$$\supset Bs_k \dots s_{i+1}Bs_iBs_iBs_{i+1} \dots s_kB$$
$$\supset Bs_k \dots s_{i+1}Bs_iBs_{i+1} \dots s_kB$$
$$\supset Bs_k \dots s_{i+1}s_is_{i+1} \dots s_kB$$

whence the result.

(v) follows immediately from (iv), which implies that $B \cup BwB$ can be a group only if $|N(w)| \leq 1$, and from (iii) which implies that $B \cup BsB$ is a group.

(vi) also follows from (iv). For $g \in BwB$ we have $^gB = B \Leftrightarrow {}^wB = B \Leftrightarrow BwBw^{-1}B = B$ which by (iv) happens only for $w = 1$. □

In a group G with a (B,N)-pair, we call **Borel subgroups** the conjugates of B and **maximal tori** the conjugates of T; this fits the terminology for algebraic groups.

Corollary 3.1.4 *In a group G with a (B,N)-pair, every pair of Borel subgroups is conjugate to a pair of the form $(B, {}^wB)$ with $w \in W$; the intersection of two Borel subgroups contains a maximal torus.*

Proof Up to conjugacy, we may assume the given pair of Borel subgroups of the form $(B, {}^gB)$. By the Bruhat decomposition we may write $g = bwb'$ where $b,b' \in B$; thus the pair is equal to $(B, {}^{bw}B)$, which is conjugate to $(B, {}^wB)$. Since B and ^wB both contain T, the intersection of every conjugate pair also contains a maximal torus. □

Example 3.1.5 For m a matrix in $\mathbf{GL}_n$, let $m_{i,j}$ be the submatrix on the last lines $i, \dots, n$ and first columns $1, \dots, j$. Let w be a permutation matrix; then

$m \in \mathbf{B}w\mathbf{B}$, where $\mathbf{B}$ is the Borel subgroup of upper triangular matrices, if and only if the matrices $m_{i,j}$ and $w_{i,j}$ have same rank for all i,j. Indeed,

- The ranks of $m_{i,j}$ are invariant by left or right multiplication of m by an upper triangular matrix.
- A permutation matrix w for the permutation σ is characterised by the ranks of $w_{i,j}$, given by $|\{k \leq j \mid \sigma(k) \geq i\}|$.

If $\{F'_i\}$ and $\{F''_i\}$ are two complete flags whose stabilisers are the Borel subgroups $\mathbf{B}'$ and $\mathbf{B}''$, then the permutation matrix w such that $(\mathbf{B}',\mathbf{B}'')$ is conjugate to $(\mathbf{B},{}^w\mathbf{B})$ (the **relative position** of the two flags) is characterised by rank $w_{i,j} = \dim \frac{F'_i \cap F''_j}{(F'_{i-1} \cap F''_j) + (F'_i \cap F''_{j-1})}$.

3.2 Parabolic Subgroups of Coxeter Groups and of (B, N)-Pairs

Lemma 3.2.1 *Let (W,S) be a Coxeter system, let I be a subset of S, and let W_I be the subgroup of W generated by I, called a* **standard parabolic subgroup** *of W. Then (W_I, I) is a Coxeter system.*

An element $w \in W$ is said to be **reduced-I** *if it satisfies one of the equivalent conditions:*

(i) *For any $v \in W_I$, we have $l(wv) = l(w) + l(v)$.*
(ii) *For any $s \in I$, we have $l(ws) > l(w)$.*
(iii) *w has minimal length in the coset wW_I.*
(iv) *$N(w) \cap I = \emptyset$.*
(v) *$N(w) \cap \mathrm{Ref}(W_I) = \emptyset$.*

There is a unique reduced-I element in wW_I.

By exchanging left and right we have the notion of *I-reduced* element which satisfies the mirror properties. A subgroup of W conjugate to a standard parabolic subgroup is called a **parabolic subgroup**.

Proof A reduced expression in W_I is reduced in W by the exchange condition and then satisfies the exchange condition in W_I, thus (W_I, I) is a Coxeter system.

(iii)$\Rightarrow$(ii) since (iii) implies $l(ws) \geq l(w)$ when $s \in I$. Let us show that "not (iii)"$\Rightarrow$ "not (ii)". If w' does not have minimal length in $w'W_I$, then $w' = wv$ with $v \in W_I$ and $l(w) < l(w')$; adding one by one the terms of a reduced expression for v to w and applying at each stage the exchange condition, we find that w' has a reduced expression of the shape $\hat{w}\hat{v}$ where $\hat{w}$ (resp. $\hat{v}$) denotes a subsequence of the chosen reduced expression. As $l(\hat{w}) \leq l(w) < l(w')$, we have $l(\hat{v}) > 0$, thus w' has a reduced expression ending by an element of I, thus w' does not satisfy (ii).

(i)$\Rightarrow$(iii) is clear. Let us show "not (i)"$\Rightarrow$ "not (iii)". If $l(wv) < l(w) + l(v)$ then a reduced expression for wv has the shape $\hat{w}\hat{v}$ where $l(\hat{w}) < l(w)$. Then $\hat{w} \in wW_I$ and has a length smaller than that of w.

By 2.1.6(ii) property (ii) is equivalent to (iv).

It is clear that (v) implies (iv), and (i) applied to $v \in \mathrm{Ref}(W)$ implies (v) by 2.1.6(ii).

Finally, an element satisfying (i) is clearly unique in wW_I. □

Lemma 3.2.2 *Let I and J be two subsets of S. An element $w \in W$ is **I-reduced-J** if it satisfies one of the equivalent properties:*

(i) *w is both I-reduced and reduced-J.*

(ii) *w has minimal length in $W_I w W_J$.*

(iii) *Every element of $W_I w W_J$ can be written uniquely xwy with $x \in W_I$, $y \in W_J$, $l(x) + l(w) + l(y) = l(xwy)$ and xw is reduced-J.*

(iii) implies that in a double coset in $W_I \backslash W / W_J$ there is a unique I-reduced-J element, which has minimal length; by symmetry we can replace in condition (iii) the assumption that xw is reduced-J by the assumption that wy is I-reduced.

Proof We first show that two elements w, w' in the same double coset and satisfying (i) have the same length. Write $w' = xwy$ with $x \in W_I$ and $y \in W_J$; then $w'y^{-1} = xw$ and $x^{-1}w' = wy$; by the defining properties of I-reduced and reduced-J and using $l(y^{-1}) = l(y)$, $l(x^{-1}) = l(x)$ we get $l(w') + l(y) = l(x) + l(w)$ and $l(x) + l(w') = l(w) + l(y)$, whence $l(x) = l(y)$ and $l(w) = l(w')$. As clearly (ii)$\Rightarrow$(i) this common length must be the minimal length, thus (i)$\Leftrightarrow$(ii).

We now show (ii)$\Rightarrow$(iii). Assume w satisfies (ii); write an element $v \in W_I w W_J$ as xwy with $x \in W_I$, $y \in W_J$ and x of minimal possible length. By the exchange lemma a reduced expression for xwy is of the form $\hat{x}\hat{w}\hat{y}$ where $\hat{x}$ (resp. $\hat{w}$, $\hat{y}$) is a subsequence of a reduced expression for x (resp. w, y). Necessarily $\hat{w} = w$ otherwise w would not be of minimal length in its double coset. Then the minimal length assumption on x implies $\hat{x} = x$, whence $\hat{y} = y$, thus $l(x) + l(w) + l(y) = l(xwy)$. The element xw is reduced-J otherwise we can write $xw = v'y'$ where $v' \in W_I w W_J$, $y' \in W_J - \{1\}$ and $l(v') + l(y') = l(xw)$. Using what we just proved on w we can write $v' = x''wy''$ with $l(x'') + l(w) + l(y'') + l(y') = l(x) + l(w)$ which implies $l(x'') < l(x)$, contradicting the minimality of $l(x)$. Finally the decomposition xwy is unique since xw is the unique J-reduced element in its coset.

Finally, (iii)$\Rightarrow$(ii) is clear. □

Note that not every decomposition xwy where w is I-reduced-J satisfies (iii); consider for instance the case $w = y = 1$, $I = J$ and x the longest element of W_I; thus the situation is not as good as in the I-reduced case.

In a group with a (B,N)-pair, we use the term **parabolic subgroups** for the subgroups containing a Borel subgroup.

Proposition 3.2.3   *Let G be a group with a (B,N)-pair. Then*

(i) *The (parabolic) subgroups containing B are the $P_I = BW_I B$ for some $I \subset S$.*
(ii) *Given two parabolic subgroups P_I and P_J, we have a **relative Bruhat decomposition** $G = \coprod_w P_I w P_J$ where w runs over the I-reduced-J elements. It follows a natural bijection $P_I \backslash G / P_J \xrightarrow{\sim} W_I \backslash W / W_J$.*

Proof   Let us show (i). Let P be a subgroup containing B and let $w \in W$ be such that $BwB \subset P$. Since P is a group we get $Bw^{-1}BwB \subset P$, thus by 3.1.3(iv) we get $BtB \subset P$ for any $t \in N(w)$. If $w = s_1 \ldots s_k$ is a reduced expression we get in particular $Bs_k B \subset P$, thus $s_1 \ldots s_{k-1} \in P$ and by induction for each i we have $s_i \in P$. It follows that $P = BW_I B$ where I is the union of the elements of S appearing in any reduced expression of any w such that $BwB \subset P$. Conversely, for any $I \subset S$, using 3.1.1(ii)(a) we see that $BW_I B$ is a group.

Let us show (ii). For any $w \in W$ we have $P_I w P_J = BW_I BwBW_J B = BW_I wW_J B$, the last equality by repeated application of 3.1.1(ii)(a) and of its right counterpart. Since, by Lemma 3.2.2 we can take I-reduced-J elements as representatives of the double cosets we see that the first assertion of (ii) is just the Bruhat decomposition. Conversely, any coset $P_I g P_J$ is of the form $P_I w P_J$ if $g \in BwB$ whence the last assertion of (ii).   □

Remark 3.2.4   Using 3.2.3 we see that in the definition 1.3.5 of a parabolic subgroup the word "closed" can be omitted. Indeed a reductive group has a (B,N) pair, hence by 3.2.3 a subgroup containing a Borel subgroup is conjugate to some $\mathbf{B}W_I\mathbf{B}$, hence it is closed. In general, if $\mathbf{G}$ is a connected group and $\mathbf{P}$ is a subgroup containing a Borel subgroup, then $\mathbf{P}/R_u(\mathbf{G})$ contains a Borel subgroup of the reductive group $\mathbf{G}/R_u(\mathbf{G})$ hence it is closed, thus $\mathbf{P}$ is closed by continuity of the quotient morphism.

Example 3.2.5   In $\mathbf{GL}_n$, the parabolic subgroup $\mathbf{P}_J$ for $J \subset S$ containing the Borel subgroup of upper triangular matrices is the subgroup of upper block-triangular matrices where the blocks correspond to maximal intervals $[i,k]$ in $[1,n]$ such that $s_i, \ldots, s_{k-1} \in J$.

Example 3.2.6   For the symplectic group $\mathbf{Sp}_{2n}$, as the stabiliser $\mathbf{B}$ of any complete isotropic flag $V_1 \subset \cdots \subset V_n$ in $\mathbf{Sp}_{2n}$ is a Borel subgroup, the stabiliser of any subflag is a parabolic subgroup. We thus get 2^n distinct parabolic subgroups containing $\mathbf{B}$. Since there are also 2^n subsets of S, they are the only parabolic subgroups containing **B**. As any isotropic flag may be completed to a complete one, we get the result that in general parabolic subgroups are the stabilisers of (complete or not) isotropic flags.

Lemma 3.2.7 (unicity in Bruhat decomposition) *Let **G** be a connected reductive group and* $\mathbf{B} = \mathbf{U} \rtimes \mathbf{T}$ *be a decomposition of a Borel subgroup **B** as in 1.3.1(ii), where* $\mathbf{U} = R_u(\mathbf{B})$. *Then* $\mathbf{B}w\mathbf{B}$ *has a direct product decomposition* $\mathbf{U} \times \mathbf{T}w \times \mathbf{U}_w$ *where* $\mathbf{U}_w := \prod_{\{\alpha \in \Phi^+ \mid w(\alpha) < 0\}} \mathbf{U}_\alpha$.

Proof Notice first that $\mathbf{U}_w$ is a group; since if in 2.3.1(vii) α and β are sent to negative roots by w, then the same holds for $\lambda\alpha + \mu\beta$. We have $\mathbf{U} = \mathbf{U}'\mathbf{U}_w$ where $\mathbf{U}' = \prod_{\{\alpha \in \Phi^+ \mid w(\alpha) > 0\}} \mathbf{U}_\alpha$ thus $^w\mathbf{U}' \subset \mathbf{U}$; thus $\mathbf{B}w\mathbf{B} = \mathbf{U}\mathbf{T}w\mathbf{U}'\mathbf{U}_w = \mathbf{U}\mathbf{T}w\mathbf{U}_w$. It remains to be shown that the decomposition is unique; that is, if $u\mathbf{T}wu' = \mathbf{T}w$ with $u \in \mathbf{U}, u' \in \mathbf{U}_w$ then $u = u' = 1$. The condition implies $^wu' \in \mathbf{B}$. But $^w\mathbf{U}_w \cap \mathbf{B} = 1$ since all $\mathbf{U}_\alpha$ in $^w\mathbf{U}_w$ are for negative α. Thus $u' = 1$, whence $u = 1$. $\qquad\square$

The next proposition says that the decomposition of **G** in **Bruhat cells** $\mathbf{B}w\mathbf{B}$ is a stratification (the closure of a stratum is a union of strata).

Proposition 3.2.8 *Let **G** be a connected reductive group and* $\mathbf{B} = \mathbf{U} \rtimes \mathbf{T}$ *be a decomposition of a Borel subgroup **B** as in 3.2.7. Then the Zariski closure of* $\mathbf{B}w\mathbf{B}$ *in **G** is given by* $\overline{\mathbf{B}w\mathbf{B}} = \coprod_{v \leq w} \mathbf{B}v\mathbf{B}$, *where* $\leq$ *is the* **Bruhat–Chevalley order** *on* w, *given by* $v \leq w$ *if a reduced expression of* v *is a subsequence of a reduced expression of* w.

Reference See Chevalley (1994, Proposition 6). $\qquad\square$

3.3 Closed Subsets of a Crystallographic Root System

In this section, Φ will be a reduced crystallographic root system in the $\mathbb{Q}$-vector space V, and Π will be a basis of Φ; we denote by Φ^+ the corresponding positive subsystem and by (W, S) the corresponding Coxeter system, where $S = \{s_\alpha\}_{\alpha \in \Pi}$.

Definition 3.3.1 A subset $\Psi \subset \Phi$ is:

(i) **closed** if $\alpha, \beta \in \Psi, \alpha + \beta \in \Phi \Rightarrow \alpha + \beta \in \Psi$.
(ii) **symmetric** if $\Psi = -\Psi$.

The intersection of two closed subsets is clearly closed.

Lemma 3.3.2 *The reduced crystallographic root systems of rank 2 are* $A_1 \times A_1$, A_2, $C_2 = B_2$ *and* G_2.

Here is a picture of their positive roots:

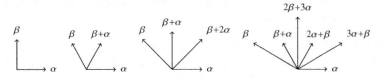

Proof   Let Φ be crystallographic of rank 2 with Weyl group W. Let $\Pi = \{\alpha, \beta\}$. Choosing a W-invariant scalar product $(.,.)$ as in 2.2.3, we have $\alpha^\vee(\beta)\beta^\vee(\alpha) = 4\frac{(\alpha,\beta)^2}{(\alpha,\alpha)(\beta,\beta)} = 4\cos^2\theta$ where θ is the angle between α and β. Since $\alpha, \beta \in \Pi$ we have $(\alpha, \beta) \le 0$ thus $\pi/2 \le \theta \le \pi$ and the integrality of $4\cos^2\theta$ implies that $4\cos^2\theta \in \{0, 1, 2, 3\}$ thus $\pi - \theta \in \{\pi/2, \pi/3, \pi/4, \pi/6\}$. Except for $A_1 \times A_1$, the ratio of the lengths of α and β is implied by the equation $\alpha^\vee(\beta)\beta^\vee(\alpha) = 4\cos^2\theta$. For instance if $4\cos^2\theta = 2$ the only integral solution, up to exchanging α and β, is $\alpha^\vee(\beta) = -1$ and $\beta^\vee(\alpha) = -2$ whence $2(\beta, \beta) = (\alpha, \alpha)$. For $A_1 \times A_1$ the ratio of the lengths is not determined, we have chosen 1 in the picture.   □

Corollary 3.3.3   *For a crystallographic root system Φ and a positive subsystem Φ^+, we have:*

  (i) *If $\alpha, \beta \in \Phi$, $\alpha \ne -\beta$ and $(\alpha, \beta) < 0$, then $\alpha + \beta \in \Phi$.*
  (ii) *If $\alpha, \beta \in \Phi$ and $\alpha + n\beta \in \Phi$ for $n \in \mathbb{N}$, then $\alpha + m\beta \in \Phi$ for all $0 \le m \le n$.*
  (iii) *If $\alpha_1, \ldots, \alpha_k \in \Phi^+$ and $\alpha = \alpha_1 + \cdots + \alpha_k \in \Phi^+$, then if $k > 1$ we have $\alpha - \alpha_i \in \Phi^+$ for some i.*
  (iv) *If $\Psi \subset \Phi$ is closed, $\alpha, \beta \in \Psi$, $\alpha \ne -\beta$ and $n\alpha + m\beta \in \Phi$ for some $n, m > 0$, then $n\alpha + m\beta \in \Psi$.*

Proof   For (i), by the argument in the proof of 3.3.2 about possible integral solutions, up to exchanging α and β we have $\alpha^\vee(\beta) = -1$, whence $\alpha + \beta = s_\alpha(\beta) \in \Phi$. For (iii) as $(\alpha, \alpha) > 0$ we must have $(\alpha, \alpha_i) > 0$ for some i thus by (i) $\alpha - \alpha_i \in \Phi$.

For (ii), by (iii) either $\alpha + (n-1)\beta$ or $n\beta$ is in Φ, and since Φ is reduced, $n\beta \notin \Phi$ if $n \ne 1$, whence the result by induction on n.

For (iv) we may assume both α and β positive (they are for some order since $\alpha \ne -\beta$), and then we apply (iii) and induction on $n + m$.   □

Corollary 3.3.4   *If $\Psi \subset \Phi$ is closed and symmetric, it is a root subsystem.*

Proof   For $\alpha, \beta \in \Psi$, we have to show that $s_\alpha(\beta) \in \Psi$. This is true if $\beta = \pm\alpha$ since Ψ is symmetric. Otherwise, replacing α by $-\alpha$ if necessary we have $s_\alpha(\beta) = \beta + n\alpha$ for some $n \in \mathbb{N}^*$; then Corollary 3.3.3(iv) gives the result.   □

Proposition 3.3.5   *If Ψ is closed and $\Psi \cap -\Psi = \emptyset$, there exists a positive subsystem Φ^+ such that $\Psi \subset \Phi^+$.*

Proof We first show by induction on $k > 0$ that 0 is not the sum of k elements of Ψ. This is clear for $k = 1$. If $0 = \alpha_1 + \cdots + \alpha_k$ then $0 < (\alpha_1, \alpha_1) = (-\alpha_1, \alpha_2 + \cdots + \alpha_k)$ thus there exists $i \neq 1$ such that $(\alpha_1, \alpha_i) < 0$. Using $\alpha_1 \neq -\alpha_i$ (since $-\alpha_i \notin \Psi$ by assumption) and 3.3.3(i) we get $\alpha_1 + \alpha_i \in \Phi$ thus $\alpha_1 + \alpha_i \in \Psi$, thus the sum is the sum of $k - 1$ elements, a contradiction.

We now build by induction on k a sequence γ_k of elements of Ψ such that $\gamma_k \in \Psi$ is the sum of k elements of Ψ. We start with γ_1 equal to an arbitrary element of Ψ. If there is $\alpha \in \Psi$ such that $(\gamma_k, \alpha) < 0$ we set $\gamma_{k+1} = \alpha + \gamma_k \in \Psi$. For $i < j$ we have $\gamma_i \neq \gamma_j$, otherwise $\gamma_j - \gamma_i$ would be a zero sum of elements of Ψ, thus by finiteness the sequence must stop on some γ_k such that $(\gamma_k, \alpha) \geq 0$ for any $\alpha \in \Psi$. The linear form $(\gamma_k, .)$ almost defines an order as in 2.2.4. We need to modify it on $\gamma_k^\perp$. But $\gamma_k^\perp \cap \Psi \subset \gamma_k^\perp \cap \Phi$ satisfies the same assumptions as the proposition and we may iterate the construction on this subspace. □

For $I \subset S$ we set $\Pi_I := \{\alpha \in \Pi \mid s_\alpha \in I\}$ and $\Phi_I = \Phi \cap \mathbb{Q}\Pi_I$; it is clearly a root subsystem with basis Π_I, since when decomposed on Π a root of Φ_I involves only elements of Π_I.

It is clear that Φ_I is closed and symmetric and that $\Phi^+ - \Phi_I$ and $\Phi^+ \cup \Phi_I$ are closed.

Example 3.3.6 There exist closed and symmetric subsystems which are not of the form Φ_I; for instance the long roots in a system B_2 form a system of type $A_1 \times A_1$, and the long roots in G_2 form a system of type A_2. See also 11.2.7.

Lemma 3.3.7 *If $s_\alpha \in W_I$ for $\alpha \in \Phi$, then $\alpha \in \Phi_I$.*

Proof Elements of W_I are the product of some s_β for $\beta \in \Pi_I$, thus they fix $\Pi_I^\perp$. Thus s_α fixes $\Pi_I^\perp$, which implies that $\alpha \in \mathbb{Q}\Pi_I \cap \Phi = \Phi_I$. □

We say that Ψ is a **parabolic** subset of Φ if Ψ is closed and $\Psi \cup -\Psi = \Phi$.

Proposition 3.3.8

(i) *A parabolic subset is conjugate to a parabolic subset containing Φ^+; such a subset is of the form $\Phi^+ \cup \Phi_I$ for some $I \subset S$.*
(ii) *A parabolic subset is a set of the form $\{\alpha \mid \lambda(\alpha) \geq 0\}$ for some linear form λ on V.*

Proof For the first part of (i) it is equivalent to show that a parabolic subset Ψ contains some positive subsystem. Choose such a positive subsystem Φ^+ such that $|\Psi \cap \Phi^+|$ is maximal. We show by contradiction that $\Phi^+ \subset \Psi$. Otherwise let Π be the basis of Φ defining Φ^+; there must exist $\alpha \in \Pi, \alpha \notin \Psi$, thus $-\alpha \in \Psi$. Since $\alpha \notin \Psi$ we have $s_\alpha(\Psi \cap \Phi^+) \subset \Phi^+$; applying s_α again we get $\Psi \cap \Phi^+ \subset s_\alpha(\Phi^+)$. But then the positive subsystem $s_\alpha(\Phi^+)$ contains $-\alpha$ thus satisfies $|\Psi \cap s_\alpha(\Phi^+)| > |\Psi \cap \Phi^+|$, a contradiction.

We now assume that $\Psi \supset \Phi^+$. Let $I = \{s_\alpha \mid -\alpha \in \{-\Pi \cap \Psi\}\}$. Let us show that $\Psi \cap \Phi^- = \Phi_I^-$.

We first show that $\Phi_I^- \subset \Psi$. Note that by 2.2.8 applied to the basis $-\Pi_I$ of Φ_I^- any root in Φ_I^- is a sum of elements of $-\Pi_I$. We show by induction on k that a root in Φ_I^- sum of k roots in $-\Pi_I$ is in Ψ. It is true by assumption when $k = 1$; in general by 3.3.3(iii) we may write the root as $\alpha + \beta$ where $\alpha \in -\Pi_I$ and $\beta \in \Phi_I^-$ sum of $k - 1$ roots in $-\Pi_I$; by induction $\beta \in \Psi$ and as $\alpha \in \Psi$ and Ψ is closed $\alpha + \beta \in \Psi$.

We finally show the reverse inclusion by induction. Let $\gamma \in \Psi \cap \Phi^-$ be the sum of k roots of $-\Pi$, and write it $\gamma = \alpha + \beta$ where $\alpha \in -\Pi$ and $\beta \in \Phi$ is the sum of $k - 1$ roots in $-\Pi$. As $-\beta \in \Phi^+ \subset \Psi$ we get $\alpha = \gamma + (-\beta) \in \Psi$ whence $\alpha \in -\Pi \cap \Psi = -\Pi_I$. Thus $-\alpha \in \Psi$ whence $\beta = \gamma + (-\alpha) \in \Psi$, and we conclude since by induction $\beta \in \Phi_I^-$.

Conversely the fact that for any $I \subset S$ the set $\Phi^+ \cup \Phi_I$ is parabolic is a consequence of the proof of (ii) below.

We now show (ii). It is clear that a subset of the form $\{\alpha \mid \lambda(\alpha) \geq 0\}$ is parabolic. It is thus sufficient to show that $\Phi^+ \cup \Phi_I$ is of this form. Take any x such that $\langle x, \alpha \rangle = 0$ if $\alpha \in \Phi_I$ and $\langle x, \alpha \rangle > 0$ if $\alpha \in \Psi - \Phi_I$. Such an x exists: the projection of Φ^+ on $\Phi_I^\perp$ lies in a half-space, and we may take x in this half-space, orthogonal to the hyperplane which delimits it. It is clear that by construction x has the required properties. $\quad\square$

A consequence of 3.3.8 is that the complement of a parabolic subset is closed.

Subgroups of Maximal Rank and Quasi-closed Sets

In the remainder of this chapter $\mathbf{G}$ is a connected reductive algebraic group, $\mathbf{T}$ is a maximal torus of $\mathbf{G}$, and Φ is the set of roots of $\mathbf{G}$ relative to $\mathbf{T}$. For $\Psi \subset \Phi$, we set $\mathbf{G}_\Psi^* := \langle \mathbf{U}_\alpha \mid \alpha \in \Psi \rangle$ and $\mathbf{G}_\Psi := \langle \mathbf{T}, \mathbf{U}_\alpha \mid \alpha \in \Psi \rangle$. These are closed connected subgroups by 1.1.3 and $\mathbf{G}_\psi^*$ is a normal subgroup of $\mathbf{G}_\Psi$.

Definition 3.3.9 A subset $\Psi \subset \Phi$ is called **quasi-closed** if $\mathbf{G}_\psi^*$ does not contain any $\mathbf{U}_\alpha$ with $\alpha \in \Phi - \Psi$.

We get an equivalent definition by replacing $\mathbf{G}_\Psi^*$ with $\mathbf{G}_\Psi$, since $\mathbf{G}_\Psi / \mathbf{G}_\Psi^*$ is a quotient of $\mathbf{T}$ and thus is a torus. Hence any $\mathbf{U}_\alpha \subset \mathbf{G}_\Psi$ is in the kernel of this quotient, and is thus in $\mathbf{G}_\psi^*$.

Proposition 3.3.10 *A closed and connected subgroup* $\mathbf{H} \subset \mathbf{G}$ *containing* $\mathbf{T}$ *is equal to* $\mathbf{G}_\Psi$ *with* $\Psi = \{\alpha \in \Phi \mid \mathbf{U}_a \subset \mathbf{H}\}$; *the set* Ψ *is quasi-closed.*

Proof By 2.3.1(iv) **H** is generated by **T** and the $\mathbf{U}_\alpha$ it contains. The subset $\Psi \subset \Phi$ of those α is quasi-closed by definition. $\square$

Let Ψ, Ψ' be quasi-closed; it is clear that $\Psi \cap \Psi'$ is quasi-closed (since $\mathbf{G}_{\Psi \cap \Psi'}$ is a subgroup of both $\mathbf{G}_\Psi$ and $\mathbf{G}_{\Psi'}$); actually one sees that $\mathbf{G}_{\Psi \cap \Psi'} = (\mathbf{G}_\Psi \cap \mathbf{G}_{\Psi'})^0$ by applying 2.3.1(iv) to the right-hand side.

Definition 3.3.11 A connected linear algebraic group **P** has a **Levi decomposition** if there is a closed subgroup $\mathbf{L} \subset \mathbf{P}$ such that $\mathbf{P} = R_u(\mathbf{P}) \rtimes \mathbf{L}$. The group **L** is called a **Levi subgroup** of **P** (or a **Levi complement**).

A Levi complement is clearly reductive.

Proposition 3.3.12 *Let* $\Psi \subset \Phi$ *be quasi-closed, and let* $\Psi_s = \{\alpha \in \Psi \mid -\alpha \in \Psi\}$ *and* $\Psi_u = \{\alpha \in \Psi \mid -\alpha \notin \Psi\}$. *Then* Ψ_s *and* Ψ_u *are quasi-closed and* $\mathbf{G}_\Psi$ *has a Levi decomposition* $\mathbf{G}_\Psi = \mathbf{G}^*_{\Psi_u} \rtimes \mathbf{G}_{\Psi_s}$ *where* $\mathbf{G}^*_{\Psi_u} = R_u(\mathbf{G}_\Psi)$. *In particular* $\mathbf{G}_\Psi$ *is reductive if and only if* Ψ *is symmetric.*

Proof We first show that Ψ_s is quasi-closed. As the intersection of two quasi-closed sets is quasi-closed, it is enough to show that $-\Psi$ is quasi-closed. This results from the existence of the opposition automorphism of **G** which acts by -1 on $X(\mathbf{T})$; see Example 2.4.9.

As a connected group normalised by **T** the group $R_u(\mathbf{G}_\Psi)$ is – by Theorem 2.3.1(v) – of the form $\mathbf{G}^*_{\Psi'}$ for a subset $\Psi' \subset \Psi$ that we may assume quasi-closed. We have $\Psi' \subset \Psi_u$, otherwise there is $\alpha \in \Psi_s \cap \Psi'$, thus $\mathbf{U}_{-\alpha} \subset \mathbf{G}_\Psi$ thus normalises $R_u(\mathbf{G}_\Psi)$ thus $[\mathbf{U}_{-\alpha}, \mathbf{U}_\alpha] \subset R_u(\mathbf{G}_\Psi)$ which is a contradiction since this commutator set contains non-unipotent elements by Theorem 2.3.1(ii).

To show $\Psi' = \Psi_u$ it is thus enough to show $\Psi - \Psi' \subset \Psi_s$. If $\alpha \in \Psi - \Psi'$, then $\mathbf{U}_\alpha \cap R_u(\mathbf{G}_\Psi) = 1$ since this intersection is normalised by **T** thus contains the whole $\mathbf{U}_\alpha$ if not trivial. Thus, in the quotient $\mathbf{G}_\Psi \to \mathbf{L}'$, where $\mathbf{L}'$ is the reductive group $\mathbf{G}_\Psi/R_u(\mathbf{G}_\Psi)$, the group $\mathbf{U}_\alpha$ maps injectively to a root subgroup of $\mathbf{L}'$. Let $\mathbf{U}'$ be the root subgroup of $\mathbf{L}'$ corresponding to the opposed root and $\mathbf{U}''$ its preimage. Any element of $\mathbf{U}''$ is unipotent since, its image being unipotent, its semi-simple part is in $R_u(\mathbf{G}_\Psi)$ so is trivial. Hence $\mathbf{U}''$ is a unipotent subgroup normalised by **T**, so is a product of certain root subgroups and must contain $\mathbf{U}_{-\alpha}$, thus $-\alpha \in \Psi$ and $\alpha \in \Psi_s$.

It also follows from the proof that $\mathbf{G}_{\Psi_s}$ maps injectively to $\mathbf{L}'$, thus $\mathbf{G}_{\Psi_s}$ is a Levi complement of $R_u(\mathbf{G}_\Psi)$. $\square$

Proposition 3.3.13 *A closed subset is quasi-closed.*

Proof Let $\Psi \subset \Phi$ be a closed subset and let Ψ_s and Ψ_u be as in 3.3.12. It is clear that Ψ_s is closed. Note that if $\alpha \in \Psi$, $\beta \in \Psi_u$ and $\alpha + \beta \in \Phi$ then $\alpha + \beta \in \Psi_u$, otherwise $\alpha + \beta \in \Psi_s$ whence $-\alpha - \beta \in \Psi_s$ thus $\alpha + (-\alpha - \beta) = -\beta \in \Psi$ which contradicts $\beta \in \Psi_u$. In particular Ψ_u is closed. By 3.3.5 there exists a positive subsystem such that $\Psi_u \subset \Phi^+$.

Lemma 3.3.14 *If Ψ is a closed subset of a positive subsystem Φ^+ of Φ, then Ψ is quasi-closed and $\mathbf{G}_\Psi^* = \prod_{\alpha \in \Psi} \mathbf{U}_\alpha$ where the product is taken in an arbitrary order.*

In the situation of the lemma we will write $\mathbf{U}_\Psi$ for $\mathbf{G}_\Psi^*$.

Proof By 2.3.1(vii) and 3.3.3(iv) $\prod_{\alpha \in \Psi} \mathbf{U}_\alpha$ is a group, thus equal to $\mathbf{G}_\Psi^*$.   □

We deduce that Ψ_u is quasi-closed. In addition, if $\alpha \in \Psi_s$, $\beta \in \Psi_u$ and $\alpha + \beta \in \Phi$, using the fact that $\alpha + \beta \in \Psi_u$ and 3.3.3(iv), we get $n\alpha + m\beta \in \Psi_u$ for $n, m \geq 1$ such that $n\alpha + m\beta \in \Phi$. Thus $\mathbf{G}_{\Psi_u}^*$ is normalised by $\mathbf{G}_{\Psi_s}$.

Since Ψ_s is closed and symmetric, it is a root subsystem by 3.3.4. Let Π_s be its basis corresponding to the positive subsystem $\Psi_s \cap \Phi^+$. Note that $\mathbf{G}_{\Psi_s}$ is already generated by $\mathbf{T}$ and $\mathbf{U}_\alpha$ such that $\alpha \in \pm\Pi_s$; indeed $\langle \mathbf{U}_\alpha, \mathbf{U}_{-\alpha} \rangle$ contains a representative of s_α by 2.3.1(iii), thus $\mathbf{G}_{\Psi_s}$ contains W_{Ψ_s}, and every root of Ψ_s is in the orbit of Π_s by 2.2.8, whence the result by the remark above 2.3.3. We show now that $\mathbf{G}_{\Psi_s} = \mathbf{U}_{\Psi_s^+} W_{\Psi_s} \mathbf{T} \mathbf{U}_{\Psi_s^+}$. For that it is enough to show that the right-hand side is a group. Since it is stable by left translation by $\mathbf{T}$ and by any $\mathbf{U}_\alpha$ for $\alpha \in \Psi_s^+$ it is enough to see it is stable by left translation by $\mathbf{U}_{-\alpha}$ for $\alpha \in \Pi_s$. Decomposing $\mathbf{U}_{\Psi_s^+} = \mathbf{U}_{\Psi_s^+ - \{\alpha\}} \mathbf{U}_\alpha$, and using that by 2.3.1(vii) $\mathbf{U}_{-\alpha}$ normalises $\mathbf{U}_{\Psi_s^+ - \{\alpha\}}$ since α is simple, it is enough to see that $\mathbf{U}_\alpha W_{\Psi_s} \mathbf{T} \mathbf{U}_{\Psi_s^+}$ is stable by left translation by $\mathbf{U}_{-\alpha}$. The Bruhat decomposition $\langle \mathbf{T}, \mathbf{U}_\alpha, \mathbf{U}_{-\alpha} \rangle = \mathbf{U}_\alpha \mathbf{T} \cup \mathbf{U}_\alpha s_\alpha \mathbf{T} \mathbf{U}_\alpha$ shows that $\mathbf{U}_{-\alpha} \mathbf{U}_\alpha \subset \mathbf{U}_\alpha \mathbf{T} \cup \mathbf{U}_\alpha s_\alpha \mathbf{T} \mathbf{U}_\alpha$. We just need to consider the second term

$$\mathbf{U}_\alpha s_\alpha \mathbf{T} \mathbf{U}_\alpha W_{\Psi_s} \mathbf{T} \mathbf{U}_{\Psi_s^+} = \bigcup_{w \in \Psi_s} \mathbf{U}_\alpha s_\alpha \mathbf{U}_\alpha w \mathbf{T} \mathbf{U}_{\Psi_s^+}.$$

If $w^{-1}(\alpha) \in \Psi^+$, then $\mathbf{U}_\alpha w \mathbf{T} = w \mathbf{T} \mathbf{U}_{w^{-1}(\alpha)}$ and the term has the right form. Otherwise, letting $\beta = -w^{-1}(\alpha) \in \Psi_s^+$ we get

$$\mathbf{U}_\alpha s_\alpha w \mathbf{T} \mathbf{U}_{w^{-1}(\alpha)} \mathbf{U}_{\Psi_s^+} = \mathbf{U}_\alpha s_\alpha w \mathbf{T} \mathbf{U}_{-\beta} \mathbf{U}_\beta \mathbf{U}_{\Psi_s^+ - \{\beta\}}$$
$$\subset \mathbf{U}_\alpha s_\alpha w \mathbf{T} (\mathbf{U}_\beta \cup \mathbf{U}_\beta s_\beta \mathbf{U}_\beta) \mathbf{U}_{\Psi_s^+ - \{\beta\}}$$
$$= \mathbf{U}_\alpha s_\alpha w \mathbf{T} \mathbf{U}_{\Psi_s^+} \cup \mathbf{U}_\alpha s_\alpha w \mathbf{T} \mathbf{U}_\beta s_\beta \mathbf{U}_{\Psi_s^+}.$$

We just need to consider the rightmost term. Since $s_\alpha w \mathbf{U}_\beta \mathbf{T} = s_\alpha \mathbf{U}_{-\alpha} w \mathbf{T} = \mathbf{U}_\alpha s_\alpha w \mathbf{T}$ we get the result.

Let us now show that Ψ_s is quasi-closed. Let γ be such that $\mathbf{U}_\gamma \subset \mathbf{G}_{\Psi_s}$; since $\Psi_s = -\Psi_s$, we may assume $\gamma \in \Phi^+$, thus $\mathbf{U}_\gamma \subset \mathbf{B}$. As each term $\mathbf{U}_{\Psi_s^+} w \mathbf{T} \mathbf{U}_{\Psi_s^+}$ is in a unique Bruhat cell of $\mathbf{G}$, we must have $\mathbf{U}_\gamma \subset \mathbf{T} \mathbf{U}_{\Psi_s^+}$. By Lemma 3.3.14 Ψ_s^+ is quasi-closed, thus $\gamma \in \Psi_s^+$.

We have seen that $\mathbf{G}_\Psi$ has a semi-direct product decomposition $\mathbf{G}^*_{\Psi_u} \rtimes \mathbf{G}_{\Psi_s}$. It follows that Ψ is quasi-closed since if $\alpha \notin \Psi_u$ and $\mathbf{U}_\alpha \subset \mathbf{G}_\Psi$ then $\mathbf{U}_\alpha$ maps isomorphically to the quotient $\mathbf{G}_{\Psi_s}$ thus $\alpha \in \Psi_s$. □

Conversely any quasi-closed subset of Φ is closed apart from some exceptions in characteristics 2 and 3; see Borel and Tits (1965, 3.8). This can be shown by proving that in other characteristics the group $\langle \mathbf{U}_\alpha, \mathbf{U}_\beta \rangle$ contains all $\mathbf{U}_{n\alpha+m\beta}$ for $n, m \in \mathbb{N}$ such that $n\alpha + m\beta \in \Phi$, using the explicit values of the coefficients in the proof of 2.3.1(vii). For a combinatorial description of quasi-closed subsets in these characteristics, see Malle and Testerman (2011, Corollary 13.7).

3.4 Parabolic Subgroups and Levi Subgroups

Proposition 3.4.1 *The parabolic subgroup* $\mathbf{P}_I = \mathbf{B} W_I \mathbf{B}$ *(see 3.2.3) has a Levi decomposition* $\mathbf{P}_I = R_u(\mathbf{P}_I) \rtimes \mathbf{L}_I$ *where* $R_u(\mathbf{P}_I) = \mathbf{U}_{\Phi^+ - \Phi_I}$ *and* $\mathbf{L}_I = \langle \mathbf{T}, \{\mathbf{U}_\alpha\}_{\alpha \in \Phi_I} \rangle$ *is reductive. We have* $\mathbf{P}_I = N_{\mathbf{G}}(R_u(\mathbf{P}_I))$.

Proof The set $\Psi = \Phi^+ \cup \Phi_I$ is quasi-closed since it is closed by Proposition 3.3.8. The proposition is then a consequence of 3.3.12 if we show that $\mathbf{P}_I = \mathbf{G}_\Psi$. We have $\mathbf{P}_I \supset \mathbf{G}_\Psi$ since $\mathbf{P}_I \supset \mathbf{U}_\alpha$ for $\alpha \in \Phi^+$ since $\mathbf{P}_I \supset \mathbf{B}$, and by 3.2.7 $\mathbf{P}_I$ contains all $\mathbf{U}_\alpha$ for $\alpha \in \Phi^-$ which change sign by some element of W_I, thus contains all $\mathbf{U}_\alpha$ for $\alpha \in \Phi_I^-$. Conversely $\mathbf{G}_\Psi$ contains $\mathbf{U}_\alpha$ and $\mathbf{U}_{-\alpha}$ for all $\alpha \in \Pi_I$, hence $\mathbf{G}_\Psi$ contains a representative of s_α in $\langle \mathbf{U}_\alpha, \mathbf{U}_{-\alpha} \rangle$ hence $\mathbf{G}_\Psi$ contains W_I, thus contains $\mathbf{P}_I$.

Finally, $N_{\mathbf{G}}(R_u(\mathbf{P}_I))$ contains $\mathbf{P}_I$ thus $\mathbf{B}$, thus is some parabolic subgroup $\mathbf{P}_J$. If $J \supsetneq I$ we have $R_u(\mathbf{P}_J) = \mathbf{U}_{\Phi^+ - \Phi_J} \subsetneq R_u(\mathbf{P}_I)$ which contradicts that $\mathbf{P}_J$ normalises $R_u(\mathbf{P}_I)$ since $R_u(\mathbf{P}_J)$ is the largest normal connected unipotent subgroup of $\mathbf{P}_J$. □

We will say that $\mathbf{P}_I$ (resp. $\mathbf{L}_I$) is a **standard** parabolic subgroup (resp. Levi subgroup) of $\mathbf{G}$.

Proposition 3.4.2 *Let* $\mathbf{P}$ *be a parabolic subgroup of* $\mathbf{G}$ *containing* $\mathbf{T}$.

(i) *There is a unique Levi subgroup of* $\mathbf{P}$ *containing* $\mathbf{T}$.
(ii) *Two Levi subgroups of* $\mathbf{P}$ *are conjugate by a unique element of* $R_u(\mathbf{P})$.

Proof The existence of a Levi subgroup containing $\mathbf{T}$ results from 3.4.1, since $\mathbf{P}$ is conjugate to some $\mathbf{P}_I$ and all maximal tori of $\mathbf{P}_I$ are conjugate in $\mathbf{P}_I$. Conversely, we may assume $\mathbf{P} = \mathbf{P}_I$; any Levi subgroup of $\mathbf{P}_I$ containing $\mathbf{T}$ is a $\mathbf{G}_\Psi$ for some $\Psi \subset \Phi^+ \cup \Phi_I$ by Proposition 3.3.10. Since any $\mathbf{U}_\alpha$ where $\alpha \in \Phi^+ - \Phi_I$ is in $R_u(\mathbf{P}_I)$, we must have $\Psi \subset \Phi_I$, thus $\mathbf{L} \subset \mathbf{L}_I$, thus there must be equality.

Two Levi subgroups $\mathbf{L}, \mathbf{L}'$ of $\mathbf{P}$ are conjugate in $\mathbf{P}$, since by (i) an element which conjugates a maximal torus $\mathbf{T}$ of $\mathbf{L}$ into $\mathbf{L}'$ conjugates $\mathbf{L}$ to $\mathbf{L}'$. Modulo $\mathbf{L}$, we can choose the conjugating element u in $R_u(\mathbf{P})$. The unicity of u is equivalent to $R_u(\mathbf{P}) \cap N_\mathbf{P}(\mathbf{L}) = 1$. Assume $u \in R_u(\mathbf{P}) \cap N_\mathbf{P}(\mathbf{L})$; then for any $l \in \mathbf{L}$ we have $[u, l] \in R_u(\mathbf{P}) \cap \mathbf{L} = 1$, thus $u \in C_\mathbf{P}(\mathbf{L})$; but $C_\mathbf{P}(\mathbf{L}) \subset C_\mathbf{G}(\mathbf{T}) = \mathbf{T}$ thus $u = 1$.   □

Proposition 3.4.3 *The $\mathbf{G}$-conjugacy classes of Levi subgroups of parabolic subgroups of $\mathbf{G}$ are in bijection with the W-orbits of subsets of S, which are themselves in bijection with the W-conjugacy classes of parabolic subgroups of W.*

Proof Since all parabolic subgroups are conjugate to a $\mathbf{P}_I$, we may assume that we consider a Levi subgroup of some $\mathbf{P}_I$. Since by 3.4.2 such a Levi subgroup is $\mathbf{G}$-conjugate to $\mathbf{L}_I$, the question becomes that of finding when $\mathbf{L}_J$ is $\mathbf{G}$-conjugate to $\mathbf{L}_I$ for two subsets I and J of S. If $\mathbf{L}_J = {}^g\mathbf{L}_I$ for some $g \in \mathbf{G}$, then, since ${}^{g^{-1}}\mathbf{T}$ and $\mathbf{T}$ are two maximal tori of $\mathbf{L}_I$, there exists $l \in \mathbf{L}_I$ such that ${}^{g^{-1}}\mathbf{T} = {}^l\mathbf{T}$ and $gl \in N_\mathbf{G}(\mathbf{T})$ also conjugates $\mathbf{L}_I$ to $\mathbf{L}_J$; so the $\mathbf{G}$-conjugacy classes of $\mathbf{L}_I$ are the same as the $W(\mathbf{T})$-conjugacy classes. Since $\mathbf{L}_I = \mathbf{G}_{\Phi_I}$, the element $w \in W$ conjugates $\mathbf{L}_I$ to $\mathbf{L}_J$ if and only if ${}^w\Phi_I = \Phi_J$. Since any two bases of Φ_I are conjugate by an element of W_I (see 2.2.6), we may assume that ${}^wI = J$ whence the first part of the statement. To see the second part it is enough to see that if $w \in N_W(W_I)$ then ${}^w\Phi_I = \Phi_I$. This results from Lemma 3.3.7.   □

The proof above shows that $N_\mathbf{G}(\mathbf{L}_I)/\mathbf{L}_I$ is isomorphic to $N_W(W_I)/W_I$.

Proposition 3.4.4 *Let $\mathbf{L}$ be a Levi subgroup of a parabolic subgroup $\mathbf{P}$. Then $R(\mathbf{P}) = R_u(\mathbf{P}) \rtimes R(\mathbf{L})$.*

Proof The quotient $\mathbf{P}/(R(\mathbf{L})R_u(\mathbf{P}))$ is isomorphic to $\mathbf{L}/R(\mathbf{L})$, so is semisimple. So $R(\mathbf{P}) \subset R(\mathbf{L})R_u(\mathbf{P})$. But $R(\mathbf{L})R_u(\mathbf{P})$ is connected, solvable and normal in $\mathbf{P}$ as the inverse image of a normal subgroup of the quotient $\mathbf{P}/R_u(\mathbf{P}) \simeq \mathbf{L}$, whence the reverse inclusion.   □

We will now characterise parabolic subgroups in terms of roots.

Proposition 3.4.5 *A closed subgroup $\mathbf{P}$ of $\mathbf{G}$ containing $\mathbf{T}$ is parabolic if and only if $\mathbf{P} = \mathbf{G}_\Psi$ for some parabolic subset Ψ.*

Proof We have seen that a parabolic subset Ψ is conjugate under W to $\Phi^+ \cup \Phi_I$; thus $\mathbf{G}_\Psi$ is conjugate under W to $\mathbf{P}_I$. Conversely, assume that $\mathbf{P}$ is a parabolic subgroup. It contains a Borel subgroup containing $\mathbf{T}$, thus up to conjugacy by W it contains $\mathbf{B}$ (see Proposition 2.3.3), thus is of the form $\mathbf{P}_I$. □

We now give an important property of Levi subgroups.

Proposition 3.4.6 *Let $\mathbf{L}$ be a Levi subgroup of a parabolic subgroup of $\mathbf{G}$; then $\mathbf{L} = C_\mathbf{G}(Z(\mathbf{L})^0)$.*

Proof We may assume that $\mathbf{L} = \mathbf{L}_I$. Then by 2.3.4(i) the group $Z(\mathbf{L})$ is the intersection of the kernels of the roots in Φ_I. The group $C_\mathbf{G}(Z(\mathbf{L})^0)$ is connected as it is the centraliser of a torus – $Z(\mathbf{L})^0$ is diagonalisable by 1.2.1(ii) and is a torus by 1.2.3(i) thus 1.3.3(iii) applies. It is normalised by $\mathbf{T}$ because it contains $\mathbf{T}$, hence by 2.3.1(iv) it is generated by $\mathbf{T}$ and the $\mathbf{U}_\alpha$ it contains. If $\mathbf{U}_\alpha \subset C_\mathbf{G}(Z(\mathbf{L})^0)$ then α is trivial on $(\bigcap_{\alpha \in \Phi_I} \operatorname{Ker} \alpha)^0$. This identity component has finite index in $\bigcap_{\alpha \in \Phi_I} \operatorname{Ker} \alpha$, hence some multiple $n\alpha$ of α is trivial on $\bigcap_{\alpha \in \Phi_I} \operatorname{Ker} \alpha$. With the notation of 1.2.12, this can be rewritten as $n\alpha \in ((\langle \Phi_I \rangle_\mathbf{T}^\perp)_{X(\mathbf{T})}^\perp$. But $((\langle \Phi_I \rangle_\mathbf{T}^\perp)_{X(\mathbf{T})}^\perp / \langle \Phi_I \rangle$ is a torsion group (see 1.2.13). This implies that $\alpha \in \langle \Phi_I \rangle \otimes \mathbb{Q}$, which in turn yields $\alpha \in \Phi_I$ by the definition of Φ_I. This proves that $C_\mathbf{G}(Z(\mathbf{L})^0) \subset \mathbf{L}$. The reverse inclusion is obvious. □

The next proposition is a kind of converse.

Proposition 3.4.7 *For any torus $\mathbf{S}$, the group $C_\mathbf{G}(\mathbf{S})$ is a Levi subgroup of some parabolic subgroup of $\mathbf{G}$.*

Proof Let $\mathbf{T}$ be a maximal torus containing $\mathbf{S}$. As the group $C_\mathbf{G}(\mathbf{S})$ is connected by 1.3.3(iii) and contains $\mathbf{T}$, by 2.3.1(iv) we have $C_\mathbf{G}(\mathbf{S}) = \langle \mathbf{T}, \mathbf{U}_\alpha \mid \mathbf{U}_\alpha \subset C_\mathbf{G}(\mathbf{S}) \rangle$. As $\mathbf{S}$ acts by α on $\mathbf{U}_\alpha$ (see 2.3.1(i)), we have

$$\mathbf{U}_\alpha \subset C_\mathbf{G}(\mathbf{S}) \Leftrightarrow \alpha|_\mathbf{S} = 0,$$

where 0 is the trivial element of $X(\mathbf{S})$. Let us choose a total order on $X(\mathbf{S})$; that is, a structure of ordered $\mathbb{Z}$-module. As $X(\mathbf{S})$ is a quotient of $X(\mathbf{T})$ (see 1.2.4) there exists a total order on $X(\mathbf{T})$ compatible with the chosen order on $X(\mathbf{S})$; that is, such that for $x \in X(\mathbf{T})$ we have $x \geq 0 \Rightarrow x|_\mathbf{S} \geq 0$. This implies that the set $\Psi = \{\alpha \in \Phi \mid \alpha > 0 \text{ or } \alpha|_\mathbf{S} = 0\}$ is also equal to $\{\alpha \in \Phi \mid \alpha|_\mathbf{S} \geq 0\}$. This last definition implies that Ψ is closed, so (see 3.3.13 and 3.3.10) Ψ is also the set of α such that $\mathbf{U}_\alpha \subset \mathbf{G}_\Psi$. Since Ψ is parabolic, it follows then from 3.4.5 that $\mathbf{G}_\Psi$ is a parabolic subgroup, of which $C_\mathbf{G}(\mathbf{S})$ is a Levi complement. □

We now study the intersection of two parabolic subgroups. First note that by 3.1.4 the intersection of two parabolic subgroups always contains some maximal torus of $\mathbf{G}$.

Proposition 3.4.8 *Let $\mathbf{P}$ and $\mathbf{P}'$ be two parabolic subgroups of $\mathbf{G}$ and let $\mathbf{L}$ and $\mathbf{L}'$ be respective Levi subgroups of $\mathbf{P}$ and $\mathbf{P}'$ containing the same maximal torus $\mathbf{T}$ of $\mathbf{G}$. Let $\mathbf{U} = R_u(\mathbf{P})$ and $\mathbf{U}' = R_u(\mathbf{P}')$. Then*

  (i) *The group $(\mathbf{P} \cap \mathbf{P}').\mathbf{U}$ is a parabolic subgroup of $\mathbf{G}$ which has the same intersection as $\mathbf{P}'$ with $\mathbf{L}$, and it has $\mathbf{L} \cap \mathbf{L}'$ as a Levi subgroup.*

  (ii) *The group $\mathbf{P} \cap \mathbf{P}'$ is connected, as well as $\mathbf{L} \cap \mathbf{L}'$ and we have the Levi decomposition*

$$\mathbf{P} \cap \mathbf{P}' = ((\mathbf{L} \cap \mathbf{U}').(\mathbf{L}' \cap \mathbf{U}).(\mathbf{U} \cap \mathbf{U}')) \rtimes (\mathbf{L} \cap \mathbf{L}')$$

where the right-hand side is a direct product of varieties – the decomposition of an element of $\mathbf{P} \cap \mathbf{P}'$ as a product of four terms is unique. On the right-hand side the last factor is a Levi subgroup of $\mathbf{P} \cap \mathbf{P}'$ and the first 3 factors form a decomposition of $R_u(\mathbf{P} \cap \mathbf{P}')$.

Proof Let Φ be the roots of $\mathbf{G}$ with respect to $\mathbf{T}$ and define subsets $\Psi, \Psi' \subset \Phi$ by $\mathbf{P} = \mathbf{G}_\Psi$ and $\mathbf{P}' = \mathbf{G}_{\Psi'}$.

Let us show first that for any $\alpha \in \Phi$, either $\mathbf{U}_\alpha$ or $\mathbf{U}_{-\alpha}$ is in the group $(\mathbf{P} \cap \mathbf{P}') \cdot \mathbf{U}$ (it is a group since $\mathbf{P}$ normalises $\mathbf{U}$). By the remarks before 3.3.11 and by 3.3.12 we have $(\mathbf{P} \cap \mathbf{P}')^0 \cdot \mathbf{U} = \mathbf{G}_{\Psi \cap \Psi'} \cdot \mathbf{G}^*_{\Psi_u}$, with the notation of 3.3.12. If neither α nor $-\alpha$ is in Ψ_u, they are both in Ψ in which case since one of them is in Ψ', one of them is in $\Psi \cap \Psi'$. Hence $(\Psi \cap \Psi') \cup \Psi_u$ is a parabolic set; indeed, this set is closed as the sum of an element of Ψ and an element of Ψ_u which is a root is in Ψ_u – see the beginning of the proof of 3.3.13. Proposition 3.4.5 then shows that $(\mathbf{P} \cap \mathbf{P}')^0.\mathbf{U}$ is a parabolic subgroup of $\mathbf{G}$, equal to $\mathbf{G}_{(\Psi \cap \Psi') \cup \Psi_u}$. Then $(\mathbf{P} \cap \mathbf{P}')\mathbf{U}$, containing a parabolic subgroup is connected, hence equal to $(\mathbf{P} \cap \mathbf{P}')^0.\mathbf{U}$.

Now $(\mathbf{P} \cap \mathbf{P}').\mathbf{U} = (\mathbf{P} \cap \mathbf{P}').\mathbf{G}^*_{\Psi_u - \Psi'}$ since $\Psi' \cap \Psi_u \subset \Psi \cap \Psi'$. The set $\Psi_u - \Psi'$ is closed as the intersection of the closed subsets Ψ_u and the complement $-\Psi'_u$ of Ψ', hence the product $(\mathbf{P} \cap \mathbf{P}').\mathbf{G}^*_{\Psi_u - \Psi'}$ is a direct product of varieties as the intersection is a unipotent subgroup normalised by $\mathbf{T}$ containing no $\mathbf{U}_\alpha$, see 2.3.1(v). As the product is connected, each term is. Thus $\mathbf{P} \cap \mathbf{P}'$ is connected equal to $\mathbf{G}_{\Psi \cap \Psi'}$. The groups $(\mathbf{P} \cap \mathbf{P}').\mathbf{U}$ and $\mathbf{P} \cap \mathbf{P}'$ have both $(\mathbf{L} \cap \mathbf{L}')^0$ as a Levi subgroup since $((\Psi \cap \Psi') \cup \Psi_u)_s = (\Psi \cap \Psi')_s = \Psi_s \cap \Psi'_s$ – indeed if $\alpha \in \Psi_u$ then $-\alpha \notin \Psi$ thus $-\alpha \notin (\Psi \cap \Psi') \cup \Psi_u$.

The decomposition $\Psi \cap \Psi' = (\Psi_s \cap \Psi'_s) \coprod (\Psi_s \cap \Psi'_u) \coprod (\Psi_u \cap \Psi'_s) \coprod (\Psi_u \cap \Psi'_u)$ shows that $\mathbf{P} \cap \mathbf{P}' = \langle \mathbf{L} \cap \mathbf{L}', \mathbf{L} \cap \mathbf{U}', \mathbf{L}' \cap \mathbf{U}, \mathbf{U} \cap \mathbf{U}' \rangle$. Using that $\mathbf{U} \cap \mathbf{U}'$ is normal in $\mathbf{P} \cap \mathbf{P}'$, then that $\mathbf{L} \cap \mathbf{L}'$ normalises $\mathbf{L} \cap \mathbf{U}'$ and $\mathbf{L}' \cap \mathbf{U}$, we get

$$\mathbf{P} \cap \mathbf{P}' = (\mathbf{L} \cap \mathbf{L}').\langle \mathbf{L} \cap \mathbf{U}', \mathbf{L}' \cap \mathbf{U} \rangle.(\mathbf{U} \cap \mathbf{U}').$$

Further, the commutator of an element of $\mathbf{L} \cap \mathbf{U}'$ with an element of $\mathbf{L}' \cap \mathbf{U}$ is in $\mathbf{U} \cap \mathbf{U}'$, thus

$$\mathbf{P} \cap \mathbf{P}' = (\mathbf{L} \cap \mathbf{L}').(\mathbf{L} \cap \mathbf{U}').(\mathbf{L}' \cap \mathbf{U}).(\mathbf{U} \cap \mathbf{U}'),$$

Write now $x = lu'uv \in \mathbf{P} \cap \mathbf{P}'$, where $l \in \mathbf{L} \cap \mathbf{L}'$, $u' \in \mathbf{L} \cap \mathbf{U}'$, $u \in \mathbf{L}' \cap \mathbf{U}$, $v \in \mathbf{U} \cap \mathbf{U}'$. Then lu' is the image of x by the projection $\mathbf{P} \to \mathbf{L}$ and l (resp. u) is the image of lu' (resp. uv) by the morphism $\mathbf{P}' \to \mathbf{L}'$. Thus the decomposition of x is unique, and the product map $(\mathbf{L} \cap \mathbf{L}') \times (\mathbf{L} \cap \mathbf{U}') \times (\mathbf{L}' \cap \mathbf{U}) \times (\mathbf{U} \cap \mathbf{U}') \to \mathbf{P} \cap \mathbf{P}'$ is an isomorphism of varieties; the four terms are connected since the product is. In particular $\mathbf{L} \cap \mathbf{L}'$ is connected. □

Proposition 3.4.9

(i) *Let* $\mathbf{P}$ *and* $\mathbf{P}'$ *be two parabolic subgroups of* $\mathbf{G}$ *such that* $\mathbf{P}' \subset \mathbf{P}$, *then* $R_u(\mathbf{P}') \supset R_u(\mathbf{P})$ *and for any Levi subgroup* $\mathbf{L}'$ *of* $\mathbf{P}'$, *there exists a unique Levi subgroup* $\mathbf{L}$ *of* $\mathbf{P}$ *such that* $\mathbf{L} \supset \mathbf{L}'$.

(ii) *Let* $\mathbf{L}$ *be a Levi subgroup of a parabolic subgroup* $\mathbf{P}$ *of* $\mathbf{G}$ *and* $\mathbf{L}'$ *be a closed subgroup of* $\mathbf{L}$. *Then the following are equivalent:*

(a) $\mathbf{L}'$ *is a Levi subgroup of a parabolic subgroup of* $\mathbf{L}$.

(b) $\mathbf{L}'$ *is a Levi subgroup of a parabolic subgroup of* $\mathbf{G}$.

Proof Let us prove (i); given a maximal torus $\mathbf{T}$ of $\mathbf{L}'$ there is by 3.4.2(i) a unique Levi subgroup $\mathbf{L}$ of $\mathbf{P}$ containing $\mathbf{T}$. Then by 3.4.8(ii) the group $\mathbf{L}' \cap \mathbf{L}$ is a Levi subgroup of $\mathbf{P}' = \mathbf{P} \cap \mathbf{P}'$ thus $\mathbf{L} \cap \mathbf{L}' = \mathbf{L}'$. Also $R_u(\mathbf{P})$ is contained in all Borel subgroups of $\mathbf{P}$, thus in $\mathbf{P}'$, whence $R_u(\mathbf{P}) \subset R_u(\mathbf{P}')$.

Let us show (ii); if $\mathbf{L}'$ is a Levi subgroup of the parabolic subgroup $\mathbf{P_L}$ of $\mathbf{L}$, then $\mathbf{P_L}R_u(\mathbf{P})$ is a parabolic subgroup of $\mathbf{G}$; indeed it is a group since $\mathbf{L}$, thus $\mathbf{P_L}$, normalises $R_u(\mathbf{P})$ and it clearly contains either $\mathbf{U}_\alpha$ or $\mathbf{U}_{-\alpha}$ for any $\alpha \in \Phi$. Thus $\mathbf{L}'$ is a Levi subgroup of $\mathbf{P_L}R_u(\mathbf{P})$, since $R_u(\mathbf{P_L}).R_u(\mathbf{P})$ is unipotent normal in $\mathbf{P_L}.R_u(\mathbf{P})$. We have shown that (a) implies (b).

Conversely, let $\mathbf{P}'$ be a parabolic subgroup of $\mathbf{G}$ with $\mathbf{L}'$ as a Levi subgroup. By 3.4.8(ii) we have $\mathbf{P} \cap \mathbf{P}' = \mathbf{L}'.(\mathbf{L} \cap \mathbf{U}').(\mathbf{U} \cap \mathbf{U}')$ thus $(\mathbf{L} \cap \mathbf{U}') \rtimes \mathbf{L}'$ is a Levi decomposition of $\mathbf{L} \cap \mathbf{P}'$, and this last group is a parabolic subgroup of $\mathbf{L}$ by 3.4.5. □

When $\mathbf{L}$ is a Levi subgroup of some parabolic subgroup of $\mathbf{G}$ we will say (improperly) "$\mathbf{L}$ is a **Levi subgroup of** $\mathbf{G}$" which is justified by statement (ii) of 3.4.9.

Proposition 3.4.10 *Let* $\mathbf{H}$ *be a closed connected reductive subgroup of* $\mathbf{G}$ *of maximal rank. Then:*

(i) *The Borel subgroups of* **H** *are the* **B** ∩ **H** *where* **B** *is a Borel subgroup of* **G** *containing a maximal torus of* **H**.

(ii) *The parabolic subgroups of* **H** *are the* **P** ∩ **H**, *where* **P** *is a parabolic subgroup of* **G** *containing a maximal torus of* **H**.

(iii) *If* **P** *is a parabolic subgroup of* **G** *containing a maximal torus of* **H**, *the Levi subgroups of* **P** ∩ **H** *are the* **L** ∩ **H** *where* **L** *is a Levi subgroup of* **P** *containing a maximal torus of* **H**.

Proof Let **T** be a maximal torus of **H**; by assumption, it is also a maximal torus of **G**. Let **B** be a Borel subgroup of **G** containing **T**, and let **B** = **U.T** be the corresponding semi-direct product decomposition. The Borel subgroup **B** defines an order on the root system Φ (resp. $\Phi_{\mathbf{H}}$) of **G** (resp. **H**) with respect to **T**. The group **U** ∩ **H** is normalised by **T**, so is connected and equal to the product of the $\mathbf{U}_\alpha$ it contains, that is those $\mathbf{U}_\alpha$ such that α is positive and in $\Phi_{\mathbf{H}}$, so (**U** ∩ **H**).**T** = **B** ∩ **H** is a Borel subgroup of **H**. This gives (i) since all Borel subgroups of **H** are conjugate under **H**.

Let us prove (ii). If **P** is a parabolic subgroup of **G** containing **T**, it contains a Borel subgroup **B** containing **T**, so its intersection with **H** contains the Borel subgroup **B** ∩ **H** of **H** and thus is a parabolic subgroup. Conversely, let **Q** be a parabolic subgroup of **H** containing **T**, and let x be a vector of $X(\mathbf{T}) \otimes \mathbb{Q}$ defining **Q** as in 3.3.8(ii). Then x defines a parabolic subgroup **P** of **G**. It remains to show that **P** ∩ **H** = **Q**. The group **P** ∩ **H** is a parabolic subgroup of **H** by the first part. It is generated by **T** and the $\mathbf{U}_\alpha$ it contains. But $\mathbf{U}_\alpha \subset$ **P** ∩ **H** if and only if $\alpha \in \Phi_{\mathbf{H}}$ and $\langle \alpha, x \rangle \geq 0$; that is, if and only if $\mathbf{U}_\alpha \subset$ **Q** by definition of x.

Similarly, the Levi subgroup of **Q** containing **T** is the intersection of the Levi subgroup of **P** containing **T** with **H**, as it is generated by **T** and the $\mathbf{U}_\alpha$ with $\alpha \in \Phi_{\mathbf{H}}$ orthogonal to x, whence (iii).   □

3.5 Centralisers of Semi-Simple Elements

Proposition 3.5.1 *Let* $s \in$ **G** *be a semi-simple element, and let* **T** *be a maximal torus containing s and* Φ *be the set of roots of* **G** *relative to* **T**; *then*

(i) *The identity component* $C_{\mathbf{G}}(s)^0$ *is generated by* **T** *and the* $\mathbf{U}_\alpha$ *for* $\alpha \in \Phi$ *such that* $\alpha(s) = 1$. *It is a connected reductive subgroup of* **G** *of maximal rank*.

(ii) $C_{\mathbf{G}}(s)$ *is generated by* $C_{\mathbf{G}}(s)^0$ *and the elements* $n \in N_{\mathbf{G}}(\mathbf{T})$ *such that* $^n s = s$.

Proof (i) is an immediate consequence of 2.3.1(iv), and of the fact that the corresponding set of α is closed and symmetric – see 3.3.12 and 3.3.13.

Let us prove (ii). Let $\mathbf{B} = \mathbf{U} \rtimes \mathbf{T}$ be the Levi decomposition of a Borel subgroup of $\mathbf{G}$ and let $g \in C_\mathbf{G}(s)$; by Lemma 3.2.7 the element g has a unique decomposition $g = unv$ with $n \in N_\mathbf{G}(\mathbf{T})$, $u \in \mathbf{U}$ and $v \in \mathbf{U}_w$ where w is the image of n in $W(\mathbf{T})$. As s normalises $\mathbf{U}$, $\mathbf{U}_w$ and $N_\mathbf{G}(\mathbf{T})$, this decomposition is invariant under conjugation by s, so each of u, n and v also centralises s. Writing again a unique decomposition of the form $u = \prod_{\alpha > 0} u_\alpha$ we see that the α must satisfy $\alpha(s) = 1$ so $u \in C_\mathbf{G}(s)^0$, and the same argument applies to v. Thus we get (ii). □

Remark 3.5.2 The Weyl group $W^0(s)$ of $C_\mathbf{G}(s)^0$ is thus the group generated by the reflections s_α for which $\alpha(s) = 1$. It is a normal subgroup of the Weyl group of $C_\mathbf{G}(s)$ which is $W(s) = \{w \in W(\mathbf{T}) \mid {}^w s = s\}$. The quotient $W(s)/W^0(s)$ is isomorphic to the quotient $C_\mathbf{G}(s)/C_\mathbf{G}(s)^0$.

Proposition 3.5.3 *If $x = su$ is the Jordan decomposition of an element of $\mathbf{G}$, where s is semi-simple and u unipotent, then $x \in C_\mathbf{G}(s)^0$.*

Proof Let $\mathbf{B} = \mathbf{U} \rtimes \mathbf{T}$ be a Levi decomposition of a Borel subgroup containing x, where $\mathbf{T}$ is a maximal torus of $\mathbf{B}$ containing s, and write $u = \prod u_\alpha$ (with $u_\alpha \in \mathbf{U}_\alpha$ where $\mathbf{U} = \prod \mathbf{U}_\alpha$). Then for any root α such that $u_\alpha \neq 1$, we have $\alpha(s) = 1$ which implies that $\mathbf{U}_\alpha \subset C_\mathbf{G}(s)^0$, whence the result as $s \in \mathbf{T} \subset C_\mathbf{G}(s)^0$. □

Examples 3.5.4

(i) All centralisers in $\mathbf{GL}_n$ are connected. Indeed, the centraliser in the variety of all matrices is an affine space, thus its intersection with $\mathbf{GL}_n$ is an open subspace of an affine space, which is always connected.

(ii) In the group $\mathbf{SL}_n$, centralisers of semi-simple elements are connected. Indeed such an element is conjugate to an element $s = \mathrm{diag}(t_1, \ldots, t_n)$ of the torus $\mathbf{T}$ of diagonal matrices where we may assume, in addition, that equal t_i are grouped in consecutive blocks, thereby defining a partition π of n. The elements of $W(\mathbf{T})$ (permutation matrices) which centralise s are products of generating reflections s_α which centralise s; that is, $W(s) = W^0(s)$ showing that the centraliser of s is connected.

(iii) We finish with an example of a semi-simple element whose centraliser is not connected. Let $s \in \mathbf{PGL}_2$ be the image of $\begin{pmatrix} 1 & 0 \\ 0 & -1 \end{pmatrix}$; in characteristic different from 2, $C_{\mathbf{PGL}_2}(s)$ has two connected components, consisting respectively of the images of the matrices of the form $\begin{pmatrix} a & 0 \\ 0 & b \end{pmatrix}$ and of the form $\begin{pmatrix} 0 & a \\ b & 0 \end{pmatrix}$.

Notes

A classic reference about (B,N)-pairs is Bourbaki (1968, Chapter IV). A detailed study of closed and quasi-closed subsets and reductive and parabolic subgroups is in Borel and Tits (1965). A detailed study of the centralisers of semi-simple elements can be found in, for example, Deriziotis (1984).

4

Rationality, the Frobenius Endomorphism, the Lang–Steinberg Theorem

4.1 k_0-Varieties, Frobenius Endomorphisms

Let k be a (non-necessarily algebraically closed) field and let k_0 be a subfield of k.

A **k_0-structure** on a k-vector space V is a k_0-subspace V_0 such that $V = V_0 \otimes_{k_0} k$.

A k_0-structure on a finite type k-algebra A is a finite type k_0-algebra A_0 such that $A = A_0 \otimes_{k_0} k$.

A k_0-structure on an affine or projective k-variety $\mathbf{V}$ is the k_0-variety defined by A_0 where the algebra A of $\mathbf{V}$ has a k_0-structure A_0.

In general a k_0-structure on a variety is essentially given by a finite open affine covering where each open affine has a k_0-structure.

In this book, all the varieties we need to consider are quasi-projective varieties. We assume all varieties are quasi-projective from now on.

Definition 4.1.1   An algebraic variety $\mathbf{V}$ over k is said to be **defined over** k_0, if it has a k_0-structure $\mathbf{V}_0$. In this case we write $\mathbf{V} = \mathbf{V}_0 \otimes_{k_0} k$.

In the literature it is also sometimes said in the same circumstances that $\mathbf{V}$ is **rational over** k_0, or even when the context of k and k_0 is unambiguous that $\mathbf{V}$ is rational.

If the variety $\mathbf{V}$ has a k_0-structure, an element $\sigma \in \mathrm{Aut}(k/k_0)$ acts on $\mathbf{V}$ by $x \otimes \lambda \mapsto x \otimes \sigma(\lambda)$ on the corresponding coordinates for the variety. We assume now that k/k_0 is a Galois extension; then one can see $\mathbf{V}_0$ as the fixed points of the action of $\mathrm{Gal}(k/k_0)$. This results from the following proposition.

Proposition 4.1.2 *If V is a k-vector space (resp. a k-algebra) with a continuous action of $\mathrm{Gal}(k/k_0)$ – as a profinite group, thus continuous means that $V = \cup_G V^G$ where G runs over subgroups of finite index of*

Gal(k/k_0) – (*resp. compatible to the algebra structure*), *the fixed points of the action define a k_0-structure.*

Reference   See Springer (1998, 11.1.6).   □

Example 4.1.3   When k is an algebraic closure $\overline{\mathbb{F}}_q$ of $\mathbb{F}_q$, the Galois group Gal($\overline{\mathbb{F}}_q/\mathbb{F}_q$) is the profinite completion $\hat{\mathbb{Z}}$ of $\mathbb{Z}$; an element of $\hat{\mathbb{Z}}$ is defined by a sequence $k_n \in \mathbb{Z}$ indexed by $n \in \mathbb{N}$ subject to the only condition $k_n \equiv k_m$ (mod m) if m divides n; this element acts on $\mathbb{F}_{q^n}$ by $x \mapsto x^{q^{k_n}}$. We have $\hat{\mathbb{Z}} \simeq \prod_{p \text{ prime}} \mathbb{Z}_p$.

Proposition 4.1.4   *A subvariety (resp. subalgebra, vector subspace) is defined over k_0 – equivalently has a k_0-structure which is a subvariety (resp. subalgebra, subspace) – if and only if it is stable under the action of* Gal(k/k_0).

Proof   We give the proof for vector spaces; the other cases follow. It is obvious that a vector subspace $W \subset V$ which has a k_0-structure is stable by $\Gamma =$ Gal(k/k_0). Let us show the converse: assume that $\Gamma(W) = W$. Let $W_0 = W^\Gamma$. We need to show that W_0 spans W over k. The space W_0 is a k_0-subspace of V_0, the k_0-structure of V. Let W' be a complementary k_0-subspace. Using that V_0 is a k_0-structure for V, if W were not spanned by W_0, there would be a non-zero element of W spanned by W'. Let $v = \sum_{i=1}^n a_i v_i$ be such an element with $v_i \in W'$, $a_i \in k^\times$, and n minimal. We may assume $a_1 = 1$ by dividing both sides by a_1. For $\gamma \in \Gamma$ we get $\gamma v - v = \sum_{i=2}^n (\gamma a_i - a_i) v_i$. The minimality of n implies that $\gamma v - v = 0$, otherwise it would be a non-zero element of W giving rise to a smaller n. But this is true for all $\gamma \in \Gamma$, which implies $v \in W_0$, a contradiction.   □

Example 4.1.5   The affine line is $\mathbb{A}^1 = \operatorname{Spec} k[T]$. The affine line on k_0, defined by the k_0-algebra $k_0[T]$, is a k_0-structure since $k[T] = k_0[T] \otimes_{k_0} k$. An element $\sigma \in$ Gal(k/k_0) acts as $\sum_i a_i T^i \mapsto \sum_i \sigma(a_i) T^i$. A k-point of $\mathbb{A}^1$ is given by $a \in k$ (or by the ideal which is the kernel of the morphism $P \mapsto P(a) : k[T] \to k$); this point is defined over k_0 if $a \in k_0$.

Frobenius Endomorphism

We now take $k = \overline{\mathbb{F}}_q$, where q is a power of the prime p.

Definition 4.1.6   Let $\mathbf{V}$ be an $\overline{\mathbb{F}}_q$-variety with an $\mathbb{F}_q$-structure $\mathbf{V}_0$. The associated **geometric Frobenius endomorphism** $F \colon \mathbf{V} \to \mathbf{V}$ is $F_0 \otimes_{\mathbb{F}_q} 1$ where F_0 is the endomorphism of $\mathbf{V}_0$ which raises the functions on $\mathbf{V}_0$ to the qth power.

The endomorphism Φ of $\mathbf{V}$ induced by $(\lambda \mapsto \lambda^q) \in$ Gal($\overline{\mathbb{F}}_q/\mathbb{F}_q$) is called the **arithmetic Frobenius endomorphism**.

On an affine variety Spec A the $\mathbb{F}_q$-structure is of the form $A = A_0 \otimes_{\mathbb{F}_q} \overline{\mathbb{F}}_q$ and the geometric Frobenius endomorphism corresponds to a morphism F^* : $a \otimes \lambda \mapsto a^q \otimes \lambda$ in a corresponding coordinate system for the variety – the geometric Frobenius raises each coordinate to the qth power. The arithmetic Frobenius is given by $\Phi \colon a \otimes \lambda \mapsto a \otimes \lambda^q$. The composition $F^* \circ \Phi$ raises each element of A to the qth power, which acts trivially on the $\overline{\mathbb{F}}_q$-points of Spec A.

Example 4.1.7 On $\mathbb{A}^1$, the geometric Frobenius is given by F^* : $P(T) \mapsto P(T^q)$; thus $F^* \circ \Phi$ maps $P(T)$ to $P(T)^q$. If $a \in \overline{\mathbb{F}}_q$ is an $\overline{\mathbb{F}}_q$-point of $\mathbb{A}^1$, the image of a by $F^* \circ \Phi$ is defined by the kernel of $P \mapsto P(a)^q$, which is the same as that of $P \mapsto P(a)$.

Note that the geometric Frobenius endomorphism is a morphism of $\overline{\mathbb{F}}_q$-varieties, while the arithmetic Frobenius endomorphism is only a morphism of $\mathbb{F}_q$-varieties. After this chapter, we will only consider the geometric Frobenius endomorphism and just call it "the **Frobenius endomorphism**".

Proposition 4.1.8 *Let* **V** *be an affine or projective* $\overline{\mathbb{F}}_q$-*variety with algebra* A.

(i) *A surjective morphism* $A \xrightarrow{F^*} A^{[q]}$, *where* $A^{[q]} = \{x^q \mid x \in A\}$, *is the Frobenius endomorphism attached to an* $\mathbb{F}_q$-*structure on* **V** *if and only if for any* $x \in A$ *there exists* n *such that* $F^{*n}(x) = x^{q^n}$.

(ii) *If* F *is a Frobenius endomorphism on* **V** *attached to an* $\mathbb{F}_q$-*structure, that* $\mathbb{F}_q$-*structure is given by* $A_0 = \{x \in A \mid x^q = F^*(x)\}$.

Proof If A has an $\mathbb{F}_q$-structure $A = A_0 \otimes_{\mathbb{F}_q} \overline{\mathbb{F}}_q \ni x = \sum_i x_i \otimes \lambda_i$ then $x^{q^n} = \sum_i x_i^{q^n} \otimes \lambda_i^{q^n}$ thus $x^{q^n} = F^{*n}(x)$ when n is such that all λ_i are in $\mathbb{F}_{q^n}$.

Conversely, if F^* is a surjective morphism as in the statement, since $x \mapsto x^{q^n}$ is injective then F^* must also be injective, thus bijective and we can define Φ by $\Phi(x) = F^{*-1}(x^q)$; then if we make the topological generator of $\mathrm{Gal}(\overline{\mathbb{F}}_q/\mathbb{F}_q)$ act by Φ, the assumptions of 4.1.2 are satisfied and Φ is the arithmetic Frobenius associated to F.

The fixed points of Φ form the $\mathbb{F}_q$-structure by 4.1.2 and are as described in the statement. $\qquad\square$

Exercise 4.1.9 Show that any $\mathbb{F}_q$-structure on the affine line $\mathbb{A}^1$ is given by a Frobenius endomorphism on $\overline{\mathbb{F}}_q[T]$ of the form $T \mapsto aT^q$ with $a \in \overline{\mathbb{F}}_q^{\times}$ up to composition with an automorphism of $\mathbb{A}^1$.

Proposition 4.1.10 *A Frobenius morphism is a finite morphism.*

Proof Since being finite is a local property we need only to prove it for an affine variety Spec A with $A = \overline{\mathbb{F}}_q[T_1, \ldots, T_n]/I$ for some ideal I. The image of A by F is $A^{[q]} = \overline{\mathbb{F}}_q[T_1^q, \ldots, T_n^q]/(I \cap \overline{\mathbb{F}}_q[T_1^q, \ldots, T_n^q])$. We have to show that the dimension of A over $A^{[q]}$ is finite. This property is true for $\overline{\mathbb{F}}_q[T_1, \ldots, T_n]$ hence it is also true for A.   □

Proposition 4.1.11 *Let* **V** *be an* $\overline{\mathbb{F}}_q$*-variety and F be the Frobenius endomorphism corresponding to an* $\mathbb{F}_q$*-structure on* **V**.

  (i) *A subvariety of* **V** *is defined over* $\mathbb{F}_q$ *if and only if it is F-stable; the corresponding Frobenius endomorphism is the restriction of F.*

 (ii) *Let* φ *be an automorphism of* **V** *such that* $(\varphi F)^n = F^n$ *for some positive integer n; then* φF *is the Frobenius endomorphism attached to another* $\mathbb{F}_q$*-structure on* **V**.

(iii) *If F' is a Frobenius endomorphism attached to another* $\mathbb{F}_q$*-structure on* **V**, *there exists an integer* $n > 0$ *such that* $F^n = F'^n$.

 (iv) F^n *is the Frobenius endomorphism attached to an* $\mathbb{F}_{q^n}$*-structure on* **V**.

  (v) *Every closed subvariety of a variety defined over* $\mathbb{F}_q$ *is defined over a finite extension of* $\mathbb{F}_q$. *Every morphism between varieties defined over* $\mathbb{F}_q$ *is defined over a finite extension of* $\mathbb{F}_q$.

 (vi) *The orbits of F on the set of points of* **V** *are finite, as well as the set* $\mathbf{V}^F$, *also denoted by* $\mathbf{V}(\mathbb{F}_q)$, *which consists of the points of* **V** *defined over* $\mathbb{F}_q$.

Proof

  (i) reflects 4.1.4.

 (ii) results from the fact that φF still satisfies 4.1.8(i).

(iii) by considering an affine open covering it is sufficient to deal with the case $\mathbf{V} = \operatorname{Spec} A$. Then we use that A is of finite type, thus there exists n such that $F^{*n}(x) = F'^{*n}(x) = x^{q^n}$ for every generator x of A.

 (iv) results from 4.1.8(i).

  (v) has a proof similar to that of (iii): there exists n such that for any element a in a finite set of generators of the ideal I defining the subvariety (resp. any coefficient a of an equation of the morphism) we have $F^{*n}a = a^{q^n}$, thus $I \subset \sqrt{F^{*n}(I)}$.

Let us show (vi). As in (iii) we may assume $\mathbf{V} = \operatorname{Spec} A$. Let A_0 define the $\mathbb{F}_q$-structure on A and let $\{a_1, \ldots, a_m\}$ be a generating set for A_0. A point $x \in \mathbf{V}$ is given by a morphism $x : A \to \overline{\mathbb{F}}_q$. It is F^{*n}-stable if and only if for any i we have $x(a_i) \in \mathbb{F}_{q^n}$, which happens for a sufficiently large n. It is F^*-stable if and only if $x(a_i) \in \mathbb{F}_q$, or equivalently if we are given a morphism $A(\mathbb{F}_q) \to \mathbb{F}_q$; there is a finite number of such morphisms.   □

Proposition 4.1.12 *Let* $\mathbf{V} \simeq \mathbb{A}^n$ *as an* $\overline{\mathbb{F}}_q$*-variety. Then* $|\mathbf{V}^F| = q^n$ *for any* $\mathbb{F}_q$*-structure on* **V**.

Proof   See Geck (2003, 4.2.4) for a (complicated) elementary proof in the case where **V** is a unipotent group. The proposition is an immediate consequence of the Lefschetz theorem in ℓ-adic cohomology, see 8.1.11(ii).   □

4.2 The Lang–Steinberg Theorem; Galois Cohomology

In the following, when we consider an algebraic group **defined over** $\mathbb{F}_q$ or endowed with a Frobenius endomorphism F, it always means that F is a group morphism.

Example 4.2.1   Let us consider the group $\mathbf{GL}_n$ over $\overline{\mathbb{F}}_q$. It is defined over $\mathbb{F}_q$ as its algebra is $\overline{\mathbb{F}}_q[t_{i,j}, \det(t_{i,j})^{-1}]$, which is isomorphic to

$$\mathbb{F}_q[t_{i,j}, \det(t_{i,j})^{-1}] \otimes \overline{\mathbb{F}}_q.$$

Its points over $\mathbb{F}_q$ form the group $\mathbf{GL}_n(\mathbb{F}_q)$. On the matrices representing $\mathbf{GL}_n(\overline{\mathbb{F}}_q)$, the endomorphism F raises the entries to the qth power.

Notation 4.2.2   *Let* **T** *be a torus with a surjective algebraic endomorphism* F. *We still denote by* F *the group endomorphism of* $X(\mathbf{T})$ *given on* $\alpha \in X(\mathbf{T})$ *by* $\alpha \mapsto \alpha \circ F$.

Proposition 4.2.3   *If* F *is the Frobenius endomorphism on the torus* **T** *attached to an* $\mathbb{F}_q$-*structure, there exists* $\tau \in \mathbf{GL}(X(\mathbf{T}))$ *of finite order such that* $F = q\tau$ *on* $X(\mathbf{T})$.

Proof   If $A = \overline{\mathbb{F}}_q[X(\mathbf{T})]$ is the algebra of **T** and Φ is the arithmetic Frobenius associated to F on A, and Φ_0 is the arithmetic Frobenius on $\overline{\mathbb{F}}_q[x, x^{-1}]$ associated to the geometric Frobenius $x \mapsto x^q$ on $\mathbb{G}_m$, using the invertibility of the arithmetic Frobenius as a semilinear morphism of algebras we can define a group automorphism τ on $X(\mathbf{T})$ by the commutative diagram

$$
\begin{array}{ccc}
\overline{\mathbb{F}}_q[x, x^{-1}] & \xrightarrow{\Phi_0} & \overline{\mathbb{F}}_q[x, x^{-1}] \\
\downarrow{\scriptstyle \tau(\alpha)^*} & & \downarrow{\scriptstyle \alpha^*} \\
A & \xrightarrow{\Phi} & A
\end{array}
$$

where α^* is the algebra morphism corresponding to $\alpha : \mathbf{T} \to \mathbb{G}_m$. We can extend this diagram to the following commutative diagram

$$
\begin{array}{ccccc}
\overline{\mathbb{F}}_q[x, x^{-1}] & \xrightarrow{\Phi_0} & \overline{\mathbb{F}}_q[x, x^{-1}] & \xrightarrow{F^*} & \overline{\mathbb{F}}_q[x, x^{-1}] \\
\downarrow{\scriptstyle \tau(\alpha)^*} & & \downarrow{\scriptstyle \alpha^*} & & \downarrow{\scriptstyle \tau(\alpha)^*} \\
A & \xrightarrow{\Phi} & A & \xrightarrow{F^*} & A
\end{array}
$$

where the downward map on the right is $\tau(\alpha)^*$ since $F^* \circ \Phi$ and $F \circ \Phi_0$ are the algebra morphisms $a \mapsto a^q$. The commutativity of the rightmost square diagram gives $F = q\tau$. By 4.1.8(i), there is a power F^a which is just multiplication by q^a on $X(\mathbf{T})$, thus the order of τ is finite.   □

A Frobenius endomorphism on a reductive group – having trivial kernel and being bijective on points – is an isogeny; but it is not an isomorphism, since it is not invertible as a morphism of varieties. Specifically, let **G** be a reductive group over $\overline{\mathbb{F}}_q$, let F be the Frobenius endomorphism attached to an $\mathbb{F}_q$-structure and let **T** be an F-stable maximal torus – we will see below that such a torus always exists. As we saw in 4.2.3, F induces a morphism of the form $q\tau$ on $X(\mathbf{T})$. By 2.4.8 the isogeny F maps a root subgroup $\mathbf{U}_\alpha$ to $\mathbf{U}_{\tau(\alpha)}$. Thus F is an isogeny associated to $\sigma = \tau$ and such that $q_\alpha = q$ for any α; the p-morphism is $q\tau$.

Definition 4.2.4

(i) Let **X** be an algebraic variety over $\overline{\mathbb{F}}_p$ and let $F : \mathbf{X} \to \mathbf{X}$ be an endomorphism such that some power is a Frobenius endomorphism. Then we call F a **Frobenius root**.

(ii) If F is a Frobenius root and F^m is the Frobenius endomorphism attached to an $\mathbb{F}_{p^n}$-structure we define a positive real number q attached to F by $q^m = p^n$.

When **X** is a torus **T**, we have the following generalisation of 4.2.3, where F on $X(\mathbf{T})$ is defined as in 4.2.2.

Lemma 4.2.5   *If F is a Frobenius root on* **T** *and q is the attached real number, then there exists a finite order $\tau \in \mathbf{GL}(X(\mathbf{T}) \otimes \mathbb{R})$ such that $F = q\tau$ on $X(\mathbf{T})$.*

Proof   Define an element $\tau \in \mathrm{End}(X(\mathbf{T}) \otimes \mathbb{R})$ by $\tau := q^{-1}F$. By 4.1.8(i) there exists a power a of F such that $F^a = q^a$ on $X(\mathbf{T})$. It follows that τ is of finite order.   □

Definition 4.2.6   Let **G** be a reductive group over $\overline{\mathbb{F}}_p$; we call Frobenius root on **G** a Frobenius root F on the variety **G** which is a group morphism. The group of fixed points $\mathbf{G}^F$ is called a **finite group of Lie type**.

Proposition 4.2.7   *A Frobenius root on a reductive group is an isogeny.*

Proof   A Frobenius root has trivial kernel since some power has trivial kernel, and is surjective since some power is surjective.   □

Lemma 4.2.8   *Let* **G** *be an affine algebraic group over $\overline{\mathbb{F}}_p$ and F be a Frobenius root. Then for $g \in G$ the map $\mathrm{ad}\, g \circ F$ is again a Frobenius root.*

Proof   It is enough to check that some power of $\mathrm{ad}\, g \circ F$ is equal to the same power of F. We have $(\mathrm{ad}\, g \circ F)^n = \mathrm{ad}\, y \circ F^n$ with $y = g^F g \ldots {}^{F^{n-1}} g$; if n is such that g is F^n-stable then y is also F^n-stable and if y has order e then $(\mathrm{ad}\, g \circ F)^{ne} = F^{ne}$.   □

The fundamental theorem on connected algebraic groups over $\overline{\mathbb{F}}_p$ – here connectedness is essential – follows next.

Theorem 4.2.9 (Lang–Steinberg) *Let* **G** *be a connected affine algebraic group over* $\overline{\mathbb{F}}_p$, *and F be a Frobenius root. Then the* **Lang map** $\mathcal{L} : g \mapsto g^{-1}.^Fg$ *is a surjective endomorphism of* **G**.

Proof The morphism $\mathcal{L}$ has fibres isomorphic to $\mathbf{G}^F$, thus finite, thus dim Im $\mathcal{L} = \dim \mathbf{G}$; as **G** is irreducible $\mathcal{L}$ is dominant (which means **G** is the closure of Im $\mathcal{L}$), thus Im $\mathcal{L}$ contains a nonempty open subset of **G**.

For a given x, the morphism $g \mapsto g^{-1}.x.^Fg$ has also finite fibres: indeed, a fibre has cardinality the number of solutions of $g^{-1}x^Fg = x$, that is $g = {}^{xF}g$ and xF still has finitely many fixed points by Lemma 4.2.8. Thus the image of $g \mapsto g^{-1}.x.^Fg$ contains also a nonempty open subset of **G**, thus meets the image of $\mathcal{L}$, using again the irreducibility of **G**. Thus there exist g and h such that $g^{-1}.^Fg = h^{-1}.x.^Fh$, thus $x = \mathcal{L}(gh^{-1})$. □

Our statement is intermediate in strength between the original theorem of Lang, which gives the same statement assuming that F is a Frobenius endomorphism, and the stronger statement of Steinberg (1968), which uses only the assumption that F is a surjective morphism with $\mathbf{G}^F$ finite. Some authors – see for instance Malle and Testerman (2011) – call **Steinberg morphisms** what we call Frobenius roots.

A consequence of the Lang–Steinberg theorem is that for $g \in \mathbf{G}$ the group $\mathbf{G}^{gF}$ is isomorphic to $\mathbf{G}^F$. Indeed, write $g = h^{-1}.^Fh$ then $\mathbf{G}^{gF} = h^{-1}\mathbf{G}^F h$.

Galois Cohomology Exact Sequence

In this subsection we give a series of consequences of the Lang–Steinberg theorem, with corresponding elementary proofs. However these results can be understood in the more general context of Galois cohomology to which we first give an introduction. The impatient reader can skip this background information and use directly the elementary proofs we give.

We follow Serre (1994, §5). If G is a profinite group acting continuously on a set E, we set $H^0(G, E) := E^G$. If E is a group on which G acts as group endomorphisms, then Serre defines a set $H^1(G, E)$ for which we do not give the general definition, but rather give it below when $G = \hat{\mathbb{Z}}$. If $A \subset B$ is a group inclusion we have the "Galois cohomology exact sequence"

$$1 \to H^0(G, A) \to H^0(G, B) \to H^0(G, B/A) \xrightarrow{p} H^1(G, A) \xrightarrow{i} H^1(G, B).$$

When F is a topological generator of $G = \hat{\mathbb{Z}}$, we will write $H^i(F, E)$ for $H^i(\hat{\mathbb{Z}}, E)$; in this case $H^1(F, E)$ is the set of **F-conjugacy classes** of E, equal by definition to the E-conjugacy orbits in the coset $E.F$, or the classes of E under the twisted conjugacy $e \mapsto e'e.^F e'^{-1}$. In this book we will use the Galois cohomology exact sequence in the form shown in the following proposition.

Proposition 4.2.10   *Let $A \subset B$ be an inclusion of algebraic groups, and F be a Frobenius root, then there is an exact sequence*

$$1 \to A^F \to B^F \to (B/A)^F \xrightarrow{p} H^1(F,A) \xrightarrow{i} H^1(F,B)$$

and when A is a normal subgroup of B, the sequence can be extended on the right to one more term $H^1(F,B/A)$.

Reference and details   For the last sentence see Serre (1994, §5, Proposition 38). The maps in 4.2.10 are the obvious ones except perhaps p which maps an F-stable coset bA to the F-conjugacy class of $b^{-1}.^F b$, an element of A since bA is F-stable. The first three terms form an exact sequence of groups. The exactness of the sequence at $(B/A)^F$ means that B^F acts naturally on $(B/A)^F$ and that two elements of $(B/A)^F$ have the same image in $H^1(F,A)$ if and only if they are in the same B^F-orbit. The exactness at $H^1(F,A)$ means that the image of p is the preimage by i of the F-conjugacy class of 1 and similarly for the exactitude at $H^1(F,B)$ when A is a normal subgroup of B: this is an "exact sequence of pointed sets".   □

In the rest of this subsection F will be a Frobenius root. The Lang–Steinberg theorem can be rephrased as:

Proposition 4.2.11   *Let $\mathbf{G}$ be a connected algebraic group over $\overline{\mathbb{F}}_p$ and F be a Frobenius root. Then: $|H^1(F,\mathbf{G})| = 1$.*

Proposition 4.2.12   *Let $\mathbf{G}$ be a connected algebraic group over $\overline{\mathbb{F}}_p$, let F be a Frobenius root and let $\mathbf{V}$ be a non-empty variety with an action of F on which $\mathbf{G}$ acts transitively and compatibly with F. Then $\mathbf{V}^F \neq \emptyset$.*

Proof   Since the action is transitive, given $v \in \mathbf{V}$, there exists $g \in \mathbf{G}$ such that $^F v = gv$. Write $g^{-1} = h^{-1\,F}h$, then $^F(hv) = {}^F hgv = hv$.   □

Lemma 4.2.13   *Let $\mathbf{G}$ be a connected algebraic group over $\overline{\mathbb{F}}_p$ and F be a Frobenius root. Let $A \subset B$ be two closed and F-stable subgroups of $\mathbf{G}$, with A connected; then:*

(i) *We have $(B/A)^F = B^F/A^F$.*

(ii) *If, in addition, A is normal in B, the quotient map induces a bijection $H^1(F,B) \to H^1(F,B/A)$.*

Proof (i) and the injectivity in (ii) are 4.2.10 since $|H^1(F,A)| = 1$ but let us give a naive proof.

For (i), by 4.2.12, any F-stable coset bA contains an F-stable element, thus the natural map $B^F/A^F \to (B/A)^F$ is surjective. It is injective since if $x, y \in B^F$ are in the same A-coset, then $x^{-1}y \in A^F$.

Let us show (ii). Surjectivity is clear. Conversely, if $b, b' \in B$ are F-conjugate modulo A, we have $ab = xb'^F x^{-1}$, with $x \in B$ and $a \in A$. We must see that ab is F-conjugate to b, that is there exists $y \in B$ such that $yab^F y^{-1} = b$ or equivalently $a = y^{-1}b^F y$. This comes from 4.2.8 which shows that we may still apply the Lang–Steinberg theorem to ad $b \circ F$. $\qquad\qquad$ □

Proposition 4.2.14 *Let* **G** *be a connected algebraic group over* $\overline{\mathbb{F}}_p$, *let* F *be a Frobenius root and let* **V** *be a variety with an action of* F *on which* **G** *acts transitively and compatibly with* F. *Let* $x \in \mathbf{V}^F$ *and* $g \in \mathbf{G}$. *Then:*

(i) *We have* $gx \in \mathbf{V}^F$ *if and only if* $g^{-1}{}^F g \in C_{\mathbf{G}}(x)$.
(ii) *The map which sends the* $\mathbf{G}^F$-*orbit of* $gx \in \mathbf{V}^F$ *to the* F-*conjugacy class of the image of* $g^{-1}{}^F g$ *in* $C_{\mathbf{G}}(x)/C_{\mathbf{G}}(x)^0$ *is well-defined and bijective from the* $\mathbf{G}^F$-*orbits in* O^F, *where* O *is the* **G**-*orbit of* x, *to* $H^1(F, C_{\mathbf{G}}(x)) = H^1(F, C_{\mathbf{G}}(x)/C_{\mathbf{G}}(x)^0)$.

Proof The proposition translates 4.2.10 applied to the inclusion $C_{\mathbf{G}}(x) \subset \mathbf{G}$, which gives $1 \to C_{\mathbf{G}}(x)^F \to \mathbf{G}^F \to \mathbf{V}^F \to H^1(F, C_{\mathbf{G}}(x)) \to 1$ since **G** is connected and since $H^1(F, C_{\mathbf{G}}(x)) = H^1(F, C_{\mathbf{G}}(x)/C_{\mathbf{G}}(x)^0)$ by 4.2.13(ii). Again we will give a naive proof.

(i) is an immediate computation. Let us show (ii). Let $x \in \mathbf{V}^F, h, g \in \mathbf{G}$ be such that $hx, gx \in \mathbf{V}^F$. Note that $hx = gx$ if and only if h and g differ by an element of $C_{\mathbf{G}}(x)$, and then $h^{-1}{}^F h$ and $g^{-1}{}^F g$ are F-conjugate in $C_{\mathbf{G}}(x)$. We have thus a well-defined map from $\mathbf{V}^F$ to the F-conjugacy classes of $C_{\mathbf{G}}(x)$. On the other hand, if $h \in \mathbf{G}^F$, then gx and hgx have the same image $g^{-1}{}^F g = (hg)^{-1}{}^F(hg)$. Thus the map goes from the $\mathbf{G}^F$-orbits in $\mathbf{V}^F$ to the F-conjugacy classes of $C_{\mathbf{G}}(x)$. If $g^{-1}{}^F g$ and $h^{-1}{}^F h$ are F-conjugate by $n \in C_{\mathbf{G}}(x)$, then $gnh^{-1} \in \mathbf{G}^F$ and sends hx to gx. The map is thus injective. By the Lang–Steinberg theorem any element of $C_{\mathbf{G}}(x)$ is of the form $g^{-1}{}^F g$ with $g \in \mathbf{G}$, which shows the surjectivity of the map. We finish the proof using 4.2.13(ii). $\qquad\qquad$ □

Corollary 4.2.15 *Let* **G** *be a connected algebraic group over* $\overline{\mathbb{F}}_p$ *and* F *be a Frobenius root.*

(i) F-*stable Borel subgroups exist and form a single* $\mathbf{G}^F$-*conjugacy class.*
(ii) *Let us say that two elements of* $\mathbf{G}^F$ *are* **geometrically conjugate** *if they are conjugate under* **G**, *and define a* **geometric conjugacy class** *of* $\mathbf{G}^F$ *as the*

intersection with $\mathbf{G}^F$ of an F-stable $\mathbf{G}$-conjugacy class. Then geometric conjugacy classes are non-empty, and if x is an element of such a class, the class splits into $\mathbf{G}^F$-conjugacy classes parametrised by $H^1(F, C_\mathbf{G}(x)/ C_\mathbf{G}(x)^0)$.

(iii) *In an F-stable Borel subgroup $\mathbf{B}$ of $\mathbf{G}$, F-stable maximal tori exist and are $\mathbf{B}^F$-conjugate. Thus there exist F-stable pairs $\mathbf{T} \subset \mathbf{B}$ of an F-stable maximal torus of $\mathbf{G}$ and an F-stable Borel subgroup of $\mathbf{G}$ containing it; any two such pairs are $\mathbf{G}^F$-conjugate.*

(iv) *If $\mathbf{G}$ is reductive, any F-stable parabolic subgroup $\mathbf{P}$ has an F-stable Levi decomposition and two F-stable Levi subgroups of $\mathbf{P}$ are conjugate by an element of $\mathbf{U}^F$ where $\mathbf{U}$ is the unipotent radical of $\mathbf{P}$.*

Proof (i) comes from 4.2.14 applied with $\mathbf{V}$ the variety of Borel subgroups, using that for $\mathbf{B}$ a Borel subgroup $N_\mathbf{G}(\mathbf{B}) = \mathbf{B}$ is connected.

For (ii) we apply 4.2.14 with $\mathbf{V}$ the $\mathbf{G}$-conjugacy class (and the action of $\mathbf{G}$ by conjugacy).

For (iii) we first use the same argument as (i) using that for a torus $\mathbf{T}$ the group $N_\mathbf{B}(\mathbf{T}) = C_\mathbf{B}(\mathbf{T})$ (by 1.3.1(iii)) is connected by 1.3.3(iii) (if $\mathbf{G}$ is reductive it is equal to $\mathbf{T}$ by 2.3.1(iii)).

Finally we use that all pairs $\mathbf{T} \subset \mathbf{B}$ are $\mathbf{G}$-conjugate and the normaliser of such a pair is $C_\mathbf{B}(\mathbf{T}) = \mathbf{T}$.

For (iv) we use that by 3.4.2(ii) $\mathbf{U}$ acts transitively on the Levi decompositions, and the normaliser of a given Levi subgroup in the unipotent radical is trivial. $\square$

Example 4.2.16   As we saw in 3.5.4(i), all centralisers in $\mathbf{GL}_n$ are connected, thus geometric conjugacy classes do not split.

Example 4.2.17   Let $\mathbf{G} = \mathbf{PGL}_2(\overline{\mathbb{F}}_q)$ with q odd and let F be the "split" Frobenius morphism which raises the matrix entries to the qth power. If $s = \begin{pmatrix} 1 & 0 \\ 0 & -1 \end{pmatrix} \in \mathbf{GL}_2(\overline{\mathbb{F}}_q)$ and $\bar{s}$ is its image in $\mathbf{G}$, then $C_\mathbf{G}(\bar{s})$ is disconnected – see 3.5.4(iii). For $\lambda \in \mathbb{F}_{q^2}$ with $\lambda^{q-1} = -1$ the image $\bar{m}$ of $m = \begin{pmatrix} 0 & \lambda^{-1} \\ \lambda & 0 \end{pmatrix}$ is in $\mathbf{G}^F$ (since $^F m = -m$ in $\mathbf{GL}_2$) and if $x = \begin{pmatrix} 1 & 1 \\ \lambda & -\lambda \end{pmatrix}$ then $xsx^{-1} = m \pmod{Z(\mathbf{GL}_2)}$ and $x^{-1}\,{}^F x = \begin{pmatrix} 0 & 1 \\ 1 & 0 \end{pmatrix} \pmod{Z(\mathbf{GL}_2)}$ which is in $C_\mathbf{G}(\bar{s}) - C_\mathbf{G}(\bar{s})^0$ thus $\bar{m}$ is geometrically conjugate but not $\mathbf{G}^F$-conjugate to $\bar{s}$.

Example 4.2.18   Two elements of $\mathbf{SL}_n(\overline{\mathbb{F}}_q)$ which are conjugate in $\mathbf{GL}_n(\overline{\mathbb{F}}_q)$ are conjugate in $\mathbf{SL}_n(\overline{\mathbb{F}}_q)$. Indeed, if $x = yx'y^{-1}$, then we also have $x = y'x'y'^{-1}$

with $y' = (\det y)^{-1/n}y \in \mathbf{SL}_n$. But $x,x' \in \mathbf{SL}_n^F$ which are conjugate under $\mathbf{GL}_n^F$ are not necessarily conjugate under $\mathbf{SL}_n^F$, where F is the "split" Frobenius morphism which raises the matrix entries to the qth power. Let us consider, for instance the unipotent element $u = \begin{pmatrix} 1 & 1 \\ 0 & 1 \end{pmatrix}$ of $\mathbf{SL}_2^F$. Its centraliser in $\mathbf{GL}_2$ consists of all the matrices $\begin{pmatrix} a & b \\ 0 & a \end{pmatrix}$ with $a \in \overline{\mathbb{F}}_q^\times$ and $b \in \overline{\mathbb{F}}_q$, so it is connected. But its intersection with $\mathbf{SL}_2$ is not connected if the characteristic is different from 2: it has two connected components corresponding respectively to $a = 1$ and to $a = -1$; so we have $C_{\mathbf{G}}(u)/C_{\mathbf{G}}(u)^0 \simeq \mathbb{Z}/2\mathbb{Z}$ and the intersection of the geometric conjugacy class of u with $\mathbf{SL}_2^F$ splits into two conjugacy classes. By 4.2.14(ii) and an easy computation, one sees that the class in $\mathbf{SL}_2^F$ of the matrix $\begin{pmatrix} 1 & a \\ 0 & 1 \end{pmatrix}$ with $a \in \mathbb{F}_q^\times$ depends only on the image of a in $\mathbb{F}_q^\times/(\mathbb{F}_q^\times)^2$.

Definition 4.2.19   An F-stable maximal torus of the algebraic group $\mathbf{G}$ which is contained in an F-stable Borel subgroup of $\mathbf{G}$ is called a **quasi-split** torus.

By 4.2.15(i) and (iii) quasi-split tori are $\mathbf{G}^F$-conjugate.

Definition 4.2.20   A torus $\mathbf{T}$ defined over $\mathbb{F}_q$ is **split** if there is an isomorphism $\mathbf{T} \xrightarrow{\sim} \mathbb{G}_m^n$ defined over $\mathbb{F}_q$.

We will see in 7.1.3 that $\mathbf{T}$ is split if and only if the τ of 4.2.5 is the identity, and in 7.1.7 that a split torus is quasi-split, which is a kind of justification of the name "quasi-split".

Lemma 4.2.21   *Let $(X,Y,\Phi,\Phi^\vee)$ be the root datum of a connected reductive group $\mathbf{G}$ attached to a quasi-split torus $\mathbf{T}$, and write $F = q\tau$ on $X \otimes \mathbb{R}$ as in 4.2.5. Let Φ^+ and Π be the positive roots and root basis defined by an F-stable Borel subgroup $\mathbf{B}$ containing $\mathbf{T}$. Then τ permutes Φ^+ and Π up to a positive scalar, and induces an automorphism of the Coxeter system (W,S) attached to the choice of $\mathbf{B}$.*

Proof   The p-morphism defined by F induces a permutation σ of the roots which differs from τ by positive scalars. By the definition of the system Φ^+ attached to $\mathbf{B}$ – see 2.3.1(vi) – the permutation σ preserves Φ^+, thus τ does up to positive scalars. For any $\alpha \in \Phi^+$ we have $\tau^{-1}(\alpha) = b\beta$ for $b \in \mathbb{R}_{\geq 0}$ and $\beta \in \Phi^+$, thus $\tau^{-1}(\alpha) = \sum_{\gamma \in \Pi} g_\gamma \gamma$ with $g_\gamma \in \mathbb{R}_{\geq 0}$. Applying τ to this equality, we find that every element of Φ^+ is a positive linear combination of elements of $\tau(\Pi)$. From the minimality of Π, see 2.2.4, we deduce that τ preserves Π up to scalars. Since F, thus τ, permutes W it follows that τ permutes S. $\quad\square$

Proposition 4.2.22   *Let* **G** *be a connected reductive group over* $\overline{\mathbb{F}}_p$, *let F be a Frobenius root and let* **T** *be a quasi-split torus of* **G**. *Then the* $\mathbf{G}^F$-*conjugacy classes of F-stable maximal tori of* **G** *are parametrised by* $H^1(F, W_{\mathbf{G}}(\mathbf{T}))$, *by mapping an F-stable maximal torus* ${}^g\mathbf{T}$ *where* $g \in \mathbf{G}$ *to the F-conjugacy class of w, the image in* $W_{\mathbf{G}}(\mathbf{T})$ *of* $g^{-1}{}^Fg \in N_{\mathbf{G}}(\mathbf{T})$.

Proof   We apply 4.2.14 with **V** the variety of maximal tori of **G**, on which **G** acts by conjugacy.   □

In the situation of 4.2.22 we call **type** of ${}^g\mathbf{T}$ (with respect to **T**) the F-conjugacy class of w, or by abuse of notation w itself. Note that if ${}^g\mathbf{T}$ is of type w, the pair $({}^g\mathbf{T}, F)$ is sent by g^{-1}-conjugation to the pair $(\mathbf{T}, wF)$. Since quasi-split tori are $\mathbf{G}^F$-conjugate, and such a conjugation induces an F-equivariant isomorphism of Weyl groups, the set of types with respect to different quasi-split tori are in canonical bijection.

Proposition 4.2.23   *Let* **G** *be as in 4.2.9. Then every F-stable semi-simple element lies in some F-stable maximal torus of* **G**.

Proof   If $s \in \mathbf{G}$ is semi-simple then $s \in C_{\mathbf{G}}(s)^0$ by 3.5.1(ii), and s being central in this group is in all its maximal tori, thus in particular in its F-stable maximal tori which are also maximal in **G**.   □

4.3 Classification of Finite Groups of Lie Type

Let **G** be a connected reductive group over $\overline{\mathbb{F}}_p$.

Proposition 4.3.1   *Let F be an isogeny, defined by the p-endomorphism f of the root datum* $(X(\mathbf{T}), Y(\mathbf{T}), \Phi, \Phi^\vee)$ *of* **G**; *then F is the Frobenius endomorphism for an* $\mathbb{F}_q$-*structure on* **G** *(where q is a power of p) if and only if* $F = q\tau$ *on* $X(\mathbf{T})$ *where* τ *is an automorphism of finite order of* $(X(\mathbf{T}), \Phi)$.

Proof   We have seen that a Frobenius endomorphism is an isogeny of the above form in 4.2.3.

Conversely, when $\tau = \mathrm{Id}$ the fact that such an isogeny is a Frobenius endomorphism is a consequence, for instance of Springer (1998, Theorem 16.3.3). The case of an arbitrary τ follows then by 4.1.11(ii).   □

For a p-morphism as in 4.3.1 where $\tau = \mathrm{Id}$, we have $\sigma = \mathrm{Id}$ and $q_\alpha = q$ for every α. The torus **T** is then split over $\mathbb{F}_q$ (by for example 7.1.3, since $\tau = \mathrm{Id}$).

Definition 4.3.2   If F is as in 4.3.1 with $\tau = \mathrm{Id}$, that is F acts as q on $X(\mathbf{T})$ for some maximal F-stable torus, we call F a **split** Frobenius endomorphism on **G**, or say that $(\mathbf{G}, F)$ is **split**.

It follows from 4.3.1 that the converse of 4.2.5 holds, that is a Frobenius root is the same thing as an isogeny such that a power is a split Frobenius endomorphism.

Example 4.3.3   An example of non-split Frobenius endomorphism is the Frobenius F' which defines the unitary group $\mathbf{GL}_n^{F'}$, which is defined as $F'(x) = F({}^t x^{-1})$, where F is the split Frobenius endomorphism which maps the matrix entries to their qth power and ${}^t g$ denotes the transposed matrix of g. For the torus $\mathbf{T}$ of diagonal matrices, F' acts by $-q$ on $X(\mathbf{T})$; the associated p-morphism has $\sigma(\alpha) = -\alpha$. The group $\mathbf{GL}_n^{F'}$ is the group of unitary transformations of $(\mathbb{F}_{q^2})^n$. The group $\mathbf{GL}_n$ with the Frobenius endomorphism F' is called the **unitary group**, denoted by $\mathbf{U}_n$. This notation is somewhat inconsistent as it is not a different group but only a different rational form of $\mathbf{GL}_n$.

In general if we want to classify isogenies which are Frobenius roots, note that q_α and σ of the corresponding p-morphism determine completely the isogeny, thus the pair $(\mathbf{G}, F)$ and $\mathbf{G}^F$ up to isomorphism, if and only if Φ spans $X(\mathbf{T}) \otimes \mathbb{Q}$, that is if and only if $\mathbf{G}$ is semi-simple. We will classify now the finite groups of Lie type coming from semi-simple reductive groups, in particular those which correspond to simple finite groups. By 2.3.9 $\mathbf{G}$ is then the almost-direct product of quasi-simple groups; the endomorphism F permutes these quasi-simple components, and the orbits of F on the quasi-simple components give rise to F-stable normal subgroups of which $\mathbf{G}$ is the almost direct product. So we may assume there is only one orbit. We may assume $\mathbf{G}$ quasi-simple, since if $\mathbf{G} = \mathbf{G}_1 \times {}^F \mathbf{G}_1 \times \cdots \times {}^{F^{n-1}} \mathbf{G}_1$ with ${}^{F^n}(\mathbf{G}_1) = \mathbf{G}_1$, we have $\mathbf{G}^F \simeq \mathbf{G}_1^{F^n}$. The Coxeter diagram being then connected, we may assume up to multiplying F by an element of the Weyl group, that F preserves a Coxeter system corresponding to an F-stable Borel subgroup containing $\mathbf{T}$ – see 4.2.21.

We first consider the case where F is a Frobenius. Then, on $X(\mathbf{T})$ we have $F = q\tau$; the σ and q_α of the corresponding p-morphism are τ and q. The possibilities for τ correspond to the automorphisms of the Coxeter diagram which preserve the length of the roots. The possibilities for such a non-trivial automorphism τ on an irreducible root system are denoted by ${}^2 A_n$, $(n \geq 2)$, ${}^2 D_n$, ${}^3 D_4$ and ${}^2 E_6$, where the exponent on the left is the order of τ.

Remark 4.3.4   Note that by 2.4.8, any automorphism of the root datum of a reductive group gives rise to an isogeny which is an automorphism of the algebraic group, in particular the automorphisms defined above. We will use the same notation as for τ and say that the corresponding automorphism is of type ${}^2 A_n$, ${}^2 D_n$, ${}^3 D_4$ or ${}^2 E_6$.

Tits (1964) has shown, using the relative (B, N)-pair (see 4.4.8) that starting from a quasi-simple algebraic group $\mathbf{G}$, to get a finite simple group we must take $\mathrm{Der}(\mathbf{G}^F)$ divided by its centre, and apart from a few exceptions, $\mathrm{Der}(\mathbf{G}^F)$ is generated by the unipotent elements of $\mathbf{G}^F$. According to Steinberg (1968, Theorem 12.4), if $\mathbf{G}$ is simply connected, then $\mathbf{G}^F$ is generated by its unipotent elements. It follows that, apart from the exceptions, we get a simple finite group by taking $\mathbf{G}^F/Z(\mathbf{G}^F)$ where $\mathbf{G}$ is quasi-simple simply connected.

We may compare this to $(\mathbf{G}/Z(\mathbf{G}))^F$ using the following lemma.

Lemma 4.3.5   *We have*

$$1 \to Z(\mathbf{G}^F) \to \mathbf{G}^F \to (\mathbf{G}/Z(\mathbf{G}))^F \to Z(\mathbf{G})/(F-1)Z(\mathbf{G}) \to 1.$$

Proof   This results from the Galois cohomology exact sequence 4.2.10 associated to the exact sequence $1 \to Z(\mathbf{G}) \to \mathbf{G} \to \mathbf{G}/Z(\mathbf{G}) \to 1$, and from $Z(\mathbf{G})^F = Z(\mathbf{G}^F)$ (see 12.2.17) and $H^1(F, Z(\mathbf{G})) = Z(\mathbf{G})/(F-1)Z(\mathbf{G})$.   □

By 4.3.5, we see that $\mathbf{G}^F/Z(\mathbf{G}^F) \neq (\mathbf{G}/Z(\mathbf{G}))^F$ if $F - 1$ is not surjective on $Z(\mathbf{G})$.

Classification 4.3.6   We now describe the groups corresponding to each case, and we list with each case the exceptions, which occur when the abelianised of $\mathbf{G}^F$ is not formed of semi-simple elements, in which case the abelianised of $\mathbf{G}^F/Z(\mathbf{G}^F)$ is unipotent.

- $A_{n-1}(n \geq 2)$ – The simple algebraic group is $\mathbf{G} = \mathbf{PGL}_n \simeq \mathbf{PSL}_n$ with split Frobenius endomorphism. The simple finite group is $\mathbf{SL}_n^F/Z(\mathbf{SL}_n^F)$; the centre $Z(\mathbf{SL}_n)$ identifies with the group $\mu_n(\overline{\mathbb{F}}_q)$ of n-th roots of unity in $\overline{\mathbb{F}}_q$, on which $F - 1$ acts by raising to the power $q - 1$; thus $\mathbf{SL}_n^F/Z(\mathbf{SL}_n^F) \neq \mathbf{PSL}_n^F$ if $q - 1$ is not prime to n.

  The exceptions to simplicity are:

  – For $q = 2$, the group $\mathbf{SL}_2^F/Z(\mathbf{SL}_2^F) = \mathbf{PSL}_2^F = \mathfrak{S}_3$.
  – For $q = 3$, the group $\mathbf{SL}_2^F/Z(\mathbf{SL}_2^F) = \mathfrak{A}_4$.

- $^2A_{n-1}(n \geq 3)$ – The simple algebraic group is the special projective unitary group $\mathbf{PSU}_n \simeq \mathbf{PU}_n = \mathbf{U}_n/Z(\mathbf{U}_n)$. A similar computation as in the split case applies to compute the simple finite group $\mathbf{G}^F/Z(\mathbf{G}^F)$ for $\mathbf{G} = \mathbf{SU}_n$ (the elements of $\mathbf{U}_n$ of determinant 1), where $F - 1$ this time raises the elements of $Z(\mathbf{G})$ to the power $-q - 1$.

  The exception to simplicity is:

  – For $q = 2$, the group $\mathbf{SU}_3^F/Z(\mathbf{SU}_3^F) \simeq (\mathbb{Z}/3\mathbb{Z})^2 \rtimes W(B_2)$.

- $C_n(n \geq 2)$ – The simple finite group is $\mathbf{Sp}_{2n}/Z(\mathbf{Sp}_{2n})^F$ where F is a split Frobenius endomorphism. It is equal to $\mathbf{PSp}_{2n}^F$ only in characteristic 2.

The exception to simplicity is:

– For $q = 2$, the group $\mathbf{Sp}_4^F/Z(\mathbf{Sp}_4^F) = \mathbf{PSp}_4^F = \mathfrak{S}_6$.

• $B_n (n \geq 2)$ – The simple finite group is $\mathbf{SO}_{2n+1}^F/Z(\mathbf{SO}_{2n+1})^F$ where F is a split Frobenius endomorphism. It is equal to $\mathbf{PSO}_{2n}^F$ only in characteristic 2, in which characteristic B_n and C_n give isomorphic groups.

• D_n $(n \geq 4)$ – In characteristic 2 the group $\mathbf{SO}_{2n}^F$ defined as $(\mathbf{O}_{2n}^0)^F$ (F split) is simple. In odd characteristic, the simple algebraic group is $\mathbf{PSO}_{2n} = \mathbf{SO}_{2n}/\{\pm 1\}$. To describe the finite simple group, we should start from the simply connected group $\mathbf{Spin}_{2n}$, but we will not describe it here. We will rather describe the simple finite group as $\mathrm{Der}(\mathbf{SO}_{2n}^F)/Z(\mathrm{Der}(\mathbf{SO}_{2n}^F))$ (F split) where $Z(\mathrm{Der}(\mathbf{SO}_{2n}^F)) = \{\pm 1\} \cap \mathbf{SO}_{2n}^F$. The group $\mathrm{Der}(\mathbf{SO}_{2n}^F)$ is always of index 2 in $\mathbf{SO}_{2n}^F$, but -1 is not always in $\mathrm{Der}(\mathbf{SO}_{2n}^F)$: it is in $\mathrm{Der}(\mathbf{SO}_{2n}^F)$ if either n is even or n is odd and $q \equiv 1 \pmod 4$.

• ^2D_n – Similarly to the definition of $\mathbf{U}_n$, we will denote by $\mathbf{O}_{2n}^-$ the orthogonal group with a Frobenius endomorphism of the form εF composed of the split Frobenius endomorphism F on $\mathbf{O}_{2n}$ with ad ε for some F-stable element $\varepsilon \in \mathbf{O}_{2n} - \mathbf{SO}_{2n}$. As two such endomorphisms differ by an element of $\mathbf{SO}_{2n}$, the $\mathbb{F}_q$-structures they define are isomorphic. They are conjugate by an inverse image under the Lang map of the element by which the Frobenius endomorphisms differ. If q is odd we may take

$$
\varepsilon = \begin{pmatrix} I_{n-1} & & \\ & \begin{matrix} 0 & 1 \\ 1 & 0 \end{matrix} & \\ & & I_{n-1} \end{pmatrix}.
$$

This element normalises both the maximal torus and the Borel subgroup of respectively diagonal and upper triangular matrices. It induces the non-trivial automorphism of the root system of type D_n. Hence we get a group of type ^2D_n.

We show that $\mathbf{O}_{2n}^{\varepsilon F}$ is the orthogonal group over $\mathbb{F}_q$ of a symmetric bilinear form which is not equivalent to the form given by the matrix J of Example 1.5.5. Let $h \in \mathbf{GL}_{2n}$ be such that $h^{-1}.^F h = \varepsilon$; then ad h maps the action of εF to that of F – in particular, it maps $\mathbf{O}_{2n}^{\varepsilon F}$ into $\mathbf{GL}_{2n}^F$. The element g preserves the form of matrix J if and only if $^h g$ preserves the form of matrix $J_1 = {}^t h^{-1} J h^{-1}$. As J is F-stable and $\varepsilon \in \mathbf{O}_{2n}$, the matrix J_1 is F-stable. Thus ad h is an isomorphism from $\mathbf{O}_{2n}^{\varepsilon F}$ to the orthogonal group over $\mathbb{F}_q$ of J_1. Let us show that the bilinear forms given by J and J_1 are not equivalent under $\mathbf{GL}_{2n}^F$. If $a \in \mathbf{GL}_{2n}^F$ satisfies $^t a J a = J_1$, then ah is in the orthogonal group of J. But $(ah)^{-1}.^F(ah) = \varepsilon$, so this would imply $\det(\varepsilon) = \det(ah)^{q-1} = (\pm 1)^{q-1} = 1$, which is false, since $\det(\varepsilon) = -1$, whence the result.

The corresponding finite simple group is $\mathbf{SO}_{2n}^{\varepsilon F}/\{\pm 1\}$.

- A similar analysis can be done for $^3D_4$, the triality group, and E_6, $^2E_6$, E_7, E_8, F_4 and G_2. Using the root datum one sees that a semi-simple group of type E_8, F_4 or G_2 is both adjoint and simply connected. A semi-simple group of type E_6, $^2E_6$ or E_7 can only be either adjoint or simply connected. A semi-simple root datum of type D_4 has an automorphism of order 3 if and only if it is adjoint or simply connected (see Exercise 4.3.8), hence a semi-simple group of type $^3D_4$ is either adjoint or simply connected.

There is one exception to simplicity:

- For $q = 2$, the group $\mathbf{G}^F$ where $\mathbf{G}$ is of type G_2 whose derived subgroup, of index 2, is simple isomorphic to $\mathbf{SU}_3^F$ for $q = 3$.

In addition to the case where F is a Frobenius there are "exceptional" isogenies which on $X(\mathbf{T}) \otimes \mathbb{R}$ are of the form $q\tau$ where τ is an automorphism of finite order and q is a non-rational scalar. In each case τ induces an automorphism of the Coxeter diagram which does not extend to the root system because it does not preserve the length of the roots. Such automorphisms are of type $^2B_2$, $^2F_4$ (resp. $^2G_2$). To make τ map the root lattice into itself, we have to scale it by an odd power q of $\sqrt{2}$ (resp. $\sqrt{3}$). After scaling, we get a p-morphism defining an isogeny whose square is a Frobenius on the field $\mathbb{F}_{q^2}$.

Example 4.3.7   We assume $\mathbf{G}$ has a root system Φ of type B_2, with basis $\Pi = \{\alpha = e_1 - e_2, \beta = 2e_2\}$. If $\operatorname{char} k = 2$ the formulae $F(\mathbf{u}_\alpha(x)) = \mathbf{u}_\beta(x^2)$, $F(\mathbf{u}_{\alpha+\beta}(x)) = \mathbf{u}_{2\alpha+\beta}(x^2), F(\mathbf{u}_\beta(x)) = \mathbf{u}_\alpha(x), F(\mathbf{u}_{2\alpha+\beta(x)}) = \mathbf{u}_{\alpha+\beta}(x)$ define an isogeny. If $t = \operatorname{diag}(t_1, t_2, t_2^{-1}, t_1^{-1}) \in \mathbf{T}$ we have $F(t) = \operatorname{diag}(t_1 t_2, t_1 t_2^{-1}, t_1^{-1} t_2, t_1^{-1} t_2^{-1})$. The associated p-morphism is $\alpha \mapsto 2\beta, \beta \mapsto \alpha$. We have $F = q\tau$ where $q = \sqrt{2}$ and $\tau^2 = 1$. Then F^2 squares all coordinates, and is thus the split Frobenius over $\mathbb{F}_2$. For any r the isogeny F^{2r+1} has $2^{2r+1} - 1$ fixed points on the torus; the group of fixed points $\mathbf{G}^{F^{2r+1}}$ is the **Suzuki group** $^2B_2(2^{2r+1})$.

We can construct similarly the **Ree groups** $^2G_2(3^{2r+1})$ and $^2F_4(2^{2r+1})$. In each case the simply connected algebraic group $\mathbf{G}$ has a trivial centre, thus the complications described above do not occur. The groups $\mathbf{G}^F$ are simple apart from the exceptions:

- $^2B_2(2)$, which is the Frobenius group of order 20.
- $^2G_2(3)$, whose derived subgroup, of index 3, is $\mathbf{SL}_2(\mathbb{F}_8)$.
- $^2F_4(2)$ whose derived subgroup, of index 2, is **Tits' simple group**.

Adding to the above list the alternating groups, we have all the non-sporadic non-abelian finite simple groups.

Exercise 4.3.8

(i) Show that in a semi-simple root datum $(X, Y, \Phi, \Phi^\vee)$, one has $\langle \Phi \rangle \subset X \subset P$ where $P := \{x \in \langle \Phi \rangle \otimes \mathbb{Q} \mid \alpha^\vee(x) \in \mathbb{Z} \text{ for all } \alpha \in \Phi\}$.

(ii) With the notation of 2.2.15, show that if Φ is a root system of type D_4 then $\{e_1, e_1 + e_2, \frac{1}{2}(e_1 + e_2 + e_3 - e_4), \frac{1}{2}(e_1 + e_2 + e_3 + e_4)\}$ is a basis of P which is dual to the basis of simple coroots.

(iii) Show that $P/\langle \Phi \rangle \simeq \mathbb{Z}/2\mathbb{Z} \times \mathbb{Z}/2\mathbb{Z}$. Deduce that a semi-simple root datum of type D_4 can have an automorphism of order 3 only if $X = P$ or $X = \langle \Phi \rangle$.

4.4 The Relative (B, N)-Pair

Proposition 4.4.1 *Let* $\mathbf{G}$ *be a connected reductive group over* $\overline{\mathbb{F}}_p$ *and let* F *be a Frobenius root.*

(i) *Let* $\mathbf{T}$ *be an F-stable maximal torus of* $\mathbf{G}$. *Then* $W(\mathbf{T})^F = N_{\mathbf{G}}(\mathbf{T})^F/\mathbf{T}^F$.
   Let $\mathbf{T} \subset \mathbf{B}$ *be an F-stable pair of a maximal torus and a Borel subgroup. Then:*

(ii) $\mathbf{G}^F = \coprod_{w \in W^F} \mathbf{B}^F w \mathbf{B}^F$.

(iii) *If $q \in \mathbb{R}_{>0}$ is attached to F as in 4.2.4(ii), then*

$$|\mathbf{G}^F| = q^{l(w_0)} |\mathbf{T}^F| \sum_{w \in W^F} q^{l(w)}.$$

(iv) $R_u(\mathbf{B})^F$ *is a Sylow p-subgroup of* $\mathbf{G}^F$.

We will recognise (ii) as the Bruhat decomposition attached to a "relative" (B,N)-pair in 4.4.8 below.

Proof (i) comes from 4.2.13(i).

For (ii) we use the "unique Bruhat decomposition" 3.2.7 which implies that $\mathbf{B}w\mathbf{B}$ is F-stable if and only if $w \in W^F$ and that an F-stable element of $\mathbf{B}w\mathbf{B}$ is in $\mathbf{B}^F n \mathbf{B}^F = \mathbf{B}^F n \mathbf{U}_w^F$ where $n \in N_{\mathbf{G}}(\mathbf{T})^F$ is a representative of w.

Let us show (iii). By the proof of (ii) $|\mathbf{G}^F/\mathbf{B}^F| = \sum_{w \in W^F} |\mathbf{U}_w^F|$, and using $|\mathbf{B}^F| = |\mathbf{T}^F||\mathbf{U}^F|$ we get the stated formula if we show $|\mathbf{U}_w^F| = q^{l(w)}$, since $\mathbf{U} = \mathbf{U}_{w_0}$. If F is a Frobenius attached to an $\mathbb{F}_q$-structure, this results from 4.1.12 and the fact that $\mathbf{U}_w$ is an affine space of dimension $l(w)$. It is still true in general, as seen in Lemma 4.4.2.

Lemma 4.4.2 *Let F be a Frobenius root on* $\mathbf{G}$ *and* $\mathbf{B} = \mathbf{U} \rtimes \mathbf{T}$ *an F-stable Levi decomposition of a Borel subgroup of* $\mathbf{G}$. *Then for any F-stable subgroup* $\mathbf{U}_1 \subset \mathbf{U}$ *normalised by* $\mathbf{T}$, *we have* $|\mathbf{U}_1^F| = q^{\dim \mathbf{U}_1}$ *where q is associated to F as in 4.2.4(ii).*

Proof We first show that $|\mathbf{U}_1^F| = \prod_{\{\alpha \mid \mathbf{U}_\alpha \subset \mathbf{U}_1\}} q_\alpha$ where the q_α define the p-morphism corresponding to the isogeny F, see 2.4.8. By 2.3.1(v) we have $\mathbf{U}_1 = \prod_{\alpha \in \Psi} \mathbf{U}_\alpha$ and $\mathrm{Der}(\mathbf{U}_1) = \prod_{\alpha \in \Psi'} \mathbf{U}_\alpha$ for some sets Ψ and Ψ' of roots, in particular these groups are connected. Thus we have $\mathbf{U}_1 / \mathrm{Der}(\mathbf{U}_1) \simeq \prod_{\alpha \in \Psi - \Psi'} \mathbf{U}_\alpha$ and $|\mathbf{U}_1^F| = |(\mathbf{U}_1 / \mathrm{Der}(\mathbf{U}_1))^F|.|\,\mathrm{Der}(\mathbf{U}_1^F)|$, since $\mathrm{Der}(\mathbf{U}_1)$ is connected. From 2.3.1(vii) we have $\Psi' \subsetneq \Psi$ which allows us to argue by induction on Ψ. An element of $\mathbf{U}_1 / \mathrm{Der}(\mathbf{U}_1)$ can be written $u = \prod_{\alpha \in \Psi - \Psi'} \mathbf{u}_\alpha(x_\alpha)$. Since $F(\mathbf{u}_\alpha(x_\alpha)) = \mathbf{u}_{\sigma(\alpha)}(\lambda_\alpha x_\alpha^{q_\alpha})$ for some non-zero constants λ_α (see 2.4.8), the element u is F-stable if and only if for each $\alpha \in \Psi - \Psi'$ we have $x_{\sigma(\alpha)} = \lambda_\alpha x_\alpha^{q_\alpha}$. Solving these equations amounts, for each orbit o of σ in $\Psi - \Psi'$, to solving an equation $x_\alpha = \lambda x_\alpha^{q_o}$ for one chosen $\alpha \in o$, where $q_o = \prod_{\beta \in o} q_\beta$ and λ is a product of the λ_β to some powers. Such an equation has q_o solutions, whence $|(\mathbf{U}_1 / \mathrm{Der}(\mathbf{U}_1))^F| = \prod_{\alpha \in \Psi - \Psi'} q_\alpha$. Since by induction $|\mathrm{Der}(\mathbf{U}_1)^F| = \prod_{\{\alpha \in \Phi^+ \setminus \Psi - \Psi'\}} q_\alpha$ we get the formula for $|\mathbf{U}_1^F|$ claimed at the beginning of the proof.

Let now o be a σ-orbit of roots in Ψ. To prove the lemma it suffices to show that $q_o := \prod_{\alpha \in o} q_\alpha = q^{|o|}$. For $\alpha \in o$ we have $F^{|o|}(\alpha) = q_o \alpha$. Let m be such that F^m is a split Frobenius; we have $F^m(\alpha) = q^m \alpha$. Hence $F^{m|o|}\alpha = q^{m|o|}\alpha = q_o^m \alpha$, whence the result. $\qquad\square$

We show (iv). As $|\mathbf{G}^F/\mathbf{U}^F| = |\mathbf{T}^F|(\sum_{w \in W^F} q^{l(w)})$ is prime to p – since $\mathbf{T}$ is a p'-group and $\sum_{w \in W^F} q^{l(w)} \equiv 1 \pmod p$ – we see that $\mathbf{U}^F$ is a Sylow p-subgroup of $\mathbf{G}^F$. $\qquad\square$

Note that by an argument similar to the proof of the last statement of 3.4.1, see Borel and Tits (1965, 5.19), we have $N_{\mathbf{G}^F}(\mathbf{U}^F) = \mathbf{B}^F$, thus $\sum_{w \in W^F} q^{l(w)}$ is the number of Sylow p-subgroups of $\mathbf{G}^F$.

The fact that the fixed points of the unipotent radical of a Borel subgroup form a Sylow p-subgroup extends to non-reductive (but connected) groups. Since $R_u(\mathbf{G})$ is a p-group, it is in all unipotent radicals of Borel subgroups, and is connected, whence $|\mathbf{G}^F| = |(\mathbf{G}/R_u(\mathbf{G}))^F||R_u(\mathbf{G})^F|$.

Corollary 4.4.3 *Let $\mathbf{G}$ be a connected algebraic group over $\overline{\mathbb{F}}_p$ and let F be a Frobenius root; any unipotent element of $\mathbf{G}^F$ lies in an F-stable Borel subgroup.*

Proof Any F-stable unipotent element is in a Sylow p-subgroup by 1.1.5, so by 4.4.1(iv) (and the remarks above) is in the unipotent radical of an F-stable Borel subgroup. $\qquad\square$

Proposition 4.4.4 *Let Γ be a group of automorphisms of the Coxeter system (W, S), that is of automorphisms of W preserving S. Let $(S/\Gamma)_{<\infty}$ the set of orbits I of Γ on S such that the subgroup W_I is finite. Then $(W^\Gamma, \{w_I\}_{I \in (S/\Gamma)_{<\infty}})$ is a Coxeter system, where W^Γ is the subgroup of Γ-fixed elements of W, and* *where w_I denotes the longest element of W_I, see 2.1.5. Further, if $w_{I_1} \ldots w_{I_k}$ is a reduced expression of some $w \in W^\Gamma$ in the above Coxeter system, we have $l(w) = \sum_{i=1}^{i=k} l(w_{I_i})$, where l is the length on W with respect to S.*

Proof We first prove Lemma 4.4.5.

Lemma 4.4.5 *If for $w \in W^\Gamma$ and $s \in S$ we have $l(ws) < l(w)$ then the Γ-orbit I of s is finite, and $l(ww_I) = l(w) - l(w_I)$. In particular $W_I^\Gamma = \{1, w_I\}$.*

Proof Since Γ is an automorphism of (W, S), for any element $t \in I$ we have $l(wt) < l(w)$. Write $w = w'v$ with w' reduced-I and $v \in W_I$; then $l(vt) < l(v)$ for any $t \in I$ which is possible only if W_I is finite and $v = w_I$, see 2.1.5. The final remark is obtained by applying the result to $w \in W_I^\Gamma$. □

Let S_Γ be the set $\{w_I \mid I \in (S/\Gamma)_{<\infty}\}$; applying inductively the lemma, we find that any $w \in W^\Gamma$ can be written $w = w_{I_1} \ldots w_{I_k}$ where $I_j \in S_\Gamma$ and $l(w) = \sum_{i=1}^{i=k} l(w_{I_i})$, in particular S_Γ generates W^Γ.

We will use the characterisation 2.2.9 to see that (W^Γ, S_Γ) is a Coxeter system, but exchanging right and left. For $w_I \in S_\Gamma$, let $D_{w_I} = \{w \in W^\Gamma \mid w$ is reduced-$I\}$. We clearly have $D_{w_I} \ni 1$ and $D_{w_I} \cap D_{w_I} w_I = \emptyset$. It remains to show that if $w \in D_{w_I}$, $w_J \in S_\Gamma$, and $w_J w \notin D_{w_I}$, then $w_J w = w w_I$. We will use the function N of 2.1.2(ii). For $I \in (S/\Gamma)_{<\infty}$ and $w \in W^\Gamma$, if $W_I \cap N(w) \neq \emptyset$, then $N(w_I) \subset N(w)$: indeed then w is not reduced-I so that by 4.4.5 it can be written $w_{I_1} \ldots w_{I_k}$ with the lengths adding and $I_k = I$.

Hence in our situation $w \in D_{w_I}$ implies $N(w) \cap N(w_I) = \emptyset$ and $w_J w \notin D_{w_I}$ implies $N(w_J w) \supset N(w_I)$; since $N(w_J w) = {}^{w^{-1}} N(w_J) \dot{+} N(w)$ it follows that ${}^{w^{-1}} N(w_J) \supset N(w_I)$. If we show that this implies $w_J = {}^w w_I$ we are done. Since 2.1.5(i) implies that $N(w)$ generates W if w is the longest element of W, we get that ${}^w N(w_I)$ and $N(w_J)$ generate respectively ${}^w W_I$ and W_J, thus ${}^w W_I \subset W_J$, thus ${}^w w_I = w_J$ since by 4.4.5 there is only one non-trivial Γ-stable element in W_I, hence in ${}^w W_I$.

We now prove the final remark in the statement of 4.4.4. If $w_{I_1} \ldots w_{I_k}$ is a reduced expression in (W^Γ, S_Γ) then by the property of $D_{w_{I_k}}$ (see 2.2.9) we have $w \notin D_{w_{I_k}}$; thus by Lemma 4.4.5 we have $w = w' w_{I_k}$ where the lengths add. This proves the result by induction on k. □

Exercise 4.4.6 For $w_I, w_J \in S_\Gamma$, the length of the braid relation between w_I and w_J (the order of $w_I w_J$) is equal to $\dfrac{2|\operatorname{Ref}(W_{I \cup J})|}{|\operatorname{Ref}(W_I)| + |\operatorname{Ref}(W_J)|}$.

Remark 4.4.7 Note that it may happen that W is a Weyl group, but W^Γ is not: if γ is the automorphism of type ${}^2 F_4$, we have $W(F_4)^\gamma = W(I_2(8))$, as can be checked using Exercise 4.4.6 – this is the only example of an irreducible Weyl group where W^γ is not a Weyl group.

Corollary 4.4.8   *Let* **G**, **T** $\subset$ **B** *and F be as in 4.4.1. Then* $(\mathbf{B}^F, N_{\mathbf{G}}(\mathbf{T})^F)$ *is a* (B,N)*-pair – the* **relative** (B,N)**-pair** *– for* $\mathbf{G}^F$ *with Weyl group* W^F. *By 4.4.4* $(W^F, \{w_I\}_{I \in S/F})$ *is a Coxeter system where I runs over the F-orbits in S and where* w_I *is the longest element in* W_I.

Proof   Here the action of F on S is as defined in 4.2.21. The corollary follows from the definition of (B,N)-pairs, from 4.4.1(ii) and from 4.4.4: it remains to check that for $I \in S/F$ and $w \in W^F$, then $\mathbf{B}^F w \mathbf{B}^F w_I \mathbf{B}^F \subset \mathbf{B}^F w \mathbf{B}^F \cup \mathbf{B}^F w w_I \mathbf{B}^F$. We use that either $l(w) + l(w_I) = l(w w_I)$, in which case $\mathbf{B}^F w \mathbf{B}^F w_I \mathbf{B}^F = \mathbf{B}^F w w_I$ $\mathbf{B}^F$, or $w = w' w_I$ with $l(w') + l(w_I) = l(w' w_I)$ in which case $\mathbf{B}^F w \mathbf{B}^F w_I \mathbf{B}^F \subset \mathbf{B}^F w' \mathbf{B}^F w_I \mathbf{B}^F w_I \mathbf{B}^F$, and $\mathbf{B}^F w_I \mathbf{B}^F w_I \mathbf{B}^F \subset \mathbf{B}^F \cup \mathbf{B}^F w_I \mathbf{B}^F$ since 1 and w_I are the only F-stable elements of W_I by 4.4.5. $\square$

Proposition 4.4.9   *Let* $\mathbf{T}_w$ *be a torus of type w with respect to* **T**. *Then* $|\mathbf{T}_w^F| = (-1)^{l(w)} \det \tau \det(wF - 1 \mid X(\mathbf{T}))$ *where* τ *is as in 4.2.5.*

Proof   It is equivalent to prove this formula for the pair $(\mathbf{T}, wF)$. Applying the exact contravariant functor X to the exact sequence $1 \to \mathbf{T}^{wF} \to \mathbf{T} \xrightarrow{wF-1} \mathbf{T} \to 1$, where the surjectivity on the right is the Lang–Steinberg theorem, we get $1 \to X(\mathbf{T}) \xrightarrow{wF-1} X(\mathbf{T}) \xrightarrow{\pi} \mathrm{Hom}(\mathbf{T}^{wF}, \mathbb{G}_m) \to 1$; by a general property of cokernels of maps between lattices we get $|\mathrm{Hom}(\mathbf{T}^{wF}, \mathbb{G}_m)| = |\mathbf{T}^{wF}| = |\det(wF - 1)|$. The statement is thus equivalent to $\mathrm{sgn}(\det(wF - 1)) = (-1)^{l(w)} \det \tau$. Using $F = q\tau$ we have $\det(wF - 1) = \det(w) \det(\tau) \det(q - (w\tau)^{-1})$. We get the statement since $\det(w) = (-1)^{l(w)}$ because w is the product of $l(w)$ reflections, since $\det \tau$ is a sign because it is a real number and a root of unity, and since $\det(q - (w\tau)^{-1}) > 0$ whenever $q > 1$ because $(w\tau)^{-1}$ is a real map of finite order: this determinant is a product of terms of the form $q - \varepsilon$, where ε is a root of unity; such a term is positive if $\varepsilon = \pm 1$, and otherwise one can pair terms $(q - \varepsilon)(q - \overline{\varepsilon}) = q^2 + 1 - q(\varepsilon + \overline{\varepsilon})$ and the positivity of this product comes from $q + q^{-1} > 2$ if $q > 1$. $\square$

Notes

General results about $\mathbb{F}_q$-structures can be found for example in Deligne (1980). The groups $\mathbf{G}^F$ of fixed points of an isogeny which has the property that $\mathbf{G}^F$ is finite were first studied in Steinberg (1968); rationality properties over an arbitrary field are studied in Borel and Tits (1965).

5

Harish-Chandra Theory

A heuristic approach to building the irreducible representations of a finite group G is to find a family of subgroups of G of the same type as G which generate G, and then build representations of G by induction from these subgroups. A typical example is the construction of the irreducible representations of the symmetric groups $\mathfrak{S}_n$ as $\mathbb{Z}$-linear combinations of $\mathrm{Ind}_{\mathfrak{S}_{n_1} \times \cdots \times \mathfrak{S}_{n_k}}^{\mathfrak{S}_n} \mathbf{1}$ with $n_1 + \cdots + n_k = n$.

In the case of a finite group of Lie type $G = \mathbf{G}^F$, a suitable family of subgroups consists of the groups $\mathbf{L}^F$ where $\mathbf{L}$ runs over the F-stable Levi subgroups of F-stable parabolic subgroups of $\mathbf{G}$. We could, as our introduction suggests, build representations of $\mathbf{G}^F$ from those of $\mathbf{L}^F$ using $\mathrm{Ind}_{\mathbf{L}^F}^{\mathbf{G}^F}$. But, actually the representations thus obtained would not have the right properties; in particular, their decomposition into irreducible representations would be rather intractable. The right construction is to use the following Harish-Chandra induction, defined as a generalised induction associated to a bimodule.

5.1 Harish-Chandra Induction and Restriction

Let A be a commutative ring with unit and let H and K be two finite groups. We will call "AH-module-AK" an A-module M endowed with a left H-action and a right K-action. We may also view it as an $AH \times AK^{\mathrm{opp}}$-module where K^{opp} denotes the opposed group to K.

Definition 5.1.1   Given M, an AH-module-AK, we define two functors:

(i) R_K^H is a functor from AK-modules to AH-modules which maps the left AK-module N to $R_K^H(N) := M \otimes_{AK} N = (M \otimes_A N)_K$, which is the coinvariant space for the diagonal action of K given by $m \otimes n \mapsto mk^{-1} \otimes kn$. The action of H on $M \otimes_{AK} N$ comes from its action on M. A morphism $f : N \to N'$ of AK-modules is mapped by R_K^H to the morphism $\mathrm{Id}_M \otimes_{AK} f : M \otimes_{AK} N \to M \otimes_{AK} N'$ of AH-modules.

79

(ii) $^*R_K^H$ is a functor from AH-modules to AK-modules which maps the AH-module P to $\mathrm{Hom}_{AH}(M,P) = \mathrm{Hom}_A(M,P)^H$, which is the invariant space for the action of H on $\mathrm{Hom}_A(M,P)$ given by $\varphi \mapsto h\varphi(h^{-1}\,.)$; here $k \in K$ acts by $\varphi \mapsto \varphi(.k)$. A morphism $f : P \to P'$ of AH-modules is mapped by $^*R_K^H$ to the morphism $\mathrm{Hom}_{AH}(M,P) \to \mathrm{Hom}_{AH}(M,P')$ of AK-modules given by composition with f.

Remark 5.1.2   If M is projective over AH then $\mathrm{Hom}_{AH}(M,.)$ is isomorphic to $M^\vee \otimes_{AH} .$ where $^\vee$ means dual over AH – see Bourbaki (1971, II §4 no 2, Corollaire); that is, $^*R_K^H$ is the functor R_K^H associated with $M^\vee$, an AK-module-AH.

Proposition 5.1.3 R_K^H *is left-adjoint to* $^*R_K^H$.

Proof   This is the classical "tensor-hom" adjunction, see Bourbaki (1971, II §4 no 1 Proposition 1 (a)).   □

Proposition 5.1.4 (Transitivity)   *Let G, H, and K be finite groups; let M be an AG-module-AH and N an AH-module-AK, then:*

(i) *The functor $R_H^G \circ R_K^H$ is equal to the functor R_K^G defined by the AG-module-AK given by $M \otimes_{AH} N$.*
(ii) *The functor $^*R_K^H \circ {}^*R_H^G$ is equal to the functor $^*R_K^G$ defined by the AG-module-AK given by $M \otimes_{AH} N$.*

Proof   The first item comes from the associativity of the tensor product. We get the second item from the first by adjunction.   □

Proposition 5.1.5   *Assume $|K|$ invertible in A. Let M be an AH-module-AK and N be an AK-module; then for $h \in H$ we have*

$$\mathrm{Trace}(h \mid R_K^H N) = |K|^{-1} \sum_{k \in K} \mathrm{Trace}((h,k^{-1}) \mid M)\,\mathrm{Trace}(k \mid N).$$

Proof   The element $e_K := |K|^{-1} \sum_k k^{-1} \otimes k \in AK \otimes_A AK$ is the projector on the K-invariants in $M \otimes_A N$ which are isomorphic to the K-coinvariants since the kernel of e_K is the image of $1 - e_K$. Hence for $h \in H$ we have

$$\mathrm{Trace}(h \mid M \otimes_{AK} N) = \mathrm{Trace}(he \mid M \otimes_A N)$$

$$= |K|^{-1} \sum_{k \in K} \mathrm{Trace}((h,k^{-1}) \otimes k \mid M \otimes_A N)$$

$$= |K|^{-1} \sum_{k \in K} \mathrm{Trace}((h,k^{-1}) \mid M)\,\mathrm{Trace}(k \mid N).   \square$$

Examples 5.1.6 (induction and restriction)   We take for K a subgroup of H and for M the group algebra AH on which H acts by left multiplication and K by right multiplication. The above construction gives the induction functor $AH \otimes_{AK} \cdot$ whose right-adjoint is $\mathrm{Hom}_{AH}(AH,.)$. Since for any AH-module P we have an isomorphism $\mathrm{Hom}_{AH}(AH,P) \simeq P$ – where the action of K on the left-hand side becomes its action by restriction on P, we get that the right-adjoint of the induction is the restriction.

We can also consider AH as an AK-module-AH where H acts by right multiplication and K by left multiplication. This defines a functor $AH \otimes_{AH} \cdot$ whose right-adjoint is $\mathrm{Hom}_{AK}(AH,.)$. Since $AH \otimes_{AH} P \simeq P$, where the action of K becomes the restriction of the action of H, we see that $AH \otimes_{AH} \cdot$ is the restriction functor. We leave as an exercise to show, using the fact that AH is free over AK, that $\mathrm{Hom}_{AK}(AH,.)$ is isomorphic to $AH \otimes_{AK} \cdot$ and deduce that restriction and induction are biadjoint.

The next example will play an important part in this book. It was first considered by Harish-Chandra (1970).

Definition 5.1.7   Let $\mathbf{G}$ be a connected reductive group over $\overline{\mathbb{F}}_p$ and F be a Frobenius root; let $\mathbf{P}$ be an F-stable parabolic subgroup of $\mathbf{G}$ and $\mathbf{L}$ an F-stable Levi subgroup of $\mathbf{P}$, so that we have an F-stable Levi decomposition $\mathbf{P} = \mathbf{U} \rtimes \mathbf{L}$.

(i) the **Harish-Chandra induction** – denoted by $R_{\mathbf{L} \subset \mathbf{P}}^{\mathbf{G}}$ – is the functor obtained by taking M in 5.1.1(i) to be the $A\mathbf{G}^F$-module-$A\mathbf{L}^F$ given by $A(\mathbf{G}^F/\mathbf{U}^F)$ where $\mathbf{G}^F$ acts by left translations and $\mathbf{L}^F$ by right translations. This is possible since $\mathbf{L}$ normalises $\mathbf{U}$.

(ii) the **Harish-Chandra restriction** – would be denoted by $^*R_{\mathbf{L} \subset \mathbf{P}}^{\mathbf{G}}$ – is the functor obtained by taking M in 5.1.1(i) to be the $A\mathbf{L}^F$-module-$A\mathbf{G}^F$ given by $A(\mathbf{U}^F \backslash \mathbf{G}^F)$ where $\mathbf{G}^F$ acts by right translations and $\mathbf{L}^F$ by left translations.

The right-adjoint of $R_{\mathbf{L} \subset \mathbf{P}}^{\mathbf{G}}$ is $\mathrm{Hom}_{A\mathbf{G}^F}(A(\mathbf{G}^F/\mathbf{U}^F),.)$. The right-adjoint of $^*R_{\mathbf{L} \subset \mathbf{P}}^{\mathbf{G}}$ is $\mathrm{Hom}_{A\mathbf{L}^F}(A(\mathbf{U}^F \backslash \mathbf{G}^F),.)$.

Proposition 5.1.8   *Assume that the prime p is invertible in A.*

(i) $R_{\mathbf{L} \subset \mathbf{P}}^{\mathbf{G}} = \mathrm{Ind}_{A\mathbf{P}^F}^{A\mathbf{G}^F} \circ \mathrm{Inf}_{A\mathbf{L}^F}^{A\mathbf{P}^F}$ *(where the* **inflation** $\mathrm{Inf}_{A\mathbf{L}^F}^{A\mathbf{P}^F}$ *is the natural lifting through the quotient $\mathbf{P}^F \to \mathbf{L}^F$).*

(ii) $^*R_{\mathbf{L} \subset \mathbf{P}}^{\mathbf{G}}$ *is* $\mathrm{Res}_{A\mathbf{P}^F}^{A\mathbf{G}^F}$ *followed with the taking of fixed points under $\mathbf{U}^F$.*

(iii) $R_{\mathbf{L} \subset \mathbf{P}}^{\mathbf{G}}$ *and* $^*R_{\mathbf{L} \subset \mathbf{P}}^{\mathbf{G}}$ *are exact and biadjoint.*

Proof   We prove first (ii). Since p is invertible in A we have an A-module isomorphism $A(\mathbf{U}^F \backslash \mathbf{G}^F) \simeq e_{\mathbf{U}^F}(A\mathbf{G}^F)$ where $e_{\mathbf{U}^F} = |\mathbf{U}^F|^{-1} \sum_{u \in \mathbf{U}^F} u$. For an $A\mathbf{G}^F$-module X this gives $^*R_{\mathbf{L} \subset \mathbf{P}}^{\mathbf{G}}(X) = A(\mathbf{U}^F \backslash \mathbf{G}^F) \otimes_{A\mathbf{G}^F} X \simeq e_{\mathbf{U}^F} X = X^{\mathbf{U}^F}$ as $A\mathbf{L}^F$-modules whence (ii).

Now for any AG^F-module X we have $\mathrm{Hom}_{AG^F}(A(\mathbf{G}^F/\mathbf{U}^F), X) \simeq X^{\mathbf{U}^F}$ as $A\mathbf{L}^F$-modules, hence ${}^*R^{\mathbf{G}}_{\mathbf{L}\subset\mathbf{P}} \simeq \mathrm{Hom}_{AG^F}(A(\mathbf{G}^F/\mathbf{U}^F),.)$. Hence ${}^*R^{\mathbf{G}}_{\mathbf{L}\subset\mathbf{P}}$ is right-adjoint to $R^{\mathbf{G}}_{\mathbf{L}\subset\mathbf{P}}$.

We now prove that ${}^*R^{\mathbf{G}}_{\mathbf{L}\subset\mathbf{P}}$ is also left-adjoint to $R^{\mathbf{G}}_{\mathbf{L}\subset\mathbf{P}}$. Since $A(\mathbf{G}^F/\mathbf{U}^F) \simeq AG^F e_{\mathbf{U}^F}$ which is a direct factor of AG^F and AG^F is free over $A\mathbf{L}^F$ we get that $A(\mathbf{G}^F/\mathbf{U}^F)$ is a finitely-generated projective $A\mathbf{L}^F$-module. Thus $R^{\mathbf{G}}_{\mathbf{L}\subset\mathbf{P}} = A(\mathbf{G}^F/\mathbf{U}^F) \otimes_{A\mathbf{L}^F} . \simeq \mathrm{Hom}_{A\mathbf{L}^F}(A(\mathbf{G}^F/\mathbf{U}^F)^\vee,.)$ – see Bourbaki (1971, II §4 no 2, Corollaire) – where $^\vee$ means dual over $A\mathbf{L}^F$, and this isomorphism is compatible with the $\mathbf{G}^F$-action. If we show that $A(\mathbf{G}^F/\mathbf{U}^F)^\vee$ is isomorphic to $A(\mathbf{U}^F\backslash\mathbf{G}^F)$ as right AG^F-modules, we will get the adjunction we want. We leave as an exercise to show that $(\mathbf{U}^F g, g'\mathbf{U}^F) \mapsto \begin{cases} 0 & \text{if } gg' \notin \mathbf{P}^F \\ l & \text{if } gg' \in l\mathbf{U}^F \text{ with } l \in \mathbf{L}^F \end{cases}$ defines a perfect $A\mathbf{L}^F$-duality compatible with the $\mathbf{G}^F$-actions.

Both $R^{\mathbf{G}}_{\mathbf{L}\subset\mathbf{P}}$ and ${}^*R^{\mathbf{G}}_{\mathbf{L}\subset\mathbf{P}}$ are exact since they are defined by tensor product with projective modules.

Using adjunction and the fact that taking the fixed points under $\mathbf{U}^F$ is adjoint to lifting representations from $A\mathbf{L}^F$ to $A\mathbf{P}^F$ we get (i).   □

Remark 5.1.9   5.1.8(iii) shows that when p is invertible in A, we could have used 5.1.1(ii) instead of 5.1.1(i) to define $R^{\mathbf{G}}_{\mathbf{L}}$ and ${}^*R^{\mathbf{G}}_{\mathbf{L}}$.

Remark 5.1.10   We have only defined here a functor $R^{\mathbf{G}}_{\mathbf{L}\subset\mathbf{P}}$ when $\mathbf{L}$ is an F-stable Levi subgroup of an *F-stable* parabolic subgroup $\mathbf{P}$ of $\mathbf{G}$; later in 9.1.3 when $A = \overline{\mathbb{Q}}_\ell$ we will define Deligne–Lusztig induction which generalises Harish-Chandra induction to the case of F-stable Levi subgroups of non-F-stable parabolic subgroups $\mathbf{P}$.

Proposition 5.1.11 (transitivity)   *Let $\mathbf{P}$ be an F-stable parabolic subgroup of $\mathbf{G}$, and $\mathbf{Q}$ an F-stable parabolic subgroup contained in $\mathbf{P}$. Let $\mathbf{L}$ be an F-stable Levi subgroup of $\mathbf{P}$ and $\mathbf{M}$ an F-stable Levi subgroup of $\mathbf{Q}$ contained in $\mathbf{L}$. Then*

$$R^{\mathbf{G}}_{\mathbf{L}\subset\mathbf{P}} \circ R^{\mathbf{L}}_{\mathbf{M}\subset\mathbf{L}\cap\mathbf{Q}} = R^{\mathbf{G}}_{\mathbf{M}\subset\mathbf{Q}} \text{ and } {}^*R^{\mathbf{L}}_{\mathbf{M}\subset\mathbf{L}\cap\mathbf{Q}} \circ {}^*R^{\mathbf{G}}_{\mathbf{L}\subset\mathbf{P}} = {}^*R^{\mathbf{G}}_{\mathbf{M}\subset\mathbf{Q}}.$$

Proof   We prove the statement for Harish-Chandra induction; the statement for restriction follows by adjunction.

Let $\mathbf{U}$ be the unipotent radical of $\mathbf{P}$ and $\mathbf{V}$ that of $\mathbf{Q}$. We have $\mathbf{U} \subset \mathbf{V}$, so according to 3.4.8(ii) we have $\mathbf{V} = \mathbf{U}(\mathbf{L}\cap\mathbf{V})$ and the uniqueness in that decomposition implies that $\mathbf{V}^F = \mathbf{U}^F(\mathbf{L}^F \cap \mathbf{V}^F)$. Using 5.1.4, proving the proposition is equivalent to showing that

$$A(\mathbf{G}^F/\mathbf{U}^F) \otimes_{A\mathbf{L}^F} A(\mathbf{L}^F/(\mathbf{L}\cap\mathbf{V})^F) \xrightarrow{\sim} A(\mathbf{G}^F/\mathbf{V}^F)$$

as AG^F-modules-AM^F. But this clearly results from the isomorphism of "$\mathbf{G}^F$-sets-$\mathbf{M}^F$" (with notation of 2.3.11)

$$\mathbf{G}^F/\mathbf{U}^F \times_{\mathbf{L}^F} \mathbf{L}^F/(\mathbf{L} \cap \mathbf{V})^F \xrightarrow{\sim} \mathbf{G}^F/\mathbf{V}^F$$

which is induced by the map $(g\mathbf{U}^F, l(\mathbf{L} \cap \mathbf{V})^F) \mapsto gl\mathbf{V}^F$ from $\mathbf{G}^F/\mathbf{U}^F \times \mathbf{L}^F/(\mathbf{L} \cap \mathbf{V})^F$ to $\mathbf{G}^F/\mathbf{V}^F$. This map is well defined since l normalises $\mathbf{U}$ and $\mathbf{U} \subset \mathbf{V}$. Using the decomposition $\mathbf{V}^F = \mathbf{U}^F(\mathbf{L} \cap \mathbf{V})^F$, it is easy to see that it factors through the amalgamated product and defines an isomorphism. □

We have the following character formula:

Proposition 5.1.12 *Assume the prime p invertible in A. If X is an AG^F-module, for $l \in \mathbf{L}^F$ we have*

$$\text{Trace}(l \mid {}^*R^{\mathbf{G}}_{\mathbf{L}\subset\mathbf{P}}(X)(l)) = |\mathbf{U}^F|^{-1} \sum_{u\in\mathbf{U}^F} \text{Trace}(lu \mid X).$$

Proof This formula comes from the description of ${}^*R^{\mathbf{G}}_{\mathbf{L}\subset\mathbf{P}}$ as the action of $\mathbf{L}^F$ on $X^{\mathbf{U}^F}$ using that $e_{\mathbf{U}^F} = |\mathbf{U}^F|^{-1} \sum_{u\in\mathbf{U}^F} u$ is the projector on that fixed point subspace. □

5.2 The Mackey Formula

A fundamental property of Harish-Chandra induction and restriction is given in Theorem 5.2.1 below. This is an analogue of the usual Mackey formula for the composition of restriction and induction. A corollary will be that in characteristic 0 Harish-Chandra induction and restriction are independent of the parabolic subgroup used in their definition.

The statement of the **Mackey formula** below remains valid for the Lusztig induction and restriction functors (dropping the assumption that the parabolic subgroups involved are F-stable) in the cases that we will be able to handle (see Chapter 9). This statement is due to Deligne but was first published in Lusztig and Spaltenstein (1985, 2.5).

Theorem 5.2.1 (Mackey formula) *Let $\mathbf{P}$ and $\mathbf{Q}$ be two F-stable parabolic subgroups of $\mathbf{G}$, and $\mathbf{L}$ (resp. $\mathbf{M}$) be an F-stable Levi subgroup of $\mathbf{P}$ (resp. $\mathbf{Q}$). Then*

$$ {}^*R^{\mathbf{G}}_{\mathbf{L}\subset\mathbf{P}} \circ R^{\mathbf{G}}_{\mathbf{M}\subset\mathbf{Q}} = \bigoplus_x R^{\mathbf{L}}_{\mathbf{L}\cap{}^x\mathbf{M}\subset\mathbf{L}\cap{}^x\mathbf{Q}} \circ {}^*R^{{}^x\mathbf{M}}_{\mathbf{L}\cap{}^x\mathbf{M}\subset\mathbf{P}\cap{}^x\mathbf{M}} \circ \text{ad}\,x, $$

where $\text{ad}\,x$ is the conjugation from AM^F-modules to $A{}^x M^F$-modules, and x runs over a set of representatives of $\mathbf{L}^F \backslash \mathcal{S}(\mathbf{L},\mathbf{M})^F / \mathbf{M}^F$, with

$$\mathcal{S}(\mathbf{L},\mathbf{M}) = \{x \in \mathbf{G} \mid \mathbf{L} \cap {}^x\mathbf{M} \text{ contains a maximal torus of } \mathbf{G}\}.$$

In order to prove this theorem, we first study double coset representatives with respect to parabolic subgroups of $\mathbf{G}$, using Proposition 3.2.3.

Lemma 5.2.2 *If $S(\mathbf{L},\mathbf{M})$ is as in 5.2.1, then:*

(i) *The natural map $S(\mathbf{L},\mathbf{M}) \to \mathbf{P}\backslash\mathbf{G}/\mathbf{Q}$ induces an isomorphism*

$$\mathbf{L}\backslash S(\mathbf{L},\mathbf{M})/\mathbf{M} \xrightarrow{\sim} \mathbf{P}\backslash\mathbf{G}/\mathbf{Q}.$$

(ii) $\mathbf{P}^F\backslash\mathbf{G}^F/\mathbf{Q}^F$ *can be identified with* $(\mathbf{P}\backslash\mathbf{G}/\mathbf{Q})^F$, *and* $\mathbf{L}^F\backslash S(\mathbf{L},\mathbf{M})^F/\mathbf{M}^F$ *with* $(\mathbf{L}\backslash S(\mathbf{L},\mathbf{M})/\mathbf{M})^F$.

Proof Let us first prove (ii): by 4.2.14, since $\mathbf{P} \times \mathbf{Q}$, as well as the stabiliser isomorphic to $\mathbf{P} \cap {}^x\mathbf{Q}$ of a point $x \in \mathbf{G}$ under the action $x \xrightarrow{(p,q)} pxq$ of $\mathbf{P} \times \mathbf{Q}$ on $\mathbf{G}$, are connected, the double cosets $\mathbf{P}^F\backslash\mathbf{G}^F/\mathbf{Q}^F$ can be identified with $(\mathbf{P}\backslash\mathbf{G}/\mathbf{Q})^F$. Similarly, as $\mathbf{L} \cap {}^x\mathbf{M}$ is connected when it contains a maximal torus (see 3.4.8(ii)), the double cosets $\mathbf{L}^F\backslash S(\mathbf{L},\mathbf{M})^F/\mathbf{M}^F$ can be identified with $(\mathbf{L}\backslash S(\mathbf{L},\mathbf{M})/\mathbf{M})^F$.

Let us prove (i). We may conjugate $\mathbf{Q}$ by an element $g \in \mathbf{G}$ so that $\mathbf{P}$ and ${}^g\mathbf{Q}$ contain a common Borel subgroup and that $\mathbf{L}$ and ${}^g\mathbf{M}$ contain a common maximal torus $\mathbf{T}$. Then for any $x \in \mathbf{G}$ we have $\mathbf{P}x\mathbf{Q} = (\mathbf{P}xg^{-1}\,{}^g\mathbf{Q})g$; that is, the double cosets with respect to $\mathbf{P}$ and $\mathbf{Q}$ are translates of those with respect to $\mathbf{P}$ and ${}^g\mathbf{Q}$. Similarly $S(\mathbf{L},\mathbf{M}) = S(\mathbf{L},{}^g\mathbf{M})g$. Thus to show (i) we may replace $\mathbf{Q}$ by ${}^g\mathbf{Q}$; that is, assume that $\mathbf{P}$ and $\mathbf{Q}$ are standard and that $\mathbf{L}$ and $\mathbf{M}$ are standard Levi subgroups, that is, $\mathbf{P} = \mathbf{P}_I$, $\mathbf{Q} = \mathbf{P}_J$, $\mathbf{L} = \mathbf{L}_I$, and $\mathbf{M} = \mathbf{L}_J$. By Proposition 3.2.3(ii), there exist representatives in $N_\mathbf{G}(\mathbf{T})$ of the double cosets in $\mathbf{P}_I\backslash\mathbf{G}/\mathbf{P}_J$. Thus every double coset has a representative in $S(\mathbf{L},\mathbf{M})$ and the map of (i) is surjective. Conversely, if $x \in S(\mathbf{L},\mathbf{M})$, the group $\mathbf{L} \cap {}^x\mathbf{M}$ contains a maximal torus. This torus is of the form ${}^l\mathbf{T}$ for some $l \in \mathbf{L}$ and also of the form ${}^{xm}\mathbf{T}$ for some $m \in \mathbf{M}$, since all maximal tori in $\mathbf{L}$ or $\mathbf{M}$ are conjugate. But then $l^{-1}xm \in N_\mathbf{G}(\mathbf{T})$ and is in the double coset $\mathbf{L}x\mathbf{M}$. Now let x and y be two elements of $S(\mathbf{L},\mathbf{M})$ such that $\mathbf{P}x\mathbf{Q} = \mathbf{P}y\mathbf{Q}$. We just saw that there are elements of $N_\mathbf{G}(\mathbf{T})$ in $\mathbf{L}x\mathbf{M}$ and $\mathbf{L}y\mathbf{M}$. By Proposition 3.2.3(ii), these elements of $N_\mathbf{G}(\mathbf{T})$ are in the same double coset with respect to W_I and W_J, thus $\mathbf{L}x\mathbf{M} = \mathbf{L}y\mathbf{M}$. □

Proof of Theorem 5.2.1 Let $\mathbf{U}$ (resp. $\mathbf{V}$) denote the unipotent radical of $\mathbf{P}$ (resp. $\mathbf{Q}$). By 5.1.4, the functor in the left-hand side of the Mackey formula corresponds to the $A\mathbf{L}^F$-module-$A\mathbf{M}^F$ given by

$$A(\mathbf{U}^F\backslash\mathbf{G}^F) \otimes_{A\mathbf{G}^F} A(\mathbf{G}^F/\mathbf{V}^F).$$

This module is clearly isomorphic to $A(\mathbf{U}^F\backslash\mathbf{G}^F/\mathbf{V}^F)$. Let us decompose $\mathbf{G}^F$ into double cosets with respect to $\mathbf{P}^F$ and $\mathbf{Q}^F$. We get

$$\mathbf{U}^F\backslash\mathbf{G}^F/\mathbf{V}^F = \coprod_{x\in L^F\backslash S(\mathbf{L},\mathbf{M})^F/\mathbf{M}^F} \mathbf{U}^F\backslash\mathbf{P}^Fx\mathbf{Q}^F/\mathbf{V}^F.$$

We now use Lemma 5.2.3 to continue.

Lemma 5.2.3 *For any $x \in S(\mathbf{L},\mathbf{M})^F$ the map*

$$(l(\mathbf{L}\cap{}^x\mathbf{V})^F,({}^x\mathbf{M}\cap\mathbf{U})^F.{}^xm) \mapsto \mathbf{U}^F lxm\mathbf{V}^F$$

$$\mathbf{L}^F/(\mathbf{L}\cap{}^x\mathbf{V})^F \times_{(\mathbf{L}\cap{}^x\mathbf{M})^F} ({}^x\mathbf{M}\cap\mathbf{U})^F\backslash{}^x\mathbf{M}^F \to \mathbf{U}^F\backslash\mathbf{P}^Fx\mathbf{Q}^F/\mathbf{V}^F,$$

where the notation is as in 2.3.11, is an isomorphism.

Proof It is easy to check that the map is well defined. The stabiliser – isomorphic to $\mathbf{U}\cap{}^x\mathbf{V}$ – of a point $x \in \mathbf{G}$ under the action $x \xrightarrow{(u,v)} uxv$ of $\mathbf{U}\times\mathbf{V}$ is connected (see 2.3.1(v)). The stabiliser of a point of $\mathbf{L}^F/(\mathbf{L}\cap{}^x\mathbf{V})^F \times ({}^x\mathbf{M}\cap\mathbf{U})^F\backslash{}^x\mathbf{M}^F$ under diagonal action of $\mathbf{L}\cap{}^x\mathbf{M}$ is reduced to 1 and thus connected. Hence it is enough to show that the same map at the level of the algebraic varieties is an isomorphism, as it clearly commutes with F. The map is surjective since any element of $\mathbf{P}x\mathbf{Q}$ has a decomposition of the form $ulxmv$ (with each term in the group suggested by its name). We now prove its injectivity. If $\mathbf{U}lxm\mathbf{V} = \mathbf{U}l'xm'\mathbf{V}$, then $lxm = l'uxvm'$ for some $u \in \mathbf{U}$ and $v \in \mathbf{V}$ since l' normalises $\mathbf{U}$ and m' normalises $\mathbf{V}$. Thus

$$u^{-1}l'^{-1}l = {}^x(vm'm^{-1}) \in \mathbf{P}\cap{}^x\mathbf{Q}. \tag{5.2.1}$$

The component in $\mathbf{L}\cap{}^x\mathbf{M}$ of the left-hand side of (5.2.1) is that of $l'^{-1}l$, and the component in $\mathbf{L}\cap{}^x\mathbf{M}$ of the right-hand side of (5.2.1) is that of ${}^x(m'm^{-1})$ by the following lemma.

Lemma 5.2.4 *Let $\mathbf{P} = \mathbf{U}\rtimes\mathbf{L}$ and $\mathbf{Q} = \mathbf{V}\rtimes\mathbf{M}$ be respective Levi decompositions of two parabolic subgroups $\mathbf{P}$ and $\mathbf{Q}$, and assume $\mathbf{L}\cap\mathbf{M}$ contains a maximal torus. If $ul = vm \in \mathbf{P}\cap\mathbf{Q}$, then $u_{\mathbf{M}} = m_{\mathbf{U}}$, $l_{\mathbf{V}} = v_{\mathbf{L}}$ and $l_{\mathbf{M}} = m_{\mathbf{L}}$, with obvious notation: $u \in \mathbf{U}$, $l \in \mathbf{L}$, $v \in \mathbf{V}$, $m \in \mathbf{M}$, $u_{\mathbf{M}}$ is the component in $\mathbf{U}\cap\mathbf{M}$ of $u \in \mathbf{U}$ in the decomposition $\mathbf{P}\cap\mathbf{Q} = (\mathbf{U}\cap\mathbf{V})(\mathbf{U}\cap\mathbf{M})(\mathbf{L}\cap\mathbf{V})(\mathbf{L}\cap\mathbf{M})$ of 3.4.8(ii), ...*

Proof We have

$$u_{\mathbf{V}}u_{\mathbf{M}}l_{\mathbf{V}}l_{\mathbf{M}} = v_{\mathbf{U}}v_{\mathbf{L}}m_{\mathbf{U}}m_{\mathbf{L}} = v_{\mathbf{U}}[v_{\mathbf{L}},m_{\mathbf{U}}]m_{\mathbf{U}}v_{\mathbf{L}}m_{\mathbf{L}}$$

and the commutator which appears in this formula is in $\mathbf{V}\cap\mathbf{U}$. The uniqueness of the decomposition in 3.4.8(ii) gives the result. □

Thus $(l(\mathbf{L} \cap {}^x\mathbf{V}), ({}^x\mathbf{M} \cap \mathbf{U})^x m)$ and $(l'(\mathbf{L} \cap {}^x\mathbf{V}), ({}^x\mathbf{M} \cap \mathbf{U})^x m')$ are equal in the amalgamated product, whence the injectivity and proof of Lemma 5.2.3. $\quad\square$

Taking then the union over all x, we get

$$\mathbf{U}^F \backslash \mathbf{G}^F / \mathbf{V}^F \xrightarrow{\sim} \coprod_{x \in \mathbf{L}^F \backslash \mathcal{S}(\mathbf{L},\mathbf{M})^F / \mathbf{M}^F} \mathbf{L}^F / (\mathbf{L} \cap {}^x\mathbf{V})^F \times_{(\mathbf{L} \cap {}^x\mathbf{M})^F} ({}^x\mathbf{M} \cap \mathbf{U})^F \backslash {}^x\mathbf{M}^F,$$

and this bijection is compatible with the action of $\mathbf{L}^F$ by left multiplication, and with that of $\mathbf{M}^F$ by right multiplication on the left-hand side and by the composite of left multiplication and $\operatorname{ad} x$ on the x term of the right-hand side. We eventually get an isomorphism of $A\mathbf{L}^F$-modules-$A\mathbf{M}^F$:

$$A(\mathbf{U}^F \backslash \mathbf{G}^F / \mathbf{V}^F)$$
$$\xrightarrow{\sim} \bigoplus_{x \in \mathbf{L}^F \backslash \mathcal{S}(\mathbf{L},\mathbf{M})^F / \mathbf{M}^F} A(\mathbf{L}^F / (\mathbf{L} \cap {}^x\mathbf{V})^F) \otimes_{A((\mathbf{L} \cap {}^x\mathbf{M})^F)} A(({}^x\mathbf{M} \cap \mathbf{U})^F \backslash {}^x\mathbf{M}^F),$$

where the action of $\mathbf{M}^F$ on the right-hand side is as explained above. The $A\mathbf{L}^F$-module-$A\mathbf{M}^F$ which appears in the right-hand side is exactly the one which corresponds to the functor in the right-hand side of the Mackey formula, whence the theorem. $\quad\square$

5.3 Harish-Chandra Theory

We explain now the concept of "cuspidal representations" which is due to Harish-Chandra.

From now on A denotes a commutative ring in which p is invertible. In many proofs and statements we assume that A is a subfield K of $\mathbb{C}$. In that case, for a finite group G, we denote by $C_K(G)$ (or $C(G)$ when K is clear from the context) the space of K-valued class functions on G, which is endowed with the **scalar product** $\langle f, f' \rangle_G = |G|^{-1} \sum_{g \in G} f(g) \overline{f'(g)}$. We will use $\mathbf{1}_G$ and reg_G respectively for the trivial and the regular characters of G (or just $\mathbf{1}$ and reg when G is clear from the context), and we will denote by $\operatorname{Irr}(G)$ the set of irreducible characters of G over K.

Theorem 5.3.1 *The functors $R^{\mathbf{G}}_{\mathbf{L} \subset \mathbf{P}}$ and ${}^*R^{\mathbf{G}}_{\mathbf{L} \subset \mathbf{P}}$ are independent of the parabolic subgroup used in their definition. Thus, from now on we will use for them the notations $R^{\mathbf{G}}_{\mathbf{L}}$ and ${}^*R^{\mathbf{G}}_{\mathbf{L}}$.*

Proof The general result when p is invertible in A is due to Dipper and Du (1993) and Howlett and Lehrer (1994). We prove it only when $A = K$, thus we need only to prove the result for characters.

We will prove 5.3.1 for Harish-Chandra induction. The result for the restriction follows by adjunction. We proceed by induction on the **semi-simple rank** of $\mathbf{G}$, which is by definition the rank of $\mathbf{G}/R\mathbf{G}$.

Let us note that the semi-simple rank of a Levi subgroup $\mathbf{L}$ of a proper parabolic subgroup of $\mathbf{G}$ is strictly less than that of $\mathbf{G}$ since by 3.4.6 the radical $Z(\mathbf{L})^0$ of $\mathbf{L}$ contains strictly that of $\mathbf{G}$. If $\mathbf{L} = \mathbf{G}$ there is nothing to prove, so we may assume that the semi-simple rank of $\mathbf{L}$ is strictly less than that of $\mathbf{G}$. By the Mackey formula applied to only one Levi subgroup $\mathbf{L}$ but with two different parabolic subgroups $\mathbf{P}$ and $\mathbf{Q}$, we get for any character π of $\mathbf{L}^F$, using adjunction

$$\langle R^{\mathbf{G}}_{\mathbf{L}\subset\mathbf{P}}\pi, R^{\mathbf{G}}_{\mathbf{L}\subset\mathbf{Q}}\pi\rangle_{\mathbf{G}^F} = \sum_{x\in\mathbf{L}^F\backslash\mathcal{S}(\mathbf{L},\mathbf{L})^F/\mathbf{L}^F} \langle {}^*R^{{}^x\mathbf{L}}_{\mathbf{L}\cap{}^x\mathbf{L}\subset\mathbf{P}\cap{}^x\mathbf{L}}{}^x\pi, {}^*R^{\mathbf{L}}_{\mathbf{L}\cap{}^x\mathbf{L}\subset\mathbf{L}\cap{}^x\mathbf{Q}}\pi\rangle_{\mathbf{L}^F\cap{}^x\mathbf{L}^F}.$$

By induction, the right-hand side does not depend on the parabolic subgroups $\mathbf{P}$ and $\mathbf{Q}$, so the same is true for the left-hand side which means that, if we put $f(\mathbf{P}) = R^{\mathbf{G}}_{\mathbf{L}\subset\mathbf{P}}\pi$, we have

$$\langle f(\mathbf{P}), f(\mathbf{P})\rangle_{\mathbf{G}^F} = \langle f(\mathbf{P}), f(\mathbf{Q})\rangle_{\mathbf{G}^F}$$
$$= \langle f(\mathbf{Q}), f(\mathbf{Q})\rangle_{\mathbf{G}^F} = \langle f(\mathbf{Q}), f(\mathbf{P})\rangle_{\mathbf{G}^F}$$

whence $\langle f(\mathbf{P}) - f(\mathbf{Q}), f(\mathbf{P}) - f(\mathbf{Q})\rangle_{\mathbf{G}^F} = 0$ and so $f(\mathbf{P}) = f(\mathbf{Q})$. □

Example 5.3.2　We construct a Hopf algebra whose commutativity relies on the fact that $R^{\mathbf{G}}_{\mathbf{L}\subset\mathbf{P}}$ is independent of $\mathbf{P}$.

Let $C\mathbf{GL} = \bigoplus_{n\geq 0} C(\mathbf{GL}_n(\mathbb{F}_q))$ where here C denotes the $\mathbb{C}$-valued class functions. One can define on $C\mathbf{GL}$ a product by putting, for two class functions $\chi \in C(\mathbf{GL}_n(\mathbb{F}_q))$ and $\psi \in C(\mathbf{GL}_m(\mathbb{F}_q))$,

$$\chi \circ \psi = R^{\mathbf{GL}_{n+m}}_{\mathbf{GL}_n\times\mathbf{GL}_m}\chi \otimes \psi,$$

where $\mathbf{GL}_n\times\mathbf{GL}_m$ is embedded in $\mathbf{GL}_{n+m}$ as a group of block-diagonal matrices

$$\begin{pmatrix} \mathbf{GL}_n & 0 \\ 0 & \mathbf{GL}_m \end{pmatrix},$$

using the parabolic subgroup $\mathbf{P}_{n,m}$ of upper block-triangular matrices with the same diagonal blocks (see Example 3.2.5). We thus get an algebra structure. Since the Levi subgroups $\mathbf{GL}_n \times \mathbf{GL}_m$ and $\mathbf{GL}_m \times \mathbf{GL}_n$ are conjugate in $\mathbf{GL}_{n+m}$ (by 5.3.1) the product is commutative. Note that the parabolic subgroups $\mathbf{P}_{n,m}$ and $\mathbf{P}_{m,n}$ are not conjugate when $n \neq m$, thus the proof of commutativity relies in an essential way on 5.3.1. One can also define on $C\mathbf{GL}$ a cocommutative coproduct $C\mathbf{GL} \to C\mathbf{GL} \otimes C\mathbf{GL} : \chi \in C(\mathbf{GL}_n(\mathbb{F}_q)) \mapsto \bigoplus_{i+j=n} {}^*R^{\mathbf{GL}_n}_{\mathbf{GL}_i\times\mathbf{GL}_j}\chi$, and the Mackey formula implies that these two laws define on $C\mathbf{GL}$ a Hopf algebra structure.

The book Zhelevinski (1981) explains the theory of representations of general linear groups from the above viewpoint; this construction is originally due to Green (1955).

Because of the transitivity and exactness of $R_\mathbf{L}^\mathbf{G}$ (see 5.1.8 and 5.1.11), one can define a partial order on the set of pairs $(\mathbf{L}, M)$ consisting of an F-stable Levi subgroup of an F-stable parabolic subgroup of $\mathbf{G}$, and of M a simple $A\mathbf{L}^F$-module, by putting $(\mathbf{L}', M') \leq (\mathbf{L}, M)$ if $\mathbf{L}' \subset \mathbf{L}$ and there exists a surjective morphism $R_{\mathbf{L}'}^\mathbf{L} M' \twoheadrightarrow M$.

Proposition 5.3.3   *The following properties are equivalent:*

(i) *The pair $(\mathbf{L}, M)$ is minimal for the partial order defined above.*

(ii) *For any F-stable Levi subgroup $\mathbf{L}'$ of an F-stable parabolic subgroup of $\mathbf{L}$, we have $^*R_{\mathbf{L}'}^\mathbf{L} M = 0$.*

If these conditions are fulfilled, the $A\mathbf{L}^F$-module M is said to be **cuspidal** and the pair $(\mathbf{L}, M)$ is said to be a **cuspidal pair**.

Proof $^*R_{\mathbf{L}'}^\mathbf{L} M \neq 0$ if and only if there exists an $A\mathbf{L}'^F$-simple module M' such that $\mathrm{Hom}_{A\mathbf{L}'^F}(M', {}^*R_{\mathbf{L}'}^\mathbf{L}(M)) \neq 0$. By adjunction this is equivalent to $\mathrm{Hom}_{A\mathbf{L}^F}(R_{\mathbf{L}'}^\mathbf{L}(M'), M) \neq 0$ which is equivalent to $(\mathbf{L}, M)$ not being minimal since, as M is simple, any non-zero morphism in this last set is surjective.   □

Notation 5.3.4   *For a finite dimensional A-algebra H, we will denote by $\mathrm{Simp}(H)$ the set of simple H-modules up to isomorphism. If G is a finite group we will write $\mathrm{Simp}(G)$ for $\mathrm{Simp}(AG)$ if the base ring is clear from the context.*

We will denote by $\mathrm{Irr}(H)$ the set of irreducible characters of H, that is the characters of simple H-modules.

Definition 5.3.5   The **Harish-Chandra series** associated to a cuspidal pair $(\mathbf{L}, \Lambda)$, denoted by $\mathrm{Simp}(\mathbf{G}^F \mid (\mathbf{L}, \Lambda))$ – or $\mathrm{Simp}(A\mathbf{G}^F \mid (\mathbf{L}, \Lambda))$ is the set of simple quotients of $R_\mathbf{L}^\mathbf{G} \Lambda$.

Lemma 5.3.6   *A simple $A\mathbf{G}^F$-module M is in the Harish-Chandra series $\mathrm{Simp}(\mathbf{G}^F \mid (\mathbf{L}, \Lambda))$ if and only if Λ is a submodule of $^*R_\mathbf{L}^\mathbf{G}(M)$.*

Proof Since $R_\mathbf{L}^\mathbf{G}$ and $^*R_\mathbf{L}^\mathbf{G}$ are adjoint we have $\mathrm{Hom}_{A\mathbf{G}^F}(R_\mathbf{L}^\mathbf{G}\Lambda, M) \simeq \mathrm{Hom}_{A\mathbf{L}^F}(\Lambda, {}^*R_\mathbf{L}^\mathbf{G}M)$, whence the result: since Λ (resp. M) is simple, any non-zero morphism from Λ to $^*R_\mathbf{L}^\mathbf{G}M$ (resp. from $R_\mathbf{L}^\mathbf{G}\Lambda$ to M) is injective (resp. surjective).   □

Theorem 5.3.7   *Assume that A is a field. Two Harish-Chandra series $\mathrm{Simp}(\mathbf{G}^F \mid (\mathbf{L}, \Lambda))$ and $\mathrm{Simp}(\mathbf{G}^F \mid (\mathbf{L}', \Lambda'))$ have a non-empty intersection if and only if the cuspidal pairs $(\mathbf{L}, \Lambda)$ and $(\mathbf{L}', \Lambda')$ are $\mathbf{G}^F$-conjugate, in which case the two series are equal.*

Proof The last assertion comes the fact that for $g \in \mathbf{G}^F$ we have $R_{\mathbf{L}}^{\mathbf{G}}\Lambda \simeq$ ${}^g R_{\mathbf{L}}^{\mathbf{G}}\Lambda = R_{{}^g\mathbf{L}}^{\mathbf{G}} \, {}^g\Lambda$. Now if M is a quotient of $R_{\mathbf{L}}^{\mathbf{G}}\Lambda$ and if P_Λ and P_M are the respective projective covers of Λ and M (which exist since A is a field), we have a surjection $R_{\mathbf{L}}^{\mathbf{G}}P_\Lambda \twoheadrightarrow M$ using the exactness of $R_{\mathbf{L}}^{\mathbf{G}}$. Hence P_M must be a direct summand of $R_{\mathbf{L}}^{\mathbf{G}}P_\Lambda$. If M is a quotient of $R_{\mathbf{L}'}^{\mathbf{G}}\Lambda'$ for some cuspidal pair $(\mathbf{L}', \Lambda')$, then $\mathrm{Hom}_{A\mathbf{G}^F}(P_M, R_{\mathbf{L}'}^{\mathbf{G}}\Lambda') \neq 0$, whence $\mathrm{Hom}_{A\mathbf{G}^F}(R_{\mathbf{L}}^{\mathbf{G}}P_\Lambda, R_{\mathbf{L}'}^{\mathbf{G}}\Lambda') \neq 0$. Using the Mackey formula and adjunction, we get

$$\bigoplus_x \mathrm{Hom}_{\mathbf{L}^F \cap {}^x\mathbf{L}'^F}({}^*R_{\mathbf{L}^F \cap {}^x\mathbf{L}'^F}^{\mathbf{L}^F}P_\Lambda, {}^*R_{\mathbf{L}^F \cap {}^x\mathbf{L}'^F}^{{}^x\mathbf{L}'^F} \, {}^x\Lambda') \neq 0.$$

Since Λ' is cuspidal, the only non-zero terms in the above sum are for x such that ${}^x\mathbf{L}' \subset \mathbf{L}$. Thus there is an $x \in \mathbf{G}^F$ such that ${}^x\mathbf{L}' \subset \mathbf{L}$. Exchanging the roles of $\mathbf{L}$ and $\mathbf{L}'$ we see that there is a $y \in \mathbf{G}^F$ such that $\mathbf{L} \subset {}^y\mathbf{L}'$. This implies that $\mathbf{L}$ and $\mathbf{L}'$ have same dimension and we must have equalities. Hence there is an $x \in \mathbf{G}^F$ such that $\mathbf{L} = {}^x\mathbf{L}'$ and $\mathrm{Hom}_{\mathbf{L}}(P_\Lambda, {}^x\Lambda') \neq 0$. Since Λ' is simple there is a non-zero morphism from P_Λ to ${}^x\Lambda'$ if and only if $\Lambda \simeq {}^x\Lambda'$, whence the result. $\square$

Corollary 5.3.8 *If $(\mathbf{L}, \Lambda)$ and $(\mathbf{L}', \Lambda')$ are cuspidal pairs, and A is a field large enough so that $\mathrm{End}_{A\mathbf{L}^F}(\Lambda) = A$, then*

$$\dim(\mathrm{Hom}_{A\mathbf{G}^F}(R_{\mathbf{L}}^{\mathbf{G}}P_\Lambda, R_{\mathbf{L}'}^{\mathbf{G}}\Lambda')) = |\{x \in \mathbf{G}^F \mid {}^x\mathbf{L} = \mathbf{L}' \text{ and } {}^x\Lambda \simeq \Lambda'\}/\mathbf{L}^F|$$
$$= \dim(\mathrm{Hom}_{A\mathbf{G}^F}(R_{\mathbf{L}}^{\mathbf{G}}\Lambda, R_{\mathbf{L}'}^{\mathbf{G}}\Lambda')),$$

where P_Λ is the projective cover of Λ.

Proof The proof of 5.3.7 gives the first equality. Now by the Mackey formula, using the cuspidality of Λ and Λ' we have $\mathrm{Hom}_{A\mathbf{G}^F}(R_{\mathbf{L}}^{\mathbf{G}}\Lambda, R_{\mathbf{L}'}^{\mathbf{G}}\Lambda') = \bigoplus_x \mathrm{Hom}_{\mathbf{L}^F}(\Lambda, {}^x\Lambda')$, where x runs over $\mathbf{L}^F$-representatives of $\{x \in \mathbf{G}^F \mid {}^x\mathbf{L}' = \mathbf{L}\}$, whence the second equality. $\square$

As a consequence of Theorem 5.3.7 we have a partition of $\mathrm{Simp}(\mathbf{G}^F)$ into Harish-Chandra series. Moreover $R_{\mathbf{L}}^{\mathbf{G}}$ and ${}^*R_{\mathbf{L}}^{\mathbf{G}}$ preserve the Harish-Chandra series in the following sense:

Proposition 5.3.9 *Let $(\mathbf{L}_0, \Lambda_0)$ be a cuspidal pair with $\mathbf{L}_0$ a Levi subgroup of a Levi subgroup $\mathbf{L}$; then any simple quotient of $R_{\mathbf{L}}^{\mathbf{G}}(M)$ for $M \in$ $\mathrm{Simp}(\mathbf{L}^F \mid (\mathbf{L}_0, \Lambda_0))$ is in $\mathrm{Simp}(\mathbf{G}^F \mid (\mathbf{L}_0, \Lambda_0))$. Similarly any simple submodule of ${}^*R_{\mathbf{L}}^{\mathbf{G}}(M)$ for $M \in \mathrm{Simp}(\mathbf{G}^F \mid (\mathbf{L}_0, \Lambda_0))$ is in $\mathrm{Simp}(\mathbf{L}^F \mid (\mathbf{L}_0, \Lambda_0))$.*

Proof Both assertions are easy consequences of the transitivity and exactness of $R_{\mathbf{L}}^{\mathbf{G}}$ and ${}^*R_{\mathbf{L}}^{\mathbf{G}}$. $\square$

Remark 5.3.10   When $A = \mathbb{C}$ – since $P_\Lambda = \Lambda$ – Corollary 5.3.8 can be expressed as

$$\langle R_{\mathbf{L}}^{\mathbf{G}} \lambda, R_{\mathbf{L}'}^{\mathbf{G}} \lambda' \rangle_{\mathbf{G}^F} = |\{x \in \mathbf{G}^F \mid {}^x\mathbf{L} = \mathbf{L}' \text{ and } {}^x\lambda = \lambda'\}/\mathbf{L}^F|$$

where $\lambda \in \mathrm{Irr}(\mathbf{L}^F)$ and $\lambda' \in \mathrm{Irr}(\mathbf{L}'^F)$ are the characters of Λ and Λ'.

Remark 5.3.11   The union of the Harish-Chandra series $\mathrm{Simp}(\mathbf{G}^F \mid (\mathbf{L}, \Lambda))$ when Λ runs over the cuspidal representations of $\mathbf{L}^F$ is called the Harish-Chandra series associated to $\mathbf{L}$. When $\mathbf{L}$ is a quasi-split torus then all irreducible (thus degree one) representations of $\mathbf{L}^F$ are cuspidal. The series associated to any such torus is called the **principal series** (all quasi-split tori are $\mathbf{G}^F$-conjugate). The set of cuspidal representations of $\mathbf{G}^F$ is also the series associated to $\mathbf{G}$, which is called the **discrete series**. So the first problem of Harish-Chandra theory is to study the discrete series, not only in $\mathbf{G}^F$ but in any F-stable Levi subgroup of an F-stable parabolic subgroup of $\mathbf{G}$.

Notes

The approach to generalised induction functors using bimodules was initiated by Broué around 1980.

One of the first references to Harish-Chandra theory is Springer's article "Cusp forms for finite groups" in Borel et al. (1970).

In this chapter, as well as in the following one, we rely for the modular setting on lecture notes by Olivier Dudas, for which we thank him.

6

Iwahori–Hecke Algebras

In this chapter $\mathbf{G}$ is a connected reductive algebraic group over $\overline{\mathbb{F}}_p$, with a Frobenius root F; we consider a Harish-Chandra series $\mathrm{Simp}(\mathbf{G}^F \mid (\mathbf{L}, \Lambda))$ as in Chapter 5 and we assume that A is a field K of characteristic different from p which is large enough so that $\mathrm{End}_{K\mathbf{L}^F}(\Lambda) = K$. We make no further assumption on K until 6.1.12.

6.1 Endomorphism Algebras

We put $N_{\mathbf{G}^F}(\mathbf{L}, \Lambda) := \{ n \in N_{\mathbf{G}^F}(\mathbf{L}) \mid {}^n\Lambda \simeq \Lambda \}$ and $W_{\mathbf{G}^F}(\mathbf{L}, \Lambda) := N_{\mathbf{G}^F}(\mathbf{L}, \Lambda)/\mathbf{L}^F$. We have seen in 5.3.8 that we have $\dim \mathrm{End}_{K\mathbf{G}^F}(R_{\mathbf{L}}^{\mathbf{G}}\Lambda) = |W_{\mathbf{G}^F}(\mathbf{L}, \Lambda)|$. In this chapter we study the structure and representations of this algebra, which by the following theorem will lead to a parametrisation of $\mathrm{Simp}(\mathbf{G}^F \mid (\mathbf{L}, \Lambda))$.

Theorem 6.1.1

(i) *The functor* $\Theta : M \mapsto \mathrm{Hom}_{K\mathbf{G}^F}(R_{\mathbf{L}}^{\mathbf{G}}\Lambda, M)$ *induces a bijection from* $\mathrm{Simp}(\mathbf{G}^F|(\mathbf{L}, \Lambda))$ *to the simple* $\mathrm{End}_{K\mathbf{G}^F}(R_{\mathbf{L}}^{\mathbf{G}}\Lambda)$*-modules up to isomorphism.*

(ii) $R_{\mathbf{L}}^{\mathbf{G}}\Lambda$ *is a semi-simple module if and only if* $\mathrm{End}_{K\mathbf{G}^F}(R_{\mathbf{L}}^{\mathbf{G}}\Lambda)$ *is semi-simple. In this case* $R_{\mathbf{L}}^{\mathbf{G}}\Lambda$ *decomposes as a* $K\mathbf{G}^F \times \mathrm{End}_{K\mathbf{G}^F}(R_{\mathbf{L}}^{\mathbf{G}}\Lambda)$*-module as* $\oplus_{M \in \mathrm{Simp}(\mathbf{G}^F|(\mathbf{L}, \Lambda))} M \otimes_K \Theta(M)$. *In particular the multiplicity in* $R_{\mathbf{L}}^{\mathbf{G}}\Lambda$ *of a simple* $K\mathbf{G}^F$*-module* M *is equal to* $\dim \Theta(M)$ *and the multiplicity of* $\Theta(M)$ *is equal to* $\dim M$.

Proof We sketch the proof of (i), following Geck and Jacon (2011, 4.2.9).

Lemma 6.1.2 $R_{\mathbf{L}}^{\mathbf{G}}\Lambda$ *is a projective* $K\mathbf{G}^F/I$*-module where* I *is its annihilator.*

Proof Let P_Λ be a projective cover of Λ; then $P := R_{\mathbf{L}}^{\mathbf{G}}P_\Lambda$ is projective and we have a surjection $\pi : P \twoheadrightarrow R_{\mathbf{L}}^{\mathbf{G}}\Lambda$, since $R_{\mathbf{L}}^{\mathbf{G}}$ is exact. We thus get an injective map $\mathrm{Hom}_{K\mathbf{G}^F}(R_{\mathbf{L}}^{\mathbf{G}}\Lambda, R_{\mathbf{L}}^{\mathbf{G}}\Lambda) \hookrightarrow \mathrm{Hom}_{K\mathbf{G}^F}(P, R_{\mathbf{L}}^{\mathbf{G}}\Lambda)$. Since these two spaces have same dimension by 5.3.8, this last map is an isomorphism; any morphism φ : $P \to R_{\mathbf{L}}^{\mathbf{G}}\Lambda$ factorises through π, hence $\operatorname{Ker}\pi \subset \operatorname{Ker}\varphi$.

Now we have $IP \subset \operatorname{Ker}\pi$; we claim that $\operatorname{Ker}\pi = IP$. Since P is finitely generated, projective, we have $\operatorname{Hom}_{KG^F}(P, R_{\mathbf{L}}^{\mathbf{G}}\Lambda) \simeq P^{\vee} \otimes_{KG^F} R_{\mathbf{L}}^{\mathbf{G}}\Lambda$, see Bourbaki (1971, II §4 no 2, Corollaire), where $P^{\vee}$ is the dual of P over KG^F. This means that $\operatorname{Hom}(P, R_{\mathbf{L}}^{\mathbf{G}}\Lambda)$ is spanned by morphisms of the form $\varphi_{f,x} : p \mapsto f(p)x$ with $f \in P^{\vee}$ and $x \in R_{\mathbf{L}}^{\mathbf{G}}\Lambda$. By the first part of the proof $\operatorname{Ker}\pi \subset \cap_{x,f} \operatorname{Ker}\varphi_{f,x}$ and this last intersection consists of the elements $p \in P$ such that $f(p) \in I$ for all $f \in P^{\vee}$, that is it consists of IP, whence our claim.

We have proved that $R_{\mathbf{L}}^{\mathbf{G}}\Lambda$ is isomorphic to P/IP, hence is a projective KG^F/I-module.   □

Using that inflation is an equivalence of categories from KG^F/I-modules to KG^F-modules annihilated by I we get that for such a module M we have $\Theta(M) = \operatorname{Hom}_{KG^F/I}(R_{\mathbf{L}}^{\mathbf{G}}\Lambda, M)$. Conversely any M in the head of $R_{\mathbf{L}}^{\mathbf{G}}\Lambda$ is annihilated by I. Hence to get (i) we are reduced to KG^F/I-modules and properties of the induced functor $\overline{\Theta} : M \mapsto \operatorname{Hom}_{KG^F/I}(R_{\mathbf{L}}^{\mathbf{G}}\Lambda, M)$. Property (i) is then proved for $\overline{\Theta}$ by noticing that, since $R_{\mathbf{L}}^{\mathbf{G}}\Lambda$ is projective, applying $\overline{\Theta}$ to any surjective map $R_{\mathbf{L}}^{\mathbf{G}}\Lambda \twoheadrightarrow M$ gives a surjective map $\operatorname{End}_{KG^F/I}(R_{\mathbf{L}}^{\mathbf{G}}\Lambda) = \overline{\Theta}(R_{\mathbf{L}}^{\mathbf{G}}\Lambda) \twoheadrightarrow \overline{\Theta}(M)$ which implies that $\overline{\Theta}(M)$ is a simple $\operatorname{End}_{KG^F}(R_{\mathbf{L}}^{\mathbf{G}}\Lambda)$-module and then it remains to prove that all $\operatorname{End}_{KG^F}(R_{\mathbf{L}}^{\mathbf{G}}\Lambda)$-simple modules are obtained in this way. For the details see Geck and Jacon (2011, 4.1.3).

For (ii): the endomorphism algebra of a semi-simple module is semi-simple, see Bourbaki (1981, §5 no 4 Proposition 5 (a)). For the converse see Geck and Jacon (2011, 4.2.10). When $R_{\mathbf{L}}^{\mathbf{G}}\Lambda$ is semi-simple, it decomposes as in (ii), see for example Bourbaki (1981, §4, no 7, Proposition 8).   □

Exercise 6.1.3

(i) Let G be a finite group and P be a subgroup. Show that if a KP-module Λ is given by KPe with $e \in KP$ a primitive idempotent, then $\operatorname{Ind}_P^G \Lambda = KGe$ and for any KG-module M one has $\operatorname{Hom}_{KG}(KGe, M) \simeq eM$. Deduce that $H := \operatorname{End}_{KG}(\operatorname{Ind}_P^G \Lambda) \simeq eKGe$, acting on eM by the restriction to H of the action of KG. Show that $\dim_K(\operatorname{Hom}_{KG}(KGe, M)) = \operatorname{Trace}(e \mid M)$.

(ii) Deduce that if KG is semi-simple, the bijection on characters induced by the functor Θ of 6.1.1 is given by the restriction from KG to H.

We will now produce a basis of $\operatorname{End}_{KG^F}(R_{\mathbf{L}}^{\mathbf{G}}\Lambda)$ and determine its algebra structure.

Lemma 6.1.4 *For $n \in N_{G^F}(\mathbf{L}, \Lambda)$ there exists a bijective linear map $\gamma_n : \Lambda \to \Lambda$ such that for $l \in \mathbf{L}^F$ acting on Λ we have*

$$\gamma_n \circ l = {}^n l \circ \gamma_n.$$

Proof This results from the fact that the KL^F-modules Λ and $^n\Lambda$ are isomorphic. □

Since $\mathrm{End}_{KL^F}(\Lambda) = K$, for $n, n' \in N_{\mathbf{G}^F}(\mathbf{L}, \Lambda)$ there exist scalars $\lambda(n, n') \in K$ such that $\gamma_{nn'} = \lambda(n, n')\gamma_n\gamma_{n'}$, and λ is a 2-cocycle on $N_{\mathbf{G}^F}(\mathbf{L}, \Lambda) \times N_{\mathbf{G}^F}(\mathbf{L}, \Lambda)$ since it satisfies $\lambda(n, n')\lambda(n, n'n'')^{-1}\lambda(nn', n'')\lambda(n', n'')^{-1} = 1$, as can be seen by factorising $nn'n''$ in two different ways.

Lemma 6.1.5 *We have* $\dim(\mathrm{End}_{K\mathbf{G}^F}(R_{\mathbf{L}}^{\mathbf{G}}\Lambda)) = |W_{\mathbf{G}^F}(\mathbf{L}, \Lambda)|$ *and*

$$\mathrm{End}_{K\mathbf{G}^F}(R_{\mathbf{L}}^{\mathbf{G}}\Lambda) \simeq \bigoplus_{w \in W_{\mathbf{G}^F}(\mathbf{L}, \Lambda)} \mathrm{Hom}_{KL^F}(\Lambda, {}^{n_w}\Lambda),$$

where $n_w \in N_{\mathbf{G}^F}(\mathbf{L}, \Lambda)$ *represents* w.

Proof This results from 5.3.8 and its proof. □

Note that the vector space $\mathrm{Hom}_{KL^F}(\Lambda, {}^{n_w}\Lambda)$ is equal to $K\gamma_{n_w}$. Our aim is to describe the algebra structure on the right-hand side of the isomorphism of 6.1.5.

Recall that $R_{\mathbf{L}}^{\mathbf{G}} = K(\mathbf{G}^F/\mathbf{U}^F) \otimes_{KL^F} . = (K\mathbf{G}^F)e_{\mathbf{U}^F} \otimes_{KL^F} .$, where $e_{\mathbf{U}^F}$ is as in the proof of 5.1.8. For $n_w \in N_{\mathbf{G}^F}(\mathbf{L}, \Lambda)$ we define a linear map

$$B_{n_w} : R_{\mathbf{L}}^{\mathbf{G}}\Lambda \to R_{\mathbf{L}}^{\mathbf{G}}\Lambda$$

$$ge_{\mathbf{U}^F} \otimes_{KL^F} x \mapsto ge_{\mathbf{U}^F}n_w^{-1}e_{\mathbf{U}^F} \otimes_{KL^F} \gamma_{n_w}(x).$$

Using the equality in 6.1.4 and the fact that $\mathbf{L}^F$ centralises $e_{\mathbf{U}^F}$ we get that B_{n_w} is well defined. It clearly commutes with the action of $\mathbf{G}^F$.

Lemma 6.1.6 *By the isomorphism of 6.1.5, B_{n_w} is mapped to γ_{n_w}. In particular B_{n_w} is invertible and $\{B_{n_w} \mid w \in W_{\mathbf{G}^F}(\mathbf{L}, \Lambda)\}$ is a basis of* $\mathrm{End}_{K\mathbf{G}^F}(R_{\mathbf{L}}^{\mathbf{G}}\Lambda)$.

Proof We have to follow B_{n_w} through the isomorphism coming from the adjunction $\mathrm{Hom}_{K\mathbf{G}^F}(R_{\mathbf{L}}^{\mathbf{G}}\Lambda, R_{\mathbf{L}}^{\mathbf{G}}\Lambda) \simeq \mathrm{Hom}_{KL^F}(\Lambda, {}^*R_{\mathbf{L}}^{\mathbf{G}}R_{\mathbf{L}}^{\mathbf{G}}\Lambda)$ and then through the isomorphism coming from the Mackey formula.

The adjunction isomorphism is the tensor-hom adjunction

$$\mathrm{Hom}_{K\mathbf{G}^F}((K\mathbf{G}^F)e_{\mathbf{U}^F} \otimes_{KL^F} \Lambda, R_{\mathbf{L}}^{\mathbf{G}}\Lambda) \simeq \mathrm{Hom}_{KL^F}(\Lambda, \mathrm{Hom}_{K\mathbf{G}^F}((K\mathbf{G}^F)e_{\mathbf{U}^F}, R_{\mathbf{L}}^{\mathbf{G}}\Lambda)),$$

under which B_{n_w} corresponds to the map

$$\Lambda \to \mathrm{Hom}_{K\mathbf{G}^F}((K\mathbf{G}^F)e_{\mathbf{U}^F}, R_{\mathbf{L}}^{\mathbf{G}}\Lambda)$$

$$x \mapsto (ge_{\mathbf{U}^F} \mapsto ge_{\mathbf{U}^F}n_w^{-1}e_{\mathbf{U}^F} \otimes_{KL^F} \gamma_{n_w}(x));$$

see Bourbaki (1971, II §4 no 1 Proposition 1 (a)). Under the identification $\mathrm{Hom}_{K\mathbf{G}^F}((K\mathbf{G}^F)e_{\mathbf{U}^F}, R_{\mathbf{L}}^{\mathbf{G}}\Lambda) \simeq e_{\mathbf{U}^F}R_{\mathbf{L}}^{\mathbf{G}}\Lambda$ as in the proof of 5.1.8, the above map becomes $x \mapsto e_{\mathbf{U}^F}n_w^{-1}e_{\mathbf{U}^F} \otimes_{KL^F} \gamma_{n_w}(x)$, whence the result since $e_{\mathbf{U}^F}n_w^{-1}e_{\mathbf{U}^F} \otimes_{KL^F} \gamma_{n_w}(x)$ is $\gamma_{n_w}(x)$ in the summand indexed by w in $^*R_{\mathbf{L}}^{\mathbf{G}}R_{\mathbf{L}}^{\mathbf{G}}\Lambda = \bigoplus_{w \in W_{\mathbf{G}^F}(\mathbf{L}, \Lambda)} {}^{n_w}\Lambda$. □

To describe the algebra structure of $\text{End}_{KG^F}(R_L^G \Lambda)$ we will make a specific choice of representatives of $W_{G^F}(\mathbf{L}, \Lambda)$. First we choose a quasi-split torus $\mathbf{T}$ of $\mathbf{G}$ contained in $\mathbf{L}$ and let $W = N_{\mathbf{G}}(\mathbf{T})/\mathbf{T}$. Up to $\mathbf{G}^F$-conjugacy $\mathbf{L}$ is a standard Levi subgroup $\mathbf{L}_I$ for some F-stable $I \subset S$.

Lemma 6.1.7 *For $I \subset S$ we have $N_W(W_I) = W_I \rtimes N_I$ where*

$$N_I = \{w \in W \mid {}^wI = I \text{ and } w \text{ is reduced-}I\}$$

is a subgroup of W. If I is F-stable, this decomposition is compatible with F so that $N_W(W_I)^F = W_I^F \rtimes N_I^F$.

Proof First note that the elements of N_I can be characterised equivalently by 3.2.2 as being I-reduced, since $w \in N_I$ has minimal length in its double coset $W_I w W_I = w W_I$.

We next show that N_I is a group. By 3.2.1(iv) an element $w \in N_I$ is characterised by ${}^wI = I$ and $N(w) \cap I = \emptyset$. If v, w satisfy these conditions it is clear that vw satisfies the first one; for the second one write $N(vw) = N(w) \dot{+} w^{-1}N(v)w$: if $N(v)$ does not meet I, since w normalises I the same is true of $w^{-1}N(v)w$ and if two sets do not meet I, neither does their symmetric difference.

Using 3.2.1 we can write any $w \in N_W(W_I)$ as $w = vw_1$ where $v \in W_I$ and w_1 is I-reduced. Since w_1 normalises W_I, for $s \in I$ we have $w_1 s = t w_1$ with $t \in W_I$, and since w_1 is I-reduced we must have $l(t) = 1$ thus w_1 normalises I, whence $N_W(W_I) = W_I \rtimes N_I$. The unicity of the decomposition vw_1 gives the decomposition of $N_W(W_I)^F$. $\qquad\square$

By the remark following Proposition 3.4.3, we have $N_{\mathbf{G}}(\mathbf{L}_I)/\mathbf{L}_I \simeq N_W(W_I)/W_I$, thus if we set $W_{\mathbf{G}^F}(\mathbf{L}) := N_{\mathbf{G}^F}(\mathbf{L})/\mathbf{L}^F$ we have $W_{\mathbf{G}^F}(\mathbf{L}_I) \simeq N_I^F$, so that we can consider $W_{\mathbf{G}^F}(\mathbf{L}, \Lambda)$ as a subgroup of W^F.

By Tits (1966) – using Borel and Tits (1965, Théorème 7.2) if $(\mathbf{G}, F)$ is not split – W^F has representatives $w \mapsto \dot{w} \in N_{G^F}(\mathbf{T})$ such that $\dot{w}\dot{w}' = (ww')\dot{}$ whenever $l(w) + l(w') = l(ww')$. From now on we fix such a choice and for $n \in N_{G^F}(\mathbf{L}, \Lambda)$ we choose γ_n such that $\gamma_{l\dot{w}} = l\gamma_{\dot{w}}$ on Λ. With this choice $B_{n_w} = B_{\dot{w}}$ for any representative n_w of w. We will write B_w instead of $B_{\dot{w}}$.

Lemma 6.1.8 *If $l(w) + l(w') = l(ww')$, then $B_w B_{w'} = \lambda(\dot{w}, \dot{w}')B_{ww'}$.*

Proof $B_w B_{w'}$ maps $ge_{\mathbf{U}^F} \otimes x$ to $ge_{\mathbf{U}^F}\dot{w}'^{-1}e_{\mathbf{U}^F}\dot{w}^{-1}e_{\mathbf{U}^F} \otimes \gamma_{\dot{w}}\gamma_{\dot{w}'}x$. We claim that $e_{\mathbf{U}^F}\dot{w}'^{-1}e_{\mathbf{U}^F}\dot{w}^{-1}e_{\mathbf{U}^F} = e_{\mathbf{U}^F}\dot{w}'^{-1}\dot{w}^{-1}e_{\mathbf{U}^F}$. Indeed, using the unique Bruhat decomposition 3.2.7, we have $\mathbf{U}^F\dot{w}\mathbf{U}^F = \mathbf{U}^F\dot{w}\mathbf{U}_w^F$ with $\mathbf{U}_w = \prod_{\{\alpha \mid s_\alpha \in N(w)\}} \mathbf{U}_\alpha$. Now, since the lengths of w and w' add, by 2.1.2(ii) and the remark following it we have $N(w) \cap N(w'^{-1}) = \emptyset$, thus $\mathbf{U}_w w' \subset w'\mathbf{U}$, hence $\mathbf{U}_w^F w' \subset w'\mathbf{U}^F$ and $\mathbf{U}^F\dot{w}\mathbf{U}^F\dot{w}'\mathbf{U}^F = \mathbf{U}^F\dot{w}\dot{w}'\mathbf{U}^F$, whence our claim, taking inverses. Using $\gamma_{\dot{w}}\gamma_{\dot{w}'} = \lambda(\dot{w}, \dot{w}')\gamma_{\dot{w}\dot{w}'}$ and $\dot{w}\dot{w}' = (ww')\dot{}$ we get the lemma. $\qquad\square$

Theorem 6.1.9 *Assume $K = \mathbb{C}$. Then the cocycle λ is trivial, thus there is a choice of the $\gamma_{\bar{w}}$ in 6.1.4 ensuring that $l(w) + l(w') = l(ww')$ implies $B_w B_{w'} = B_{ww'}$.*

References This has been proven in Lusztig (1984a, 8.6) when $Z(\mathbf{G})$ is connected, then in general in Geck (1993). □

To get a complete description of the algebra structure of $\mathrm{End}_{K\mathbf{G}^F}(R_{\mathbf{L}}^{\mathbf{G}}\Lambda)$ we note the following lemma.

Lemma 6.1.10 *Let $\mathbf{M}$ be an F-stable Levi subgroup containing $\mathbf{L}$ of an F-stable parabolic subgroup such that $|W_{\mathbf{M}^F}(\mathbf{L},\Lambda)| = 2$; then B_w satisfies a quadratic relation for the non-trivial $w \in W_{\mathbf{M}^F}(\mathbf{L},\Lambda) \subset W_{\mathbf{G}^F}(\mathbf{L},\Lambda)$.*

Proof For $w \in W_{\mathbf{M}^F}(\mathbf{L},\Lambda)$ we have an operator $B_w^{\mathbf{M}} \in \mathrm{End}_{K\mathbf{M}^F}(R_{\mathbf{L}}^{\mathbf{M}}\Lambda)$ and by definition, using the transitivity of the functor $R_{\mathbf{L}}^{\mathbf{G}}$, we have $B_w = R_{\mathbf{M}}^{\mathbf{G}}(B_w^{\mathbf{M}})$. If w is the only non-trivial element of $W_{\mathbf{M}^F}(\mathbf{L},\Lambda)$, then, since $\{B_1^{\mathbf{M}}, B_w^{\mathbf{M}}\}$ is a basis of $\mathrm{End}_{K\mathbf{M}^F}(R_{\mathbf{L}}^{\mathbf{G}}\Lambda)$, we have $(B_w^{\mathbf{M}})^2 = \alpha B_w^{\mathbf{M}} + \beta B_1^{\mathbf{M}}$ for some $\alpha, \beta \in K$. Applying $R_{\mathbf{M}}^{\mathbf{G}}$ we get the result. □

Note that β above is not 0 since B_w is invertible.

To show that there exist $\mathbf{M}$ as in 6.1.10 and for computing α and β of 6.1.10 we need to know more on $W_{\mathbf{G}}(\mathbf{L},\Lambda)$. The result is known in all cases only when $K = \mathbb{C}$ (or a large enough subfield of $\mathbb{C}$).

Assumption 6.1.11 *Until 6.1.19 we will assume that K is a large enough subfield of $\mathbb{C}$.*

Proposition 6.1.12 *If $(\mathbf{L}_I,\Lambda)$ is a cuspidal pair, then $W_{\mathbf{G}^F}(\mathbf{L}_I)$ is normalised by w_J for any F-stable $J \subset S$ containing I, where w_J is the longest element of the Coxeter group W_J; it is a Coxeter group with generating simple reflections the elements $w_J w_I$ where $J \supset I$ and $J - I$ is a single orbit under F.*

Proof The first part is proved in Lusztig (1984a, 8.2) using a classification of cuspidal pairs. The second part follows from the next proposition applied to $(W^F, S/F)$. □

Proposition 6.1.13 *If (W,S) is a finite Coxeter system and $I \subset S$ is such that W_I is normalised by the $w_{I \cup \{s\}}$ for $s \in S - I$ and N_I is as in 6.1.7, then $(N_I, \{\sigma_s\}_{s \in S-I})$, with $\sigma_s := w_{I \cup \{s\}} w_I$, is a Coxeter system. Further, if $\sigma_{s_1} \ldots \sigma_{s_k}$ is a reduced expression of some $w \in N_I$ in the above Coxeter system, we have $l(w) = \sum_{i=1}^{i=k} l(\sigma_{s_i})$, where l is the length on W with respect to S.*

Proof This proposition is proved for Weyl groups in Lusztig (1976a, §5) using root systems. We give a proof in the setting of the elementary theory of Coxeter groups.

We first check that $\sigma_s \in N_I$. Since $w_{I\cup\{s\}}$ normalises W_I and induces an automorphism of $(W_{I\cup\{s\}}, I \cup \{s\})$ (see 2.1.5), it normalises I. Also by 2.1.5(ii) applied in $W_{I\cup\{s\}}$ we have $l(\sigma_s) = l(w_{I\cup\{s\}}) - l(w_I)$ which together with 3.2.1(i) implies that σ_s is shortest in the coset $\sigma_s W_I$.

We now follow the same outline as the proof of 4.4.4.

Lemma 6.1.14 *If $w \in N_I$ and $l(ws) < l(w)$ where $s \in S - I$, then $l(w\sigma_s) = l(w) - l(\sigma_s)$.*

Proof Since w normalises I it commutes with w_I and since it is reduced-I we have $l(ww_I) = l(w) + l(w_I)$. Thus $N(ww_I) = N(w_I w) = N(w) + w^{-1}N(w_I)w = N(w) + N(w_I)$; since by assumption $N(w) \ni s$ we have $N(w_I w) \ni s$ and $N(ww_I) \supset I \cup \{s\}$. Thus by 3.2.1(i) and (iv) we have $ww_I = w'w_{I\cup\{s\}}$ with $l(w') + l(w_{I\cup\{s\}}) = l(ww_I)$, that is $l(w\sigma_s) = l(ww_I) - l(w_{I\cup\{s\}})$. Whence the lemma since $l(w) = l(ww_I) - l(w_I)$ and $l(\sigma_s) = l(w_{I\cup\{s\}}) - l(w_I)$. □

Applying inductively the lemma, we get that any $w \in N_I$ can be written $\sigma_{s_1} \ldots \sigma_{s_k}$ with $l(w) = \sum_{i=1}^{i=k} l(\sigma_{s_i})$. This gives the final remark of 6.1.13 and proves in particular that the σ_s generate N_I.

We again use 2.2.9 to show that $(N_I, \{\sigma_s\}_{s\in S-I})$ is a Coxeter system. For $s \in S - I$, let $D_s = \{w \in N_I \mid s \notin N(w)\}$. We have clearly $D_s \ni 1$ and $D_s \cap D_s\sigma_s = \emptyset$, so it remains to show that if $w \in D_s$, $s' \in S - I$ are such that $\sigma_{s'}w \notin D_s$, then $\sigma_{s'}w = w\sigma_s$. Since w is in D_s, it is $I \cup \{s\}$-reduced, hence $N(w) \cap \mathrm{Ref}(W_{I\cup\{s\}}) = \emptyset$. On the other hand, we have $s \in N(\sigma_{s'}w) = {}^{w^{-1}}N(\sigma_{s'})\dotplus N(w) = ({}^{w^{-1}}\mathrm{Ref}(W_{I\cup\{s'\}}) - {}^{w^{-1}}\mathrm{Ref}(W_I))\dotplus N(w)$, and $s \notin N(w)$ and $s \notin \mathrm{Ref}(W_I) = {}^{w^{-1}}\mathrm{Ref}(W_I)$, thus $s \in {}^{w^{-1}}\mathrm{Ref}(W_{I\cup\{s'\}})$. Since $I = {}^{w^{-1}}I \subset {}^{w^{-1}}\mathrm{Ref}(W_{I\cup\{s'\}})$, we have $\{s\} \cup I \subset {}^{w^{-1}}\mathrm{Ref}(W_{I\cup\{s'\}})$, thus ${}^{w^{-1}}W_{I\cup\{s'\}} \supset W_{I\cup\{s\}}$.

As a non-trivial element $v \in W_{I\cup\{s'\}} \cap N_I$ satisfies $l(vs') < l(v)$, the lemma shows that $W_{I\cup\{s'\}} \cap N_I = \{1, \sigma_{s'}\}$ and similarly $W_{I\cup\{s\}} \cap N_I = \{1, \sigma_s\}$, whence ${}^{w^{-1}}\sigma_{s'} = \sigma_s$. □

Exercise 6.1.15 For $s, s' \in S_I$, the length of the braid relation between σ_s and $\sigma_{s'}$ (the order of $\sigma_s\sigma_{s'}$) is equal to $\dfrac{2|\mathrm{Ref}(W_{I\cup\{s,s'\}})|}{|\mathrm{Ref}(W_{I\cup\{s\}})| + |\mathrm{Ref}(W_{I\cup\{s'\}})|}$.

Proposition 6.1.16 *Assume the centre of $\mathbf{G}$ connected. If $(\mathbf{L}_I, \Lambda)$ is a cuspidal pair, the group $W_{\mathbf{G}^F}(\mathbf{L}_I, \Lambda)$ is a reflection subgroup of N_I^F with reflections $\mathrm{Ref}(N_I^F) \cap W_{\mathbf{G}^F}(\mathbf{L}_I, \Lambda)$. For $w \in W_{\mathbf{G}^F}(\mathbf{L}_I, \Lambda)$ we define $T_w := q^{l(w)}B_w$. We denote by $\tilde{l}$ the length in the Coxeter group $W_{\mathbf{G}^F}(\mathbf{L}_I, \Lambda)$. Then*

(i) $T_w T_{w'} = T_{ww'}$ if $\tilde{l}(w) + \tilde{l}(w') = \tilde{l}(ww')$.

(ii) Let ${}^n\sigma_s$ be a reflection of $W_{\mathbf{G}^F}(\mathbf{L}_I, \Lambda)$ where $n \in N_I^F$ and $s \in S/F - I/F$. Then $T_{{}^n\sigma_s}^2 = (q^{c_s} - 1)T_{{}^n\sigma_s} + q^{c_s}$ where c_s is a positive integer such that q^{c_s} is the quotient of the degrees of the two irreducible constituents of $R_{\mathbf{L}_I}^{\mathbf{L}_{I \cup s}{}^{n^{-1}}} \Lambda$.

The "two irreducible constituents" in the last sentence refers to the fact that the condition of 6.1.10 holds: $|W_{\mathbf{L}_{I \cup s}}(\mathbf{L}_I, \Lambda)| = 2$.

References   See Lusztig (1984a, 8.5.8, 8.5.13 and 8.6).   □

Remark 6.1.17   It can be shown that $l(w) + l(w') = l(ww')$ implies $\tilde{l}(w) + \tilde{l}(w') = \tilde{l}(ww')$, hence relation 6.1.16(i) is compatible with 6.1.8. The converse implication is not true.

Example 6.1.18   Consider the case where $\mathbf{L} = \mathbf{T}$ is a maximal torus contained in an F-stable Borel subgroup $\mathbf{B} = \mathbf{TU}$. Any representation of $\mathbf{T}^F$ is cuspidal in particular the trivial representation. In this example K can be any field such that $|\mathbf{B}^F|$ is invertible. The representation $R_{\mathbf{T}}^{\mathbf{G}}(1) = K(\mathbf{G}^F/\mathbf{U}^F) \otimes_{\mathbf{T}^F} K = K(\mathbf{G}^F/\mathbf{B}^F) = \text{Ind}_{\mathbf{B}^F}^{\mathbf{G}^F}(1)$ is the permutation representation of $\mathbf{G}^F$ on the F-stable Borel subgroups. We have $W_{\mathbf{G}^F}(\mathbf{T}, 1) = W^F$. The γ_n of 6.1.4 is the identity so that the cocycle is trivial. For $w \in W^F$, the operator B_w is $ge_{\mathbf{B}^F} \mapsto ge_{\mathbf{B}^F}w^{-1}e_{\mathbf{B}^F}$. The Levi subgroups $\mathbf{M}$ as in 6.1.10 are the $\mathbf{L}_J$ for $J \in S/F$, an orbit of F in S. We compute the quadratic relation satisfied by B_{w_J}. We have $B_{w_J}^2 : ge_{\mathbf{B}^F} \mapsto ge_{\mathbf{B}^F}w_J e_{\mathbf{B}^F}w_J e_{\mathbf{B}^F}$. Using the properties of the relative (B,N)-pair of 4.4.8, we have $\mathbf{B}^F w_J \mathbf{B}^F w_J \mathbf{B}^F = \mathbf{B}^F w_J \mathbf{B}^F \cup \mathbf{B}^F$. The computation gives $B_{w_J}^2 = (1 - \frac{|\mathbf{B}^F \cap {}^{w_J}\mathbf{B}^F|}{|\mathbf{B}^F|})B_{w_J} + \frac{|\mathbf{B}^F \cap {}^{w_J}\mathbf{B}^F|}{|\mathbf{B}^F|}B_1$. We have $\frac{|\mathbf{B}^F \cap {}^{w_J}\mathbf{B}^F|}{|\mathbf{B}^F|} = |\mathbf{U}_{w_J}^F|^{-1} = q^{-l(w_J)}$ (see the proof of 4.4.1). If we define $T_{w_J} := q^{l(w_J)}B_{w_J}$ we have $T_{w_J}^2 = (q^{l(w_J)} - 1)T_{w_J} + q^{l(w_J)}$. We recover in this case 6.1.16 without needing the assumption that the centre of $\mathbf{G}$ is connected.

Remark 6.1.19   It is shown in Lusztig (1978, 3.26) that for Λ a unipotent cuspidal module (see definition 11.3.4), the statements of 6.1.16 are true without any assumption on the centre of $\mathbf{G}$. In this case one has $W_{\mathbf{G}^F}(\mathbf{L}, \Lambda) = W_{\mathbf{G}^F}(\mathbf{L})$.

6.2 Iwahori–Hecke Algebras

We give now general results on algebras having the same kind of presentation as $\text{End}_{K\mathbf{G}^F}(R_{\mathbf{L}}^{\mathbf{G}}(\Lambda))$.

Definition 6.2.1   Let A be a commutative ring, (W,S) be a Coxeter system, and $(q_s, q_s')_{s \in S}$ be pairs of elements of A depending only on the W-conjugacy class of s. The **Iwahori–Hecke algebra** $H_{\{q_s,q'_s\}_{s\in S}}(W,A)$ is the A-algebra with presentation

$$\Big\langle (T_s)_{s\in S} \mid \underbrace{T_sT_{s'}T_s\ldots}_{\text{order of } ss'} = \underbrace{T_{s'}T_sT_{s'}\ldots}_{\text{order of } ss'} \text{ for } s,s' \in S,$$

$$\text{and } (T_s - q_s)(T_s - q'_s) = 0 \text{ for } s \in S\Big\rangle.$$

In the particular case when $q_s = 1$ and $q'_s = -1$, the Iwahori–Hecke algebra is the group algebra AW. For the study of $\mathrm{End}_{K\mathbf{G}^F}(R_{\mathbf{L}}^{\mathbf{G}}(\Lambda))$, it would be enough to consider the case where $q'_s = -1$, but the consideration of the general case has some advantages: it is used when studying invariants of knots, and for rings A which have a Galois automorphism interchanging q_s and q'_s, this exchange induces a duality on characters, see 6.2.3. Note that T_s is invertible if and only if q_s and q'_s are invertible.

Proposition 6.2.2 *An equivalent presentation of* $H_{\{q_s,q'_s\}_{s\in S}}(W,A)$ *is given by*

$$\Big\langle (T_w)_{w\in W} \mid \text{for } s \in S,$$

$$T_sT_w = \begin{cases} T_{sw} & \text{if } l(sw) = l(w) + 1 \\ (q_s + q'_s)T_w - q_sq'_sT_{sw} & \text{otherwise} \end{cases}\Big\rangle.$$

Proof We first show that 6.2.2 implies 6.2.1. If $s_1\ldots s_k \in W$ is a reduced expression of $w \in W$, iterating the first relation of 6.2.2 gives $T_w = T_{s_1}\ldots T_{s_k}$; thus the T_s generate the algebra. The particular case of the second relation for $w = s$ gives the second relation of 6.2.1. The first relation of 6.2.1 holds since both sides represent $T_{\Delta_{s,s'}}$ (see 2.1.1).

We show now that 6.2.1 implies 6.2.2. If $s_1\ldots s_k \in W$ is a reduced expression of $w \in W$, let $T_w := T_{s_1}\ldots T_{s_k}$; this element does not depend on the reduced expression thanks to the first relation of 6.2.1 and 2.1.2(iv). Let $s \in S$ and $w \in W$; if $l(sw) = l(w) + 1$ then we get a reduced expression of sw by prefixing by s a reduced expression of w, thus $T_{sw} = T_sT_w$; if $l(sw) = l(w) - 1$ then we have $T_w = T_sT_{sw}$, hence using the second relation of 6.2.1 we have $T_sT_w = T_s^2T_{sw} = (q_s + q'_s)T_sT_{sw} - q_sq'_sT_{sw} = (q_s + q'_s)T_w - q_sq'_sT_{sw}$. □

Theorem 6.2.3 *The set* $\{T_w|w\in W\}$ *is an A-basis of* $H_{\{q_s,q'_s\}_{s\in S}}(W,A)$.

Proof By 6.2.2 the set $\{T_w|w\in W\}$ spans the Iwahori–Hecke algebra. To prove the linear independence, for $s \in S$, we consider the operators $L_s \in \mathrm{End}_A(AW)$ defined by

$$L_s(w) = \begin{cases} sw & \text{if } l(sw) = l(w) + 1, \\ (q_s + q'_s)w - q_sq'_ssw & \text{otherwise.} \end{cases}$$

We get the theorem if we show that $L : T_s \mapsto L_s$ defines a representation $L : H_{\{q_s,q'_s\}_{s \in S}}(W,A) \to \mathrm{End}_A(AW)$ and that the operators $L(T_w)$ are linearly independent. This will be a consequence of

Lemma 6.2.4 *Let $\lambda \in \mathrm{End}_A(AW)$ be in the subalgebra generated by the L_s. Then $\lambda = 0$ if and only if $\lambda(1) = 0$.*

Proof For $t \in S$ define $R_t \in \mathrm{End}_A(AW)$ by

$$R_t(w) = \begin{cases} wt & \text{if } l(wt) = l(w) + 1, \\ (q_t + q'_t)w - q_t q'_t wt & \text{otherwise.} \end{cases}$$

We claim that L_s and R_t commute. To simplify the computation we put $L_s(w) = l_s(w)sw + l'_s(w)w$ and $R_t(w) = r_t(w)wt + r'_t(w)w$. We have

$$L_s R_t(w) = l_s(wt)r_t(w)swt + l'_s(wt)r_t(w)wt + l_s(w)r'_t(w)sw + l'_s(w)r'_t(w)w$$

and

$$R_t L_s(w) = r_t(sw)l_s(w)swt + l'_s(w)r_t(w)wt + l_s(w)r'_t(sw)sw + l'_s(w)r'_t(w)w.$$

The four basis elements involved are distinct unless $sw = wt$; in this case s and t are conjugate, hence $l_s(w) = r_t(w)$, $l'_s(wt) = r'_t(sw)$, $l'_s(w) = r'_t(w)$ and $l_s(wt) = r_t(sw)$, thus the two expressions are equal.

Assume $sw \neq wt$. The coefficients of w are the same in both expressions. The coefficients of sw are the same unless $l(wt) - l(w) \neq l(swt) - l(sw)$ which cannot happen since by the exchange property (see Exercise 2.1.4) it implies $sw = wt$. The same applies to the coefficients of wt. The coefficients of swt are different only if $l(swt) - l(sw) \neq l(wt) - l(w)$ or $l(swt) - l(wt) \neq l(sw) - l(w)$. Both cases imply $sw = wt$ hence do not happen.

We deduce the lemma: if $\lambda(1) = 0$, since λ commutes with R_s we have $\lambda(s) = \lambda(R_s(1)) = R_s(\lambda(1)) = 0$ for all $s \in S$, hence $\lambda = 0$. $\qquad\square$

The theorem follows: both $\underbrace{L_s L_{s'} \ldots}_{order(ss')}$ and $\underbrace{L_{s'} L_s \ldots}_{order(ss')}$ map 1 to $\Delta_{s,s'}$ whence the equality of these two operators, and the quadratic relations are clear by definition, so we indeed get a representation of the Iwahori–Hecke algebra. We have $L(w)(1) = w$ as can be seen by taking a reduced expression for w. If $\sum_w a_w L(w) = 0$ is a linear dependency relation we get $0 = \sum_w a_w L(w)(1) = \sum_w a_w w$ so that $a_w = 0$ for all w. $\qquad\square$

Before going on, we need to recall the following theorem proved by Dyer (1990) and Deodhar (1989).

Theorem 6.2.5 *Let (W,S) be a finite Coxeter system and let W' be a reflection subgroup of W – that is, W' is generated by $W' \cap \mathrm{Ref}\, W$.*

Let $S(W') = \{r \in \text{Ref}(W) \mid N(r) \cap W' = \{r\}\}$. Then $(W', S(W'))$ is a Coxeter system, called the canonical Coxeter system of W' with respect to W.

Proof We follow Dyer's proof. Let W_1' be the subgroup of W' generated by $S(W')$. The first thing we prove is

> Any $r \in \text{Ref}(W) \cap W'$ is W_1'-conjugate to an element of $S(W')$.   (∗)

We shall prove (∗) by induction on $l(r)$, where the induction assumption is that it holds for all $r' \in \text{Ref}(W)$ such that $l(r') < l(r)$ and all reflection subgroups of W. The start of the induction is when $l(r) = 1$ in which case $r \in S(W')$ and there is nothing to prove. For the general step of the induction, by 2.1.6(iv) there exists $s \in S$ such that $l(srs) = l(r) - 2$. By the induction assumption, there exists $r_0, \ldots, r_m \in S(^sW')$ such that $srs = r_m \ldots r_1 r_0 \ldots r_m$. There are now two cases:

- If $s \in W'$. Then $^sW' = W'$ and $S(^sW') = S(W')$ thus r is W_1'-conjugate to $r_0 \in S(W')$, as needed.
- If $s \notin W'$. We have $r = {}^s r_m \ldots {}^s r_1 {}^s r_0 \ldots {}^s r_m$ where $^s r_i \in {}^s S(^sW')$. It is thus sufficient to show that $^s S(W') = S(^sW')$. Now let $r \in \text{Ref}(W) \cap W'$; we have

$$N(^sr) \cap {}^sW' = (\{{}^{sr}s\} \dot{+} {}^sN(r) \dot{+} \{s\}) \cap {}^sW'$$
$$= {}^s((\{{}^rs\} \dot{+} N(r) \dot{+} \{s\}) \cap W')$$
$$= {}^s(N(r) \cap W'),$$

where the first equality is by the defining relation of N, and the third one since $s \notin W'$ and $^rs \notin W' = {}^rW'$. These equalities show that $N(^sr) \cap {}^sW' = \{{}^sr\}$ is equivalent to $N(r) \cap W' = \{r\}$.

We can now finish the proof of the theorem by applying criterion 2.1.2(ii) for being a Coxeter system, taking for N the function N' on W' defined by $N'(w) := N(w) \cap W'$; the property $N'(xy) = N'(y) \dot{+} y^{-1} N'(x) y$ is an immediate consequence of the same property for N.   □

The theory of Iwahori–Hecke algebras applies in particular in the two following examples. Part (i) was proved by Iwahori, whence the terminology "Iwahori–Hecke" algebra.

Proposition 6.2.6

(i) *Let $\mathbf{T}$ be an F-stable maximal torus of an F-stable Borel subgroup $\mathbf{B}$ of $\mathbf{G}$ and let K be a field in which $|\mathbf{B}^F|$ is invertible; then*

$$\text{End}_{K\mathbf{G}^F}(R_{\mathbf{T}}^{\mathbf{G}}(1)) = H_{\{q^{l(w_J)}, -1\}_{J \in S/F}}(W^F, K).$$

(ii) *Assume that the centre of* **G** *is connected; let* $(\mathbf{L}, \Lambda)$ *be a cuspidal pair over a large enough subfield K of* $\mathbb{C}$, *then* $\mathrm{End}_{K\mathbf{G}^F}(R_{\mathbf{L}}^{\mathbf{G}}\Lambda) = H_{\{q^{c_s}, -1\}_{s \in S(L, \Lambda)}}$ $(W_{\mathbf{G}^F}(\mathbf{L}, \Lambda), K)$, *where c_s is as in 6.1.16 and $S(L, \Lambda)$ is the canonical set of Coxeter generators with respect to N_I^F of the reflection subgroup* $W_{\mathbf{G}^F}(\mathbf{L}_I, \Lambda)$.

Proof (i) is a consequence of what has been seen in Example 6.1.18.

For (ii): by 6.1.16 and 6.1.6, $\mathrm{End}_{K\mathbf{G}^F}(R_{\mathbf{L}}^{\mathbf{G}}\Lambda)$ has a basis $(T_w)_{w \in W_{\mathbf{G}^F}(\mathbf{L}, \Lambda)}$ which satisfies the relations of 6.2.2 with parameters q^{c_s} and -1, whence the result. $\qquad\square$

Recall that if K is a field, a finite dimensional K-algebra H is **absolutely semi-simple** if $H \otimes_K \overline{K}$ is semi-simple where $\overline{K}$ is an algebraic closure of K, in which case $H \otimes_K \overline{K}$ is isomorphic to a product of matrix algebras $\prod_i M_{n_i}(\overline{K})$ and $H \otimes_K L$ is semi-simple for any extension L of K, see Bourbaki (1981, §13 no 3 Théorème 1). The multi-set $\{n_i\}$ is called the **numerical invariant** of H. A criterion for being absolutely semi-simple is given by the following

Proposition 6.2.7 *Let H be a finite dimensional algebra over a field K.*

(i) *If the bilinear form* $(a, b) \mapsto \mathrm{Trace}_{H/K}(ab)$ *is non-degenerate then H is absolutely semi-simple.*

(ii) *The converse of (i) is true when the characteristic of K is 0.*

Proof We recall that $\mathrm{Trace}_{H/K}(a)$ is the matrix trace of the multiplication by a expressed in a K-basis of H.

For (i): The form $\mathrm{Trace}_{H/K}(ab)$ remains non-degenerate when extended to $\overline{K}$. An element of the radical of $H \otimes_K \overline{K}$ being nilpotent has trace 0. Thus for a in the radical we have $\mathrm{Trace}_{H \otimes_K \overline{K}/\overline{K}}(ab) = 0$ for all b whence $a = 0$.

For (ii): it is sufficient to prove the result for $H = M_n(\overline{K})$. We can take the elementary matrices $(E_{i,j})_{i,j}$ (see 1.5.3) as a $\overline{K}$-basis for H. We get $\mathrm{Trace}_{H/\overline{K}}(E_{i,j}) = n\delta_{i,j}$ thus for $h \in M_n(\overline{K})$ the matrix trace of h is equal to $\frac{1}{n}\mathrm{Trace}_{H/\overline{K}}(h)$, since the characteristic is 0, hence the bilinear form is non-degenerate since if the matrix trace of aE_{ij} is zero for all i, j then $a = 0$. $\qquad\square$

Remark 6.2.8 If W is a finite group and K a field, then for $w \in W$ we have $\mathrm{Trace}_{KW/K}(w) = |W|\delta_{w,1}$. Hence if the characteristic of K does not divide $|W|$, the form $(a, b) \mapsto \mathrm{Trace}_{KW/K}(ab)$ is non-degenerate, and in particular KW is absolutely semi-simple.

The next theorem, due to Tits, appears for the first time in Exercises 26 and 27 of Bourbaki (1968, IV, §2).

Theorem 6.2.9 (Tits deformation theorem)   *Let H be a free finite dimensional A-algebra where A is an integral domain. Let K be the quotient field of A and let f : A → k be a ring morphism from A to some field k.*

   (i) *If the form $(a,b) \mapsto \mathrm{Trace}_{H \otimes_A k}(ab)$ is non-degenerate, then $H \otimes_A K$ is absolutely semi-simple.*
  (ii) *If $H \otimes k$ and $H \otimes K$ are absolutely semi-simple they have the same numerical invariant.*
 (iii) *If $\overline{A}$ is the integral closure of A in $\overline{K}$ then f can be extended to $\overline{f} : \overline{A} \to \overline{k}$; for any irreducible character χ of $H \otimes_A \overline{K}$, one has $\chi(H) \subset \overline{A}$ and $\chi \mapsto (\overline{f} \circ \chi) \otimes 1$ is a bijection from $\mathrm{Irr}(H \otimes_A \overline{K})$ to $\mathrm{Irr}(H \otimes_A \overline{k})$.*

Proof   We write $H_K, H_{\overline{K}}, H_k, \ldots$, respectively for $H \otimes_A K$ $H \otimes_A \overline{K}$, $H \otimes_A k, \ldots$. An A-basis of H specialises to a k basis of H_k and is also a K-basis of H_K. This shows that, under the assumption of (i), $(a,b) \mapsto \mathrm{Trace}_{H_K / K}(ab)$ is non-degenerate on H_K, whence (i) by 6.2.7.

For proving (ii), we can assume k and K algebraically closed. Then H_K and H_k are products of matrix algebras. Let $\{e_i \mid i = 1, \ldots, r\}$ be an A-basis of H and consider the algebra $H_{K[x_1, \ldots, x_r]}$ where x_i are indeterminates. We claim that the characteristic polynomial of the multiplication by $\sum_i x_i e_i$ is of the form $\prod_{j=1}^{s} P_j^{n_j} \in K[x_1, \ldots, x_r][t]$, where $\{n_1, \ldots, n_s\}$ is the numerical invariant of H_K and each P_j is irreducible of degree n_j in t. It is sufficient to see it when H_K is a matrix algebra $M_n(K)$. Since the degree of the characteristic polynomial does not change by a (linear) base change, we can do the computation taking as basis of $M_n(K)$ the elementary matrices $E_{i,j}$. The matrix of the multiplication by $x := (x_{i,j})$ is a block diagonal matrix with n blocks, all equal to x, thus the characteristic polynomial is P^n where P is the characteristic polynomial $\det(x - t\, \mathrm{I}_n)$ of the matrix x, an irreducible polynomial of degree n in t (since it is the "generic" characteristic polynomial of an $n \times n$ matrix), whence our claim. Now, the basis (e_i) of H specialises to a basis of H_k. The multiplication by $\sum_i x_i e_i$ specialises to the multiplication by $\sum x_i e_i$ in $H \otimes_A k[x_1, \ldots, x_n]$, so that the specialisation $\prod_j f(P_j)^{n_j}$ is the analogous polynomial over $k[x_1, \ldots, x_n]$. Since this has to be the product of irreducible polynomials, each one raised to its degree and of the same degree r as P we must have $f(P_j)$ irreducible and $\{n_1, \ldots, n_s\}$ must be the numerical invariant of H_k, whence (ii).

We prove (iii). The fact that f can be extended to $\overline{A}$ is well known; see for example, Lang (2002, VII, Proposition 3.1). Each irreducible character χ of $H_{\overline{K}[x_1, \ldots, x_r]} \simeq \oplus_{i=1}^{i=s} M_{n_i}(\overline{K}[x_1, \ldots, x_r])$ is given by the matrix trace composed with the projection onto one of the $M_{n_i}(\overline{K}[x_1, \ldots, x_r])$. Its value $\chi(h)$ is thus the coefficient of degree $n_i - 1$ of the irreducible factor $P_i \in \overline{K}[x_1, \ldots, x_r]$, of the characteristic polynomial P of the multiplication by h. For $h = \sum_i x_i e_i \in$

$H_{K[x_1,\ldots,x_r]}$ the coefficients of P are in $A[x_1,\ldots,x_r]$, hence its roots are integral over this ring, so that the coefficients of its irreducible factors are in the integral closure $\overline{A[x_1,\ldots,x_r]}$ of $A[x_1,\ldots,x_r]$ in $\overline{K}[x_1,\ldots,x_r]$. By Bourbaki (1975, V, §1 no 3, Proposition 13), we have $\overline{A[x_1,\ldots,x_r]} = \overline{A}[x_1,\ldots,x_r]$. Since $\overline{f}(\chi(h))$ is $(-1)^{n_i-1}$ times the coefficient of degree $n_i - 1$ of $\overline{f}(P_i)$ we get the result. □

Theorem 6.2.10 *Let $K \subset \mathbb{C}$ be a splitting field for the finite Coxeter group W and let q_s, q'_s be indeterminates; the field $K(\sqrt{q_s}, \sqrt{-q'_s})_{s\in S}$ is a splitting field for $H_{\{q_s,q'_s\}_{s\in S}}(W, \mathbb{Q}[q_s^{\pm 1}, q'^{\pm 1}_s]_{s\in S})$.*

Proof This can be deduced from Geck and Pfeiffer (2000, 9.3.1 and 9.3.5); see also Malle (1999, Corollary 4.8 and Proposition 5.1). □

Corollary 6.2.11 *Let (W,S) be a finite Coxeter system and K a field whose characteristic does not divide $|W|$.*

(i) *If q_s, q'_s are indeterminates, then $H_{\{q_s,q'_s\}_{s\in S}}(W, K(q_s, q'_s)_{s\in S})$ is absolutely semi-simple and has same numerical invariant as W.*

(ii) *If $a_s, a'_s \in K$ are such that $H_{\{a_s,a'_s\}_{s\in S}}(W,K)$ is absolutely semi-simple then $H_{\{a_s,a'_s\}_{s\in S}}(W,K) \simeq KW$.*

(iii) *If K is a splitting field for W, the characters of KW are specialised from the characters of $H_{\{q_s,q'_s\}_{s\in S}}(W, K[\sqrt{q_s}^{\pm 1}, \sqrt{-q'_s}^{\pm 1}]_{s\in S})$ for the specialisation $\sqrt{q_s} \mapsto 1$ and $\sqrt{-q'_s} \mapsto 1$.*

Proof (i) is a consequence of Remark 6.2.8 and Theorem 6.2.9(i). (iii) is by 6.2.9(iii), taking in account 6.2.10.

For (ii): by 6.2.9(ii) $H_{\{q_s,q'_s\}_{s\in S}}(W, K(q_s, q'_s)_{s\in S})$ has same numerical invariant as KW and as $H_{\{a_s,a'_s\}_{s\in S}}(W,K)$. Hence these last two K-algebras are isomorphic. □

Definition 6.2.12 The algebra $H_{\{q_s,q'_s\}_{s\in S}}(W, K[q_s^{\pm 1}, q'^{\pm 1}_s]_{s\in S})$, when q_s, q'_s are indeterminates, is called the **generic Iwahori–Hecke algebra** of W over K.

Remark 6.2.13 Statement 6.2.11(i) is true for variations of the generic Iwahori–Hecke algebra like the algebra $H_{\{q^{c_s}, -1\}_{s\in S}}(W, K[q^{\pm 1}])$ where q is an indeterminate and c_s are positive integers. The same proof applies.

Corollary 6.2.11 applies in particular in the two following cases:

Corollary 6.2.14

(i) *Let $\mathbf{B}$ be an F-stable Borel subgroup in a connected reductive group $\mathbf{G}$; then for any algebraically closed field K in which $|\mathbf{G}^F|$ is invertible, the irreducible constituents of $\mathrm{Ind}_{\mathbf{B}^F}^{\mathbf{G}^F}(\mathbf{1})$ are in bijection with the irreducible characters of KW^F. By this bijection the trivial representation of $\mathbf{G}^F$ corresponds to the trivial representation of W^F.*

(ii) *We assume moreover that $Z(\mathbf{G})$ is connected. Let $(\mathbf{L}, \Lambda)$ be a cuspidal pair over a large enough subfield K of $\mathbb{C}$, then there is a bijective correspondence between $\mathrm{Simp}(K\mathbf{G}^F \mid (\mathbf{L}, \Lambda))$ and $\mathrm{Simp}(KW_{\mathbf{G}^F}(\mathbf{L}, \Lambda))$.*

Proof If V is a representation of a finite group G over an algebraically closed field K whose characteristic does not divide $|G|$ then $\mathrm{End}_G(V)$ is absolutely semi-simple. Thus in view of 6.2.6 (i) and (ii) we can apply 6.2.11(iii) to both cases of the Corollary, which, using 6.1.1(i), gives (ii) and the first assertion of (i).

Now the trivial representation of $\mathbf{G}^F$ is a constituent of $\mathrm{Ind}_{\mathbf{B}^F}^{\mathbf{G}^F}(\mathbf{1})$ since this last representation is a permutation representation. We use the notation and results of 6.1.18. We have $\mathrm{Ind}_{\mathbf{B}^F}^{\mathbf{G}^F}(\mathbf{1}) = R_{\mathbf{T}}^{\mathbf{G}}(\mathbf{1})$ where $\mathbf{T}$ is an F-stable maximal torus of $\mathbf{B}$ and $\mathrm{End}_{K\mathbf{G}^F}(\mathrm{Ind}_{\mathbf{B}^F}^{\mathbf{G}^F}(\mathbf{1})) = H_{\{q^{l(w_J)}, -1\}_{J \in S/F}}(W^F, K)$. By 6.1.1(i) the representation of $\mathrm{End}_{K\mathbf{G}^F}(R_{\mathbf{T}}^{\mathbf{G}}(\mathbf{1}))$ corresponding to the trivial representation of $\mathbf{G}^F$ is $\mathrm{Hom}_{K\mathbf{G}^F}(R_{\mathbf{T}}^{\mathbf{G}}\mathbf{1}, \mathbf{1})$. A morphism in this last space maps all $ge_{\mathbf{B}^F}$ to the same element of K, and the action of T_{w_J} is by composition with $e_{\mathbf{B}^F} \mapsto T_{w_J} e_{\mathbf{B}^F} = q^{l(w_J)} e_{\mathbf{B}^F} w_J^{-1} e_{\mathbf{B}^F}$, so that this action is just multiplication by $q^{l(w_J)}$. This one dimensional representation of the Iwahori–Hecke algebra is the specialisation of the representation of the generic algebra of 6.2.13 given by the same formula where q is an indeterminate. This last representation in turn specialises to the trivial representation of W^F when the indeterminate q goes to 1, whence the last assertion of (i). $\qquad\square$

6.3 Schur Elements and Generic Degrees

We give now more properties of the characters of Iwahori–Hecke algebras.

Proposition 6.3.1 *Let A be an integral domain and let a_s, a'_s be invertible elements of A.*

(i) *Let τ be the linear map $H_{\{a_s, a'_s\}_{s \in S}}(W, A) \to A$ defined by $\tau(T_w) = \delta_{w,1}$. Then the bilinear form $(a, b) \mapsto \tau(ab)$ on $H_{\{a_s, a'_s\}_{s \in S}}(W, A)$ is non-degenerate.*

(ii) *The map $T_s \mapsto a_s$ for $s \in S$ (resp. $T_s \mapsto a'_s$ for $s \in S$) defines a linear character of $H_{\{a_s, a'_s\}_{s \in S}}(W, A)$. If we denote by a_w (resp. a'_w) the value of this character at T_w, then $\{(-1)^{l(w)} a_w^{-1} a_w'^{-1} T_{w^{-1}}\}$ is a basis dual to $\{T_w\}$ for the bilinear form of (i).*

Proof We prove first that the maps of (ii) are characters. We need to check the compatibility with the relations; this is obvious for the quadratic relations. For an even length braid relation the number of s_i equal to a given s is the same on both sides of the relation thus the product of the a_{s_i} is the same on both sides; if an odd length braid relation involves s and s', then s and s' are conjugate, hence $a_s = a_{s'}, a'_s = a'_{s'}$, and the two products are again equal.

We now prove by induction on $l(w)$ that $\tau(T_w T_{w'}) = 0$ unless $w = w'^{-1}$ and that $\tau(T_w T_{w^{-1}}) = (-1)^{l(w)} a_w a'_w$. This is true if $w = 1$. Otherwise write $w = vs$ with $l(v) = l(w) - 1$ and $s \in S$. Then $T_w T_{w'} = T_v T_s T_{w'}$. If $l(sw') = l(w') + 1$, then $T_w T_{w'} = T_v T_{sw'}$. Since $l(vs) > l(v)$ and $l(w') < l(sw')$, we cannot have $v = (sw')^{-1}$, thus by induction $\tau(T_w T_{w'}) = 0$. If $l(sw') = l(w') - 1$ then $w' = sv'$ for some $s \in S$ by the exchange property, hence $T_w T_{w'} = T_v T_s^2 T_{v'} = (a_s + a'_s) T_v T_{w'} - a_s a'_s T_v T_{v'}$. Since $l(vs) > l(v)$ and $l(sw') < l(w')$, we cannot have $v = w'^{-1}$, hence we get by induction $\tau(T_w T_{w'}) = -a_s a'_s \tau(T_v T_{v'}) = (-1)^{l(v)+1} a_s a'_s a_v a'_v = (-1)^{l(w)} a_w a'_w$ if $vv' = 1$ and 0 otherwise, which gives the result. This proves that we have dual bases, thus that the bilinear form is non-degenerate. □

Note that when $a_s = 1$ and $a'_s = -1$ for all s, the two characters of 6.3.1(ii) are the trivial and the sign character of AW and the dual basis to $\{w\}$ is $\{w^{-1}\}$.

Proposition 6.3.1 says in particular that $H_{\{a_s, a'_s\}_{s \in S}}(W, A)$ is a **symmetric algebra**. A symmetric algebra over an integral domain is a finite dimensional free algebra on which there exists a linear form τ (the symmetrising form) such that $(a, b) \mapsto \tau(ab)$ is symmetric non-degenerate. We will use general properties of symmetric algebras which we reprove here for the convenience of the reader.

Notation 6.3.2 *Until 6.3.8 H will be a symmetric algebra over an integrally closed integral domain A with symmetrising form τ. We denote by K the quotient field of A and by H_K the algebra $H \otimes_A K$.*

Proposition 6.3.3

(i) *The element $I := \sum_i e_i \otimes e'_i \in H \otimes_A H^{\mathrm{opp}}$, where (e_i) and (e'_i) are dual bases of H with respect to τ, does not depend on the choice of dual bases.*

(ii) *If M and N are H-modules, then $\varphi \mapsto I\varphi$ maps $\mathrm{Hom}_K(M, N)$ to $\mathrm{Hom}_H(M, N)$.*

Proof For any $h, h' \in H$ we have $(\tau \otimes \tau)(I(h \otimes h')) = \sum_i \tau(e_i h) \tau(h' e'_i) = \sum_i \tau(\tau(e_i h) e'_i h') = \tau(\sum_i \tau(e_i h) e'_i h') = \tau(hh')$. The last equality holds since, by definition of dual bases, we have $\sum_i \tau(he_i) e'_i = h$. This shows the independence since $\tau \otimes \tau$ is non-degenerate. Since $\tau(hh') = \tau(h'h)$, the above computation shows also $I(h \otimes h') = I(h' \otimes h)$.

In (ii) the action of $H \otimes H^{\mathrm{opp}}$ on φ is given by $((x \otimes y)\varphi)(m) = x\varphi(ym)$. We get the result from the equality $I(1 \otimes h) = I(h \otimes 1)$ seen in the proof of (i). □

Remark 6.3.4 If H^* is the dual of H, by the isomorphism $H \otimes_A H \simeq \mathrm{Hom}_A(H^*, H) \simeq \mathrm{Hom}_A(H, H)$ given by $x \otimes y \mapsto (h \mapsto \tau(xh)y)$ the element I goes to $h \mapsto \sum_i \tau(e_i h) e_i' = h$ hence to the identity, which gives another proof of the independence of I from the choice of basis.

The symmetrising form τ induces an isomorphism $f \mapsto \check{f} : H^* \to H$, by $f(h) = \tau(\check{f}h)$. By definition $\check{\tau} = 1$. Note that $f \in H^*$ is a trace on H, that is it satisfies $f(hh') = f(h'h)$ for all elements $h, h' \in H$, if and only if $\check{f}$ is in the centre of H. For any pair of dual bases $(e_i), (e_i')$ we have $\check{f} = \sum_i f(e_i) e_i'$.

Definition 6.3.5   If χ is the character of an absolutely irreducible H_K-module M, for h in the centre of H_K, we denote by $\omega_\chi(h) \in K$ the scalar by which h acts on M and we call **Schur element** of χ the scalar $S_\chi := \omega_\chi(\check{\chi}) \in A$.

The name "Schur element" and the semi-simplicity criterion of 6.3.8 are due to Geck.

Note that $\check{\chi}$ is in H since $\chi(H) \subset A$, which is true since, A being integrally closed, the character of any finite dimensional H_K-module takes values in A on H.

Proposition 6.3.6   *Assume moreover H_K split semi-simple.*

(i) *The central primitive idempotent corresponding to an irreducible character χ is $e_\chi = \frac{1}{S_\chi} \check{\chi}$.*

(ii) *We have $\tau = \sum_{\chi \in \mathrm{Irr}\, H_K} \frac{1}{S_\chi} \chi$.*

Proof   For any $h \in H$ we have $\chi(e_\chi h) = \chi(h)$, hence $\tau(\check{\chi} e_\chi h) = \tau(\check{\chi} h)$, which implies $\check{\chi} e_\chi = \check{\chi}$. Hence $\check{\chi}$ acts by 0 on any simple H-module with character different from χ. Since $\check{\chi}$ is central in H, we have $\check{\chi} = \sum_{\varphi \in \mathrm{Irr}\, H_K} \omega_\varphi(\check{\chi}) e_\varphi = S_\chi e_\chi$, whence (i) since $\check{\chi} \neq 0$, whence $S_\chi \neq 0$. From $1 = \sum_\chi e_\chi$, we get thus $1 = \sum_\chi \frac{1}{S_\chi} \check{\chi}$ which is equivalent to $\tau = \sum_\chi \frac{1}{S_\chi} \chi$ since $1 = \check{\tau}$. $\square$

We define a scalar product on functions on H_K by $\langle f, g \rangle_{H_K} := \sum (f \otimes g)(I) = \sum_i f(e_i) g(e_i')$.

Proposition 6.3.7   *Assume H_K split semi-simple; then for $\chi, \varphi \in \mathrm{Irr}(H_K)$ we have*

$$\langle \chi, \varphi \rangle = \begin{cases} \chi(1) S_\chi & \text{if } \chi = \varphi, \\ 0 & \text{if } \chi \neq \varphi. \end{cases}$$

Proof   We have $\langle \tau, \chi \rangle = \sum_i \tau(e_i) \chi(e_i') = \chi(\sum_i \tau(e_i) e_i') = \chi(1)$. By 6.3.6, the left hand side is equal to $\sum_{\varphi \in \mathrm{Irr}(H_K)} \frac{1}{S_\varphi} \langle \varphi, \chi \rangle$. But $\langle \varphi, \chi \rangle = \sum_i \varphi(e_i) \chi(e_i') = \chi(\sum_i \tau(\check{\varphi} e_i) e_i') = \chi(\check{\varphi})$. We have seen in the proof of 6.3.6 that $\check{\varphi} = \check{\varphi} e_\varphi$ so that $\chi(\check{\varphi}) = 0$ if $\varphi \neq \chi$. We get then $\langle \tau, \chi \rangle = \frac{1}{S_\chi} \langle \chi, \chi \rangle = \chi(1)$, whence the proposition. $\square$

The next result gives a criterion of semi-simplicity for a symmetric algebra. For a proof see Broué et al. (1999, 7.13) or Geck and Pfeiffer (2000, Theorem 7.4.7).

Proposition 6.3.8 *Assume H_K split semi-simple and let $A \to k$ be a surjective morphism from A onto a field k. Assume H_k split, then H_k is semi-simple if and only if for all $\chi \in \mathrm{Irr}(H_K)$ the Schur element S_χ has a non-zero image in k.*

We now apply these results to Iwahori–Hecke algebras. As before (W, S) denotes a finite Coxeter system. Let A be an integrally closed integral domain with quotient field K.

Definition 6.3.9   A representation of an Iwahori–Hecke algebra whose character vanishes on T_w for all $w \neq 1$ is called **trace-special**.

Corollary 6.3.10 *Take $H = H_{\{a_s, a'_s\}_{s \in S}}(W, A)$ and τ as in 6.3.1. Assume that H_K is split semi-simple and that H_K has a trace-special representation on a finite dimensional K-vector space V. Then the multiplicity in V of an irreducible representation of H_K with character χ is equal to $\frac{\dim V}{S_\chi}$.*

Proof   This is a consequence of 6.3.6 since the character of V is $(\dim V)\tau$.   □

We can apply 6.3.10 to the two cases of 6.2.6.

Proposition 6.3.11

(i) *If $\mathbf{B}$ is an F-stable Borel subgroup in the reductive group $\mathbf{G}$ then the representation $\mathrm{Ind}_{\mathbf{B}^F}^{\mathbf{G}^F}(1)$ over a field K whose characteristic does not divide $|\mathbf{B}^F|$ is a trace-special representation of $H_{\{q^{l(w_J)}, -1\}_{J \in S/F}}(W^F, K)$.*

(ii) *With assumption and notation as in 6.2.6(ii), the representation of $H_{\{q^{cs}, -1\}_{s \in S(L, \Lambda)}}(W_{\mathbf{G}^F}(\mathbf{L}, \Lambda), K)$ on $R_{\mathbf{L}}^{\mathbf{G}} \Lambda$ is trace-special.*

Proof   Let us prove (ii). We can take as a basis of the space $R_{\mathbf{L}}^{\mathbf{G}} \Lambda = K(\mathbf{G}^F/\mathbf{U}^F) \otimes_{K\mathbf{L}^F} \Lambda$ the elements $g e_{\mathbf{U}^F} \otimes e_i$ where g runs over a set of representatives of $\mathbf{G}^F/\mathbf{L}^F\mathbf{U}^F$ and (e_i) is a basis of Λ. The operator B_w maps $g e_{\mathbf{U}^F} \otimes e_i$ to $g e_{\mathbf{U}^F} n_w^{-1} e_{\mathbf{U}^F} \otimes \gamma_{n_w}(e_i)$. Since $(\mathbf{U}^F n_w \mathbf{U}^F) \cap \mathbf{U}^F = \emptyset$ unless $w = 1$ by 3.2.3(ii), the matrix of B_w has all diagonal entries 0, hence trace 0; whence (ii) since T_w is a scalar multiple of B_w. The same argument applies to (i).   □

Proposition 6.3.12 *Let $H = H_{\{q_s, q'_s\}_{s \in S}}(W, K[q_s^{\pm 1}, q'^{\pm 1}_s]_{s \in S})$ be the generic Hecke algebra. Assume H_L split semi-simple for some extension L of $K(q_s^{\pm 1}, q'^{\pm 1}_s)$. For any irreducible character χ of H_L and any specialisation such that $q_s \mapsto 1$ and $q'_s \mapsto -1$, the Schur element S_χ specialises to $|W|/\chi(1)$.*

Proof By 6.3.7 and 6.3.1(ii), we have

$$\sum_{w\in W} \chi(T_w)\chi((-1)^{l(w)}q_w^{-1}q_w'^{-1}T_{w^{-1}}) = \chi(1)S_\chi,$$

where q_w is defined as a_w in 6.3.1(ii). The left-hand side specialises to $|W|$, thus S_χ specialises to $|W|/\chi(1)$.   □

Definition 6.3.13 Let H and L be as in 6.3.12; For $\chi \in \mathrm{Irr}(H_L)$ the **generic degree** of χ is $d_\chi := \frac{\sum_{w\in W}(-1)^{l(w)}q_w q_w'^{-1}}{S_\chi} \in L$.

Note that $\sum_{w\in W}(-1)^{l(w)}q_w q_w'^{-1}$ is the Schur element of the linear character $T_s \mapsto q_s$.

Since $d_\chi = \frac{\sum_{w\in W}(-1)^{l(w)}q_w q_w'^{-1}}{S_\chi}$ and $\sum_{w\in W}(-1)^{l(w)}q_w q_w'^{-1}$ specialises to $|W|$ one could say that the generic degree "specialises to $\chi(1)$" although specialising from a field has no meaning. In the same vein we have

Proposition 6.3.14 *Under the assumptions of 6.3.11(i), let $\{q_J, q_J' \mid J \in S/F\}$ be indeterminates; the degree of the constituent $\gamma_\chi \in \mathrm{Irr}(\mathbf{G}^F)$ of $\mathrm{Ind}_{\mathbf{B}^F}^{\mathbf{G}^F}(\mathbf{1})$ parametrised by $\chi \in \mathrm{Irr}(W^F)$ is equal to the "specialisation" of the generic degree of the corresponding generic algebra for any specialisation such that $q_J \mapsto q^{l(w_J)}$ and $q_J' \mapsto -1$.*

Proof By 6.2.11(i) the generic algebra is absolutely semi-simple so that there exists a splitting field L as in 6.3.12. By 6.2.9(iii) $\mathrm{Irr}(LW^F)$ is in bijection with the irreducible characters of the generic algebra. By Benson and Curtis (1972, Theorem 2.6) except for Ree and Suzuki groups and by the tables in Lusztig (1984a, Appendix) for those last groups, we have $d_\chi \in \mathbb{Q}[q_J, q_J']_{J\in S/F}$ for all $\chi \in \mathrm{Irr}(W^F)$. The proposition is thus a consequence of 6.3.10 and 6.1.1(ii), since $\dim(\mathrm{Ind}_{\mathbf{B}^F}^{\mathbf{G}^F}(\mathbf{1})) = |\mathbf{G}^F/\mathbf{B}^F|$, which by 4.4.1(iii) is the specialisation of $\sum_{w\in W^F}(-1)^{l(w)}q_w q_w'^{-1}$ where q_w is defined as a_w in 6.3.1(ii).   □

6.4 The Example of G_2

We end this chapter with the example of the generic Hecke algebra of type G_2.

Example 6.4.1 Let $(W, \{s,t\})$ be a Coxeter system of type G_2. Consider the Iwahori–Hecke algebra $H := H_{(x,x'),(y,y')}(W, \mathbb{Z}[x^{\pm 1}, y^{\pm 1}, x'^{\pm 1}, y'^{\pm 1}])$ where (x,x') and (y,y') are pairs of indeterminates corresponding to s and t – which are not conjugate. By 6.2.11(i) $H_{\mathbb{Q}(x,y,x',y')}$ is absolutely semi-simple. We will compute the values of irreducible characters on the basis elements T_w and their generic degrees over a splitting field which we will show to be $\mathbb{Q}(x,y,x',y', \sqrt{xyx'y'})$ (compare with 6.2.10). These irreducible characters are in bijection with $\mathrm{Irr}(W)$ thus there are four irreducible characters of degree 1 and two characters of degree 2. The representations of dimension 1 are determined by the possible eigenvalues for T_s which must be either x or x' and the possible eigenvalues for T_t which must be y or y'. These four possibilities give the four characters of degree 1. We now compute the irreducible representations of dimension 2. Since the values of the irreducible characters of W on s and t are 0, and are specialisations when x and y go to 1 and x', y' go to -1 of the values of the corresponding character on T_s and T_t, the two eigenvalues must be distinct. Take a basis consisting of an eigenvector of T_s for the eigenvalue x' and an eigenvector of T_t for the eigenvalue y', which are not colinear since the representation is irreducible. The matrices of T_s and T_t are then of the form $\begin{pmatrix} x' & u \\ 0 & x \end{pmatrix}$ and $\begin{pmatrix} y & 0 \\ v & y' \end{pmatrix}$ for some u and v which we can determine using the braid relation between T_s and T_t. The computation gives $uv = -(xx' + yy') + \varepsilon \sqrt{xyx'y'}$ with $\varepsilon = \pm 1$. Thus a splitting field for H is $\mathbb{Q}(x,y,x',y', \sqrt{xyx'y'})$, and the two representations of dimension 2 are conjugate by $\sqrt{xyx'y'} \mapsto -\sqrt{xyx'y'}$, an element of $\mathrm{Gal}(\mathbb{Q}(x,y, \sqrt{xyx'y'})/\mathbb{Q}(x,y,x',y'))$.

We can compute thus all character values, and then the Schur elements either by 6.3.7 or by solving equation 6.3.6(ii), and finally the generic degrees.

We denote by $\mathbf{1}$, sgn, τ, and σ the four characters of degree 1 of W and by ρ and ρ' the two characters of degree 2 and we use the same notation for the corresponding characters of H. The two characters ρ and ρ' correspond to the two possible values of ε.

In Table 6.1 we factorised generic degrees in terms of cyclotomic polynomials ϕ_i to save space, and we have only given the character values on elements T_w for one element w of minimal length in each W-conjugacy class. The values at other elements can be computed thanks to Theorem 6.4.2.

Theorem 6.4.2 *Let (W,S) be a finite Coxeter system. Then*

(i) *If $w \in W$ is not of minimal length in its conjugacy class, there exists $s \in S$ such that $l(sws) = l(w) - 2$;*

(ii) *If w,w' are of minimal length in the same conjugacy class, they are conjugate by a sequence of "cyclic" conjugations of the form $w = xy$, $w' = yx$ with $l(w) = l(x) + l(y) = l(w')$.*

Reference See Geck and Pfeiffer (2000, Theorem 3.2.9). Note that (i) for $w \in \mathrm{Ref}(W)$ follows from 2.1.6(iv). $\qquad\square$

Table 6.1. *Character table of* $H_{(x,x'),(y,y')}(W(G_2), \mathbb{Z}[x^{\pm 1}, y^{\pm 1}, x'^{\pm 1}, y'^{\pm 1}])$

	Generic degree	T_1	T_s	T_t	T_{st}	T_{stst}	T_{ststst}
1	1	1	x	y	xy	$(xy)^2$	$(xy)^3$
sgn	$\left(\dfrac{xy}{x'y'}\right)^3$	1	x'	y'	$x'y'$	$(x'y')^2$	$(x'y')^3$
σ	$-\dfrac{x\phi_3(xy/x'y')}{x'\phi_3(yx'/xy')}$	1	x'	y	$x'y$	$(x'y)^2$	$(x'y)^3$
τ	$-\dfrac{y\phi_3(xy/x'y')}{y'\phi_3(xy'/x'y)}$	1	x	y'	$y'x$	$(xy')^2$	$(xy')^3$
ρ	$-\dfrac{x\phi_1(x/x')\phi_1(y/y')\phi_3(\sqrt{xy/x'y'})}{2x'\phi_3(\sqrt{xy'/x'y})}$	2	$x+x'$	$y+y'$	$\sqrt{xx'yy'}$	$-xx'yy'$	$-2\sqrt{(xx'yy')^3}$
ρ'	$-\dfrac{x\phi_1(x/x')\phi_2(y/y')\phi_6(\sqrt{xy/x'y'})}{2x'\phi_6(\sqrt{xy'/x'y})}$	2	$x+x'$	$y+y'$	$-\sqrt{xx'yy'}$	$-xx'yy'$	$2\sqrt{(xx'yy')^3}$

Table 6.2. *Character table of* $H_{(q,-1),(q,-1)}(W(G_2), BZ[q^{\pm 1}])$

χ	Degree of γ_χ	T_1	T_s	T_t	T_{st}	T_{stst}	T_{ststst}
1	1	1	q	q	q^2	q^4	q^6
sgn	q^6	1	-1	-1	1	1	1
σ	$\frac{q}{3}\phi_3(q)\phi_6(q)$	1	-1	q	$-q$	q^2	$-q^3$
τ	$\frac{q}{3}\phi_3(q)\phi_6(q)$	1	q	-1	$-q$	q^2	$-q^3$
ρ	$\frac{q}{6}\phi_2(q)^2\phi_3(q)$	2	$q-1$	$q-1$	q	$-q^2$	$-2q^3$
ρ'	$\frac{q}{2}\phi_2(q)^2\phi_6(q)$	2	$q-1$	$q-1$	$-q$	$-q^2$	$2q^3$

Since for elements as in (ii) we have $T_w = T_x T_y$ and $T_y T_x = T_{w'}$ it follows that the T_w corresponding to elements of minimal length are conjugate in H, provided the parameters are invertible, thus have same character values.

Assume now that w is not of minimal length in its conjugacy class, and let s and $w' := sws$ be as in (i) of the above theorem. Then $T_w = T_s T_{w'} T_s$ thus if $\chi \in \text{Irr}(H)$ then $\chi(T_w) = \chi(T_s T_{w'} T_s) = \chi(T_{w'} T_s^2) = (q_s + q_s')\chi(T_{w's}) - q_s q_s' \chi(T_{w'})$, reducing the computation to character values on smaller length elements.

By 6.3.14, when H is specialised to $\text{End}(\text{Ind}_{\mathbf{B}^F}^{\mathbf{G}^F}(1))$ (with $\mathbf{G}$ of type G_2) by $x \mapsto q$, $y \mapsto q$ $x' \mapsto -1$, $y' \mapsto -1$ and $\sqrt{xyx'y'} \mapsto q$, the generic degrees specialise to the dimensions of the elements of $\text{Simp}(\mathbf{G}^F \mid (\mathbf{T}, 1))$, called the **principal series unipotent** representations. We give the character table of this specialised algebra in Table 6.2. By 6.2.14(i) γ_1 is the trivial character and by 7.2.14 γ_{sgn} is the Steinberg character.

Note that the four characters of degree 1 of the generic Iwahori–Hecke algebra are permuted by a Galois action, exchanging x and x' and/or y and y'. This Galois action disappears in the specialised algebra. The two characters of degree 2 of H are interchanged by the Galois action which changes the sign of $\sqrt{xx'yy'}$. Exercise 6.4.3 shows that these properties generalise to all Hecke algebras of finite Coxeter groups.

Exercise 6.4.3

(i) Show that for any field K, the map $\varphi : T_s \mapsto q_s q_s' T_s^{-1}$ defines an automorphism of the generic Iwahori–Hecke algebra $H := H_{(q_s, q_s')_{s \in S}}(W, A)$ where $A = K[q_s^{\pm 1}, q_s'^{\pm 1}]_{s \in S}$.

(ii) Denote by $a \mapsto \bar{a}$ the automorphism of A which maps q_s and q_s' to their inverses. Show that $j(\sum_w a_w T_w) = \sum_w \overline{a_w}(q_w q_w')^{-1} T_w$ is a semilinear automorphism of H.

(iii) Let $K \subset \mathbb{C}$ be a splitting field for W and let $A' = K[\sqrt{q_s}^{\pm 1}, \sqrt{-q'_s}^{\pm 1}]_{s \in S}$. The automorphism φ extends naturally to $H \otimes_A A'$. Show that if χ_1 denotes the specialisation to KW of a character χ of $H \otimes_A A'$ for some specialisation such that $q_s \mapsto 1$ and $q'_s \mapsto -1$, then the character $\chi^\varphi := \chi \circ \varphi$ specialises to $\chi_1 \otimes \text{sgn}$.

(iv) Let K and A' be as in (iii). We denote again by j an extension of j to A'. It can be shown, by similar arguments as in Geck and Pfeiffer (2000, 9.2.5), that for any irreducible character χ of $H \otimes_A A'$ the value $\chi(T_w) \in A'$ is homogeneous of degree $l(w)$ in q_s, q'_s. Using this fact, show that if ρ is a representation of H over A' with character χ then $\rho^j(T_w) := \overline{\rho(j(T_w))}$ is a representation of H whose character is $\chi^j(T_w) = q_w q'_w \overline{\chi(T_w)}$ and specialises to $\chi_1 \otimes \text{sgn}$. Deduce that the Galois automorphism σ of A' which for all s interchanges $\sqrt{q_s}$ and $\sqrt{-q'_s}$ satisfies $\chi^\varphi = \sigma \circ \chi$ for any character χ of $H \otimes_A A'$.

We will see in 7.2.13 that the unipotent characters γ_χ and γ_{χ^φ} are dual to each other. Thus the Galois automorphism of (iv) of the generic Hecke algebra corresponds to duality.

Notes

A good reference about Iwahori–Hecke algebras is Geck and Pfeiffer (2000). The structure of the algebra $\text{End}_{K\mathbf{G}^F}(R_{\mathbf{L}}^{\mathbf{G}} \Lambda)$, in particular Lemmas 6.1.8 and 6.1.10, was first determined in Howlett and Lehrer (1980) in characteristic 0 and by Geck et al. (1996b) in general.

7

The Duality Functor
and the Steinberg Character

This chapter is devoted to the exposition of the main properties of the "duality" functor for the characters of $\mathbf{G}^F$. This duality, introduced in 1979, has an elementary definition and allows substantial simplification in some areas of the representation theory of $\mathbf{G}^F$, for instance the theory of the Steinberg character.

7.1 F-rank

First we attach signs to F-stable tori and reductive subgroups of a reductive group $\mathbf{G}$ with a Frobenius root F.

Definition 7.1.1   If F is a Frobenius root on the torus $\mathbf{T}$ acting as $q\tau$ on $E = X(\mathbf{T}) \otimes \mathbb{R}$ – see 4.2.5 – we define the **F-rank** of $\mathbf{T}$ as $\dim E^\tau$, and define a sign $\varepsilon_{\mathbf{T}} = (-1)^{F\text{-rank}(\mathbf{T})}$.

Lemma 7.1.2   *We have $\varepsilon_{\mathbf{T}} = (-1)^{\text{rank}(\mathbf{T})} \det \tau$.*

Proof   Since τ is real and of finite order, its eigenvalues are roots of unity which are 1 with multiplicity $\dim E^\tau$, -1 with a certain multiplicity n_{-1}, and non-real roots of unity which come in pairs $(\lambda, \overline{\lambda})$. Since $\lambda\overline{\lambda} = 1$, is follows that $\det \tau = (-1)^{n_{-1}}$. We have $\dim E^\tau + n_{-1} \equiv \dim E \pmod 2$, which gives the result since $\dim E = \text{rank}(\mathbf{T})$.   □

When F is the Frobenius endomorphism attached to an $\mathbb{F}_q$-structure, the F-rank is called also the **$\mathbb{F}_q$-rank**, and $\mathbb{F}_q$-rank$(\mathbf{T}) = \text{rank}(X(\mathbf{T})^\tau)$ since τ is defined on $X(\mathbf{T})$ in this case.

Lemma 7.1.3   *A torus $\mathbf{T}$ defined over $\mathbb{F}_q$ is split if and only if $\mathbb{F}_q$-rank$(\mathbf{T}) = \text{rank}(\mathbf{T})$.*

Proof Let $\{a_1, \ldots, a_n\}$ be a $\mathbb{Z}$-basis of $X(\mathbf{T})$. Using that

$$\mathbf{T} = \mathrm{Spec}(\overline{\mathbb{F}}_q X(\mathbf{T})) = \mathrm{Spec}(\overline{\mathbb{F}}_q[a_1^{\pm 1}, \ldots, a_n^{\pm 1}]),$$

we see that it is equivalent that F acts by q on all a_i or that the isomorphism $\mathbf{T} \xrightarrow{\sim} \mathbb{G}_m^n$ given by $T_i \mapsto a_i : \overline{\mathbb{F}}_q[T_1^{\pm 1}, \ldots, T_n^{\pm 1}] \to \overline{\mathbb{F}}_q[a_1^{\pm 1}, \ldots, a_n^{\pm 1}]$ be defined over $\mathbb{F}_q$.   □

We have then the following alternative definition of $\mathbb{F}_q$-rank:

Proposition 7.1.4 *A torus $\mathbf{T}$ defined over $\mathbb{F}_q$ contains a unique maximum split subtorus $\mathbf{T}_1$ and $\mathrm{rank}(\mathbf{T}_1) = \mathbb{F}_q$-$\mathrm{rank}(\mathbf{T})$.*

Proof The image of a split torus $\mathbf{S}$ by a surjective morphism of tori $\mathbf{S} \xrightarrow{f} \mathbf{S}'$ defined over $\mathbb{F}_q$ is split, as $f^* : X(\mathbf{S}') \to X(\mathbf{S})$ is injective. This implies that the product of two split subtori of $\mathbf{T}$ is again split, whence the existence of $\mathbf{T}_1$.

If we denote again by τ the transposed action of τ on $Y(\mathbf{T})$, since rank $(X(\mathbf{T})^\tau) = \mathrm{rank}(Y(\mathbf{T})^\tau)$, to prove the second part of the statement it is sufficient to prove that $\mathrm{rank}(Y(\mathbf{T})^\tau) = \mathrm{rank}(Y(\mathbf{T}_1))$. The inclusion of $\mathbf{T}_1$ in $\mathbf{T}$ already proves that the dimension of the left-hand side is not smaller than the dimension of the right-hand side. On the other hand the sublattice $Y(\mathbf{T})^\tau$ defines a split subtorus, which must thus be $\mathbf{T}_1$.   □

Definition 7.1.5 Let F be a Frobenius root on the connected algebraic group $\mathbf{G}$. We call **F-rank** of $\mathbf{G}$ the F-rank of a quasi-split torus of $\mathbf{G}$, and define a sign $\varepsilon_\mathbf{G} = (-1)^{F\text{-rank}(\mathbf{G})}$.

Note that, since all quasi-split tori of $\mathbf{G}$ are $\mathbf{G}^F$-conjugate, this definition makes sense. We will also call $\mathbb{F}_q$-rank the F-rank of $\mathbf{G}$ when F is a Frobenius endomorphism.

Note that the F-rank of a group is equal to that of its reductive quotient, as tori map isomorphically in that quotient; that is, F-rank$(\mathbf{G})$ = F-rank $(\mathbf{G}/R_u(\mathbf{G}))$.

Lemma 7.1.6 *Let $\mathbf{T}$ be an F-stable maximal torus of type w (see 4.2.22) of $\mathbf{G}$. Then $\varepsilon_\mathbf{G}\varepsilon_\mathbf{T} = (-1)^{l(w)}$.*

Proof By definition of the type, if $\mathbf{T}_0$ is quasi-split, then the pair $(\mathbf{T}, F)$ is conjugate to $(\mathbf{T}_0, wF)$. Thus if $F = q\tau$ on $X(\mathbf{T}_0) \otimes \mathbb{R}$, we have $F = wq\tau$ on $X(\mathbf{T}) \otimes \mathbb{R}$ and by 7.1.2 we have $\varepsilon_\mathbf{T}\varepsilon_\mathbf{G} = \det w$. Now w is a product of $l(w)$ reflections, of determinant -1, whence the result.   □

Proposition 7.1.7 *An F-stable torus $\mathbf{T}$ of $\mathbf{G}$ is quasi-split if and only if F-rank$(\mathbf{T}) = F$-rank$(\mathbf{G})$, which is the maximum over all F-stable maximal tori $\mathbf{T}$ of $\mathbf{G}$ of F-rank$(\mathbf{T})$. In particular, if F is a Frobenius endomorphism, an*

F-stable torus **T** *is quasi-split if and only if* **maximally split**; *that is,* $\mathbf{T}_1$ *of 7.1.4 is a maximal split torus of* **G**.

Proof Taking the quotient by $R_u(\mathbf{G})$ we can assume **G** reductive. We use again that any F-stable torus is conjugate to $(\mathbf{T}_0, wF)$ for some w, where $\mathbf{T}_0$ is quasi-split. Since F preserves a Borel subgroup **B** containing $\mathbf{T}_0$, if Π is the basis of the roots corresponding to the order defined by **B**, then by 4.2.21 τ preserves the open cone $C = \{x \in E \mid \alpha(x) > 0$ for any $\alpha \in \Pi\}$. This implies that there is a τ-fixed point in C, thus lying outside all reflecting hyperplanes (indeed, any $\alpha \in \Phi^+$ being a linear combination of elements of Π with nonnegative coefficients, we also have $\alpha(x) > 0$ for $x \in C$). This means that τ has a regular eigenvector for the eigenvalue 1, see Springer (1974, 6.5), thus by Springer (1974, 6.2) E^τ has maximal dimension among all the $E^{w\tau}$, and any other element such that $\dim E^{w\tau} = \dim E^\tau$ satisfies $w\tau = v\tau v^{-1}$; that is, $w\tau$ is conjugate to τ, or in other terms w is F-conjugate to 1. □

Example 7.1.8 Let us compute the $\mathbb{F}_q$-rank of the unitary group $\mathbf{U}_n$, where the Frobenius F' is as in 4.3.3. If **B** is the Borel subgroup of upper triangular matrices, the group $^{F'}\mathbf{B}$ is the Borel subgroup of lower triangular matrices, and is also equal to $^{w_0}\mathbf{B}$, where w_0 is the longest element of $W(\mathbf{T})$ where **T** is the torus of diagonal matrices (w_0 is the permutation $(1, 2, \ldots, n) \mapsto (n, n-1, \ldots, 1)$). Thus **T** has type w_0 with respect to some quasi-split torus, so $(\mathbf{T}, w_0 F')$ is geometrically conjugate to $(\mathbf{T}_0, F')$ where $\mathbf{T}_0$ is a quasi-split torus. The $\mathbb{F}_q$-rank of $\mathbf{U}_n$ is thus the dimension of $(X(\mathbf{T}) \otimes \mathbb{R})^{w_0 F'/q}$; that is, the number of eigenvalues q of $w_0 F'$, which is equal to $\lceil n/2 \rceil$. Thus $(\mathbf{U}_n, F')$ is not split and in particular not isomorphic to $(\mathbf{GL}_n, F)$.

The following lemma relates the $\mathbb{F}_q$-rank of a group to that of its semi-simple quotient.

Lemma 7.1.9 *With the notation of 7.1.5(ii), we have*

$$F\text{-rank}(\mathbf{G}) = F\text{-rank}(\mathbf{G}/R\mathbf{G}) + F\text{-rank}(R\mathbf{G}).$$

Proof Taking the quotient by $R_u(\mathbf{G})$ we can assume **G** reductive, then $R(\mathbf{G})$ is a torus. Let **T** be an F-stable maximal torus of **G**. Applying to the exact sequence

$$1 \to R\mathbf{G} \to \mathbf{T} \to \mathbf{T}/R\mathbf{G} \to 1$$

the exact functor X, see 1.2.4, we get

$$0 \leftarrow X(R\mathbf{G}) \leftarrow X(\mathbf{T}) \leftarrow X(\mathbf{T}/R\mathbf{G}) \leftarrow 0.$$

This sequence remains exact when tensoring with $\mathbb{R}$, and it remains so when taking the fixed points under τ, as τ, being of finite order, is semi-simple. If we take for $\mathbf{T}$ a quasi-split torus of $\mathbf{G}$, we get the result. $\quad\square$

Definition 7.1.10 Let $\mathbf{G}$ be a connected algebraic group over $\overline{\mathbb{F}}_q$, with a Frobenius root F. We call **semi-simple F-rank** of $\mathbf{G}$, denoted by $r(\mathbf{G})$, the F-rank of $\mathbf{G}/R\mathbf{G}$.

Lemma 7.1.11 *Let $\mathbf{T}$ be an F-stable maximal torus of a connected reductive group $\mathbf{G}$, and let $E = X(\mathbf{T}) \otimes \mathbb{R}$; if $\Phi^{\vee}$ is the set of coroots of $\mathbf{G}$ relative to $\mathbf{T}$, we have F-rank$(R\mathbf{G}) = \dim(\langle\Phi^{\vee}\rangle^{\perp} \cap E^{\tau})$.*

Proof We first use that $(\mathbf{T}, F)$ is $\mathbf{G}^F$-conjugate to $(\mathbf{T}_0, wF)$ for some quasi-split $\mathbf{T}_0$ and $w \in W_{\mathbf{G}}(\mathbf{T}_0)$; as W acts trivially on $\langle\Phi^{\vee}\rangle^{\perp}$, the actions of τ and $w\tau$ on $\langle\Phi^{\vee}\rangle^{\perp}$ are the same and we may assume $w = 1$. The lemma is a consequence of the isomorphism $X(R\mathbf{G}) \otimes \mathbb{R} \xrightarrow{\sim} \langle\Phi^{\vee}\rangle^{\perp}$, which itself results from the second exact sequence of 7.1.9 tensored by $\mathbb{R}$: indeed Φ is the root system of the semi-simple group $\mathbf{G}/R\mathbf{G}$ relative to the quasi-split torus $\mathbf{T}/R(\mathbf{G})$, so it spans $X(\mathbf{T}/R(\mathbf{G})) \otimes \mathbb{R}$ (see 2.3.5(iii)). $\quad\square$

Proposition 7.1.12 *Let $\mathbf{G}$ be a connected reductive group with a Frobenius root F. In the context of the (B, N) pair for $\mathbf{G}$ defined by an F-stable pair $\mathbf{T} \subset \mathbf{B}$, let $\mathbf{P}_I = \mathbf{B}W_I\mathbf{B}$ be the F-stable parabolic subgroup of $\mathbf{G}$ defined by an F-stable subset $I \subset S$. Then $r(\mathbf{P}_I) = |I/F|$.*

Proof Without changing the semi-simple rank, we may replace $\mathbf{P}_I$ by its Levi subgroup $\mathbf{L}_I$. We may then also assume that $\mathbf{L}_I = \mathbf{G}$ and is semi-simple, so that $I = S$. We have then to show that $\dim(E^{\tau}) = |S/F|$. As τ is of finite order and normalises W, it follows that the group $\langle\tau, W\rangle$ is finite, thus there is a scalar product on E which is W and τ invariant. Thus τ is defined by the permutation it induces of the unitary vectors for this scalar product colinear to the elements of Π. The proposition then results from the fact that the dimension of the 1-eigenspace of a linear map which permutes a basis is the number of orbits of the map on the basis. $\quad\square$

7.2 The Duality Functor

Here C will denote class functions over $\mathbb{C}$. In the rest of this chapter we assume that $\mathbf{G}$ is connected reductive.

Definition 7.2.1 Let F be a Frobenius root on $\mathbf{G}$ over $\overline{\mathbb{F}}_p$ and let $\mathbf{B}$ be an F-stable Borel subgroup of $\mathbf{G}$. By **duality** we mean the operator $D_{\mathbf{G}}$ on $C(\mathbf{G}^F)$ defined by

$$D_{\mathbf{G}} = \sum_{\mathbf{P} \supset \mathbf{B}} (-1)^{r(\mathbf{P})} R_{\mathbf{L}}^{\mathbf{G}} \circ {}^*R_{\mathbf{L}}^{\mathbf{G}}$$

where the summation is over the set of F-stable parabolic subgroups of $\mathbf{G}$ containing $\mathbf{B}$ and where $\mathbf{L}$ denotes an arbitrarily chosen F-stable Levi subgroup of $\mathbf{P}$.

Remark 7.2.2   This definition is consistent because of the two following properties:

(i) The operator $R_{\mathbf{L}}^{\mathbf{G}} \circ {}^*R_{\mathbf{L}}^{\mathbf{G}}$ depends only on the parabolic subgroup $\mathbf{P}$ and not on the chosen Levi subgroup.
(ii) $D_{\mathbf{G}}$ does not depend on the F-stable Borel subgroup $\mathbf{B}$.

Property (i) comes from the fact that two F-stable Levi subgroups of $\mathbf{P}$ are conjugate under $\mathbf{P}^F$ and from the equality $R_{\mathbf{L}}^{\mathbf{G}} \circ \mathrm{ad}\, x^{-1} = R_{{}^x\mathbf{L}}^{\mathbf{G}}$ along with its analogue for ${}^*R_{\mathbf{L}}^{\mathbf{G}}$. We could also deduce it from the equality $R_{\mathbf{L}}^{\mathbf{G}} \circ {}^*R_{\mathbf{L}}^{\mathbf{G}}(\chi) = \mathrm{Ind}_{\mathbf{P}^F}^{\mathbf{G}^F}(\chi')$ where $\chi'(p) := |\mathbf{U}^F|^{-1} \sum_{u \in \mathbf{U}^F} \chi(pu)$ for $p \in \mathbf{P}^F$, where $\mathbf{U} = R_u(\mathbf{P})$.

Property (ii) comes from the fact that two F-stable Borel subgroups of $\mathbf{G}$ are $\mathbf{G}^F$-conjugate.

Proposition 7.2.3   *The functor $D_{\mathbf{G}}$ is self-adjoint.*

Proof   This is straightforward from the definition of $D_{\mathbf{G}}$ and the fact that $R_{\mathbf{L}}^{\mathbf{G}}$ and ${}^*R_{\mathbf{L}}^{\mathbf{G}}$ are adjoint to each other (see 5.1.8(iii)).   □

Theorem 7.2.4 (C. Curtis)   *For any F-stable Levi subgroup $\mathbf{L}$ of an F-stable parabolic subgroup of $\mathbf{G}$ we have*

$$D_{\mathbf{G}} \circ R_{\mathbf{L}}^{\mathbf{G}} = R_{\mathbf{L}}^{\mathbf{G}} \circ D_{\mathbf{L}}.$$

The proof we shall give is valid even when the parabolic subgroup is not F-stable, in which case we have to replace $R_{\mathbf{L}}^{\mathbf{G}}$ with the Lusztig functor which is its generalisation and is again denoted by $R_{\mathbf{L}}^{\mathbf{G}}$ – this functor will be extensively studied later, starting from Chapter 9 – and the formula has then to be written

$$\varepsilon_{\mathbf{G}} D_{\mathbf{G}} \circ R_{\mathbf{L}}^{\mathbf{G}} = \varepsilon_{\mathbf{L}} R_{\mathbf{L}}^{\mathbf{G}} \circ D_{\mathbf{L}}.$$

In the case of an F-stable parabolic subgroup this formula reduces to the previous one because $\mathbf{L}$ then contains a quasi-split torus of $\mathbf{G}$, so that F-rank$(\mathbf{G}) = F$-rank$(\mathbf{L})$.

The only properties of $R_{\mathbf{L}}^{\mathbf{G}}$ that we shall use in the proof are transitivity, the formula $R_{\mathbf{L}}^{\mathbf{G}} = R_{{}^x\mathbf{L}}^{\mathbf{G}} \circ \mathrm{ad}\, x$ (for $x \in \mathbf{G}^F$) and the "Mackey formula"; these facts hold for the Lusztig functor, with some restrictions for the Mackey formula (see Chapter 9). The reader will check that in the remainder of the book we shall use 7.2.4 for the Lusztig functor only in those cases where the Mackey formula has been proved.

Proof By the Mackey formula, using the equality $(\mathbf{M}^F \backslash S(\mathbf{M},\mathbf{L})^F / \mathbf{L}^F)^{-1} = \mathbf{L}^F \backslash S(\mathbf{L},\mathbf{M})^F / \mathbf{M}^F$ we get

$$
\begin{aligned}
D_{\mathbf{G}} \circ R_{\mathbf{L}}^{\mathbf{G}} &= \sum_{\mathbf{Q} \supset \mathbf{B}} (-1)^{r(\mathbf{Q})} R_{\mathbf{M}}^{\mathbf{G}} {}^* R_{\mathbf{M}}^{\mathbf{G}} R_{\mathbf{L}}^{\mathbf{G}} \\
&= \sum_{\mathbf{Q} \supset \mathbf{B}} (-1)^{r(\mathbf{Q})} \sum_{x \in \mathbf{L}^F \backslash S(\mathbf{L},\mathbf{M})^F / \mathbf{M}^F} R_{\mathbf{M}}^{\mathbf{G}} \operatorname{ad} x^{-1} R_{^x\mathbf{M} \cap \mathbf{L}}^{^x\mathbf{M}} {}^* R_{^x\mathbf{M} \cap \mathbf{L}}^{\mathbf{L}}.
\end{aligned}
$$

We then use the equalities $R_{\mathbf{M}}^{\mathbf{G}} \operatorname{ad} x^{-1} = R_{^x\mathbf{M}}^{\mathbf{G}}$ and $R_{^x\mathbf{M}}^{\mathbf{G}} R_{^x\mathbf{M} \cap \mathbf{L}}^{^x\mathbf{M}} = R_{^x\mathbf{M} \cap \mathbf{L}}^{\mathbf{G}} = R_{\mathbf{L}}^{\mathbf{G}} R_{^x\mathbf{M} \cap \mathbf{L}}^{\mathbf{L}}$ to get

$$
D_{\mathbf{G}} \circ R_{\mathbf{L}}^{\mathbf{G}} = R_{\mathbf{L}}^{\mathbf{G}} \circ \left(\sum_{\mathbf{Q} \supset \mathbf{B}} (-1)^{r(\mathbf{Q})} \sum_{x \in \mathbf{L}^F \backslash S(\mathbf{L},\mathbf{M})^F / \mathbf{M}^F} R_{^x\mathbf{M} \cap \mathbf{L}}^{\mathbf{L}} {}^* R_{^x\mathbf{M} \cap \mathbf{L}}^{\mathbf{L}} \right).
$$

We now transform the right-hand side of the above equality. Let $\mathbf{B}_{\mathbf{L}}$ be a fixed F-stable Borel subgroup of $\mathbf{L}$. For each parabolic subgroup $\mathbf{Q} \supset \mathbf{B}$ of $\mathbf{G}$ with Levi subgroup $\mathbf{M}$ we define a bijection $\mathcal{P}$ from $\mathbf{L}^F \backslash S(\mathbf{L},\mathbf{M})^F / \mathbf{M}^F$ to the set of parabolic subgroups of $\mathbf{G}$ which are $\mathbf{G}^F$-conjugate to $\mathbf{Q}$ and contain $\mathbf{B}_{\mathbf{L}}$. If $x \in S(\mathbf{L},\mathbf{M})^F$, as $^x\mathbf{Q}$ and $\mathbf{L}$ contain a common maximal torus, the group $^x\mathbf{Q} \cap \mathbf{L}$ is a parabolic subgroup of $\mathbf{L}$ – see 3.4.8(i). As it is F-stable this parabolic subgroup contains an F-stable Borel subgroup of $\mathbf{L}$ and such a Borel subgroup is conjugate by an element $l \in \mathbf{L}^F$ to $\mathbf{B}_{\mathbf{L}}$. Then we have $\overline{lx} = \overline{x}$, where $\overline{x}$ denotes the image of x in $\mathbf{L}^F \backslash S(\mathbf{L},\mathbf{M})^F / \mathbf{M}^F$, and $^{lx}\mathbf{Q} \supset \mathbf{B}_{\mathbf{L}}$. The map $\mathcal{P} : \overline{x} \mapsto {}^{lx}\mathbf{Q}$ is well defined; that is, $^{lx}\mathbf{Q}$ does not depend on the chosen l, since if $^{l'lx}\mathbf{Q}$ also contains $\mathbf{B}_{\mathbf{L}}$ then $^{l'lx}\mathbf{Q} \cap \mathbf{L} = {}^{lx}\mathbf{Q} \cap \mathbf{L}$ by 1.3.6(ii), thus $l' \in {}^{lx}\mathbf{Q} \cap \mathbf{L}$.

The map $\mathcal{P}$ is onto. Indeed, let $^x\mathbf{Q}$ with $x \in \mathbf{G}^F$ be a parabolic subgroup in $\mathbf{G}^F$ containing $\mathbf{B}_{\mathbf{L}}$, let $\mathbf{T}_0$ be a maximal torus of $\mathbf{B}_{\mathbf{L}}$ and let $\mathbf{M}'$ be the unique Levi subgroup of $^x\mathbf{Q}$ containing $\mathbf{T}_0$ (see 3.4.2). Then $\mathbf{M}'$ and $^x\mathbf{M}$ are conjugate under $^x\mathbf{Q}^F$; that is, there exists $q \in \mathbf{Q}^F$ such that $\mathbf{M}' = {}^{xq}\mathbf{M}$. But then $^{xq}\mathbf{M} \cap \mathbf{L} \supset \mathbf{T}_0$, so xq is in $S(\mathbf{L},\mathbf{M})^F$ and satisfies $\mathcal{P}(\overline{xq}) = {}^{xq}\mathbf{Q} = {}^x\mathbf{Q}$.

We now prove the injectivity of $\mathcal{P}$. If $^x\mathbf{Q} = {}^y\mathbf{Q}$ with x and y in $S(\mathbf{L},\mathbf{M})^F$ then since $N_{\mathbf{G}^F}(\mathbf{Q}) = \mathbf{Q}^F$ there exists $q \in \mathbf{Q}^F$ such that $y = xq$. The groups $\mathbf{L} \cap {}^x\mathbf{M}$ and $\mathbf{L} \cap {}^{xq}\mathbf{M}$ are two F-stable Levi subgroups of the parabolic subgroup $\mathbf{L} \cap {}^x\mathbf{Q}$ of $\mathbf{L}$. Thus there exists $l \in \mathbf{L}^F \cap {}^x\mathbf{Q}$ such that $\mathbf{L} \cap {}^x\mathbf{M} = {}^l(\mathbf{L} \cap {}^{xq}\mathbf{M}) = \mathbf{L} \cap {}^{lxq}\mathbf{M}$, so $^x\mathbf{M}$ and $^{lxq}\mathbf{M}$ are two Levi subgroups of $^x\mathbf{Q}$ containing a common maximal torus (any maximal torus of $\mathbf{L} \cap {}^x\mathbf{M}$), so they are equal. Thus $lxqx^{-1} \in N_{\mathbf{G}^F}(^x\mathbf{M}) \cap {}^x\mathbf{Q} = {}^x\mathbf{M}$ (equality by 3.4.2), so $y = xq \in l^{-1}x\mathbf{M} \subset \mathbf{L}x\mathbf{M}$ which implies, using the Lang–Steinberg theorem in $\mathbf{L} \cap {}^x\mathbf{M}$, that $y \in \mathbf{L}^F x \mathbf{M}^F$, whence $\overline{y} = \overline{x}$.

We have noticed in 7.2.2(i) that the functor $R_{^x\mathbf{M}\cap\mathbf{L}}^{\mathbf{L}}{}^*R_{^x\mathbf{M}\cap\mathbf{L}}^{\mathbf{L}}$ depends only on $^x\mathbf{Q}\cap\mathbf{L}$. We denote by $f(^x\mathbf{Q})$ this functor; we have $f(^x\mathbf{Q}) = f(\mathcal{P}(\overline{x}))$ as $\mathcal{P}(\overline{x})$ and $^x\mathbf{Q}$ are conjugate under $\mathbf{L}^F$. Since $\mathcal{P}$ is bijective and $r(\mathbf{Q}) = r(^x\mathbf{Q}) = r(\mathcal{P}(\overline{x}))$, we get

$$D_{\mathbf{G}} \circ R_{\mathbf{L}}^{\mathbf{G}} = R_{\mathbf{L}}^{\mathbf{G}} \circ \sum_{\mathbf{Q}'} (-1)^{r(\mathbf{Q}')} f(\mathbf{Q}')$$

where $\mathbf{Q}'$ in the summation runs over all F-stable parabolic subgroups of $\mathbf{G}$ containing $\mathbf{B_L}$. The summation in the right-hand side can be written

$$\sum_{\mathbf{Q}_0 \supset \mathbf{B_L}} \left(\sum_{\{\mathbf{Q}'|\mathbf{Q}'\cap\mathbf{L}=\mathbf{Q}_0\}} (-1)^{r(\mathbf{Q}')} \right) R_{\mathbf{M}_0}^{\mathbf{L}} \circ {}^*R_{\mathbf{M}_0}^{\mathbf{L}}$$

where $\mathbf{Q}_0$ runs over F-stable parabolic subgroups of $\mathbf{L}$ containing $\mathbf{B_L}$, where $\mathbf{M}_0$ denotes an F-stable Levi subgroup of $\mathbf{Q}_0$, and $\mathbf{Q}'$ runs over F-stable parabolic subgroups of $\mathbf{G}$ such that $\mathbf{Q}' \cap \mathbf{L} = \mathbf{Q}_0$. Theorem 7.2.4 is thus a consequence of the following result.

Proposition 7.2.5 *Let $\mathbf{H}$ be an F-stable reductive subgroup of maximal rank of $\mathbf{G}$, and let $\mathbf{Q_H}$ be an F-stable parabolic subgroup of $\mathbf{H}$. Then we have*

$$\varepsilon_{\mathbf{G}} \sum_{\{\mathbf{Q}|\mathbf{Q}\cap\mathbf{H}=\mathbf{Q_H}\}} (-1)^{r(\mathbf{Q})} = (-1)^{r(\mathbf{Q_H})} \varepsilon_{\mathbf{H}}$$

where $\mathbf{Q}$ in the summation runs over F-stable parabolic subgroups of $\mathbf{G}$ such that $\mathbf{Q} \cap \mathbf{H} = \mathbf{Q_H}$.

Proof Let $\mathbf{T}$ be an F-stable maximal torus of $\mathbf{Q_H}$, and let $E = X(\mathbf{T}) \otimes \mathbb{R}$. The coroots $\Phi^\vee$ of $\mathbf{G}$ with respect to $\mathbf{T}$ define a hyperplane arrangement $\mathcal{H} = \{\alpha^\perp \mid \alpha \in \Phi^\vee\}$ on E. By 4.2.5, we have $F = q\tau$ on E and τ induces a permutation of $\mathcal{H}$ since it differs by scalars from the permutation of the coroots attached to the p-morphism defined by F.

Lemma 7.2.6 *There is a bijection $\mathbf{Q} \mapsto \mathcal{F}_{\mathbf{Q}}$ between parabolic subgroups of $\mathbf{G}$ containing $\mathbf{T}$ and facets of the hyperplane arrangement $\mathcal{H}$. The parabolic subgroup $\mathbf{Q}$ is F-stable if and only if $\mathcal{F}_{\mathbf{Q}} \cap E^\tau \neq \emptyset$, and then $r(\mathbf{Q}) = \dim E^\tau - \dim(\mathcal{F}_{\mathbf{Q}} \cap E^\tau)$.*

Proof We call here facets intersection of hyperplanes and of open half-spaces defined by hyperplanes. Thus a facet can always be written $\{x \in E \mid \langle x,\alpha \rangle = 0 \text{ for } \alpha \in \Phi_1^\vee, \langle x,\alpha \rangle > 0 \text{ for } \alpha \in \Phi_2^\vee\}$, where $\Phi_1^\vee$ is some subset of $\Phi^\vee$ and $\Phi_2^\vee$ is the half of $\Phi^\vee - \Phi_1^\vee$ for which the scalar products are positive. Then $\Phi_1^\vee \cup \Phi_2^\vee$ is clearly a parabolic subset thus defines a parabolic subgroup (see 3.4.5). Conversely, a parabolic subset $\Psi^\vee$ defines a facet where $\Phi_1^\vee = \Psi^\vee \cap -\Psi^\vee$ and $\Phi_2^\vee = \Psi^\vee - \Phi_1^\vee$.

If $\mathbf{Q}$ is F-stable, then $\mathcal{F}_{\mathbf{Q}}$ is τ-stable, and the sum of the τ-orbit of $x \in \mathcal{F}_{\mathbf{Q}}$ defines an element of E^τ, non-zero since it has a positive scalar product with some coroot (unless $\mathcal{F}_{\mathbf{Q}}$ is the zero facet, which is in E^τ). Conversely a facet which meets E^τ non-trivially must be τ-stable since τ permutes the facets.

Finally, if $\mathbf{Q}$ is F-stable, it has same F-rank as $\mathbf{G}$, thus $r(\mathbf{Q}) = F\text{-rank}(\mathbf{G}) - F\text{-rank}(R(\mathbf{Q}))$ and if $\mathbf{Q}$ corresponds to a parabolic subset of the form $\Phi_1^\vee \cup \Phi_2^\vee$ as above, since $\Phi_2^\vee$ corresponds to the unipotent radical of $\mathbf{Q}$, it follows from 7.1.11 that $r(\mathbf{Q}) = \dim E^\tau - \dim(\langle \Phi_1^\vee \rangle^\perp \cap E^\tau) = \dim E^\tau - \dim(\mathcal{F}_{\mathbf{Q}} \cap E^\tau)$.   □

In the same way we associate to $\mathbf{Q_H}$ a facet $\mathcal{F}_{\mathbf{Q_H}}$ of the set of hyperplanes $\mathcal{H}_{\mathbf{H}} = \{\alpha^\perp \mid \alpha \in \Phi_{\mathbf{H}}^\vee\}$ where $\Phi_{\mathbf{H}}^\vee$ is the set of coroots of $\mathbf{H}$ relative to $\mathbf{T}$, and the same formula applies, thus, factoring out $(-1)^{\dim E^\tau}$, the formula we have to prove can now be written

$$\sum_{\{\mathbf{Q} \mid \mathbf{Q} \cap \mathbf{H} = \mathbf{Q_H}\}} (-1)^{\dim(\mathcal{F}_{\mathbf{Q}} \cap E^\tau)} = (-1)^{\dim(\mathcal{F}_{\mathbf{Q_H}} \cap E^\tau)}.$$

Further using that a facet

$$\{x \in E \mid \langle x, \alpha \rangle = 0 \text{ for } \alpha \in \Phi_1^\vee, \langle x, \alpha \rangle > 0 \text{ for } \alpha \in \Phi_2^\vee\}$$

is in the facet of $\mathcal{H}_{\mathbf{H}}$ given by

$$\{\langle x, \alpha \rangle = 0 \text{ for } \alpha \in \Phi_1^\vee \cap \Phi_{\mathbf{H}}^\vee, \langle x, \alpha \rangle > 0 \text{ for } \alpha \in \Phi_2^\vee \cap \Phi_{\mathbf{H}}^\vee\},$$

we see that

$$\mathbf{Q} \cap \mathbf{H} = \mathbf{Q_H} \Leftrightarrow \mathcal{F}_{\mathbf{Q}} \subset \mathcal{F}_{\mathbf{Q_H}},$$

so we can rewrite the formula as

$$\sum_{\mathcal{F} \subset \mathcal{F}_{\mathbf{Q_H}} \cap E^\tau} (-1)^{\dim \mathcal{F}} = (-1)^{\dim \mathcal{F}_{\mathbf{Q_H}} \cap E^\tau}.$$

But this last formula is well-known: it expresses the fact that the Euler characteristic of the convex open set $\mathcal{F}_{\mathbf{Q_H}} \cap E^\tau$ can be computed using the subdivision into facets defined by $\mathcal{H}_{\mathbf{G}} \cap E^\tau$.   □

□

Corollary 7.2.7 (of 7.2.4) ${}^*R_{\mathbf{L}}^{\mathbf{G}} \circ D_{\mathbf{G}} = D_{\mathbf{L}} \circ {}^*R_{\mathbf{L}}^{\mathbf{G}}$.

Proof   This is just the adjoint formula to 7.2.4.   □

Corollary 7.2.8 $D_{\mathbf{G}}$ *is an isometry for the scalar product on* $C(\mathbf{G}^F)$ *and* $D_{\mathbf{G}} \circ D_{\mathbf{G}}$ *is the identity functor.*

Proof The first assertion is a consequence of the second one using the fact that $D_{\mathbf{G}}$ is self adjoint (7.2.3) which implies that it is self adjoint with respect to the scalar product on class functions. Let us prove the second assertion. Let $\mathbf{T} \subset \mathbf{B}$ be an F-stable pair of a maximal torus and a Borel subgroup of $\mathbf{G}$, and let (W, S) be the corresponding Coxeter system. The F-stable parabolic subgroups of $\mathbf{G}$ which contain $\mathbf{B}$ correspond one-to-one to the F-stable subsets of S; that is, to the subsets of the set S/F. If $\mathbf{L}_I$ denotes a Levi subgroup of the parabolic subgroup corresponding to the F-stable subset of S of image $I \subset S/F$, then $r(\mathbf{L}_I) = |I|$ by 7.1.12.

Thus we can write $D_{\mathbf{G}} = \sum_{I \subset S/F} (-1)^{|I|} R_{\mathbf{L}_I}^{\mathbf{G}} {}^* R_{\mathbf{L}_I}^{\mathbf{G}}$, whence

$$
\begin{aligned}
D_{\mathbf{G}} \circ D_{\mathbf{G}} &= \sum_{I \subset S/F} (-1)^{|I|} R_{\mathbf{L}_I}^{\mathbf{G}} \circ {}^* R_{\mathbf{L}_I}^{\mathbf{G}} \circ D_{\mathbf{G}} = \sum_{I \subset S/F} (-1)^{|I|} R_{\mathbf{L}_I}^{\mathbf{G}} \circ D_{\mathbf{L}_I} \circ {}^* R_{\mathbf{L}_I}^{\mathbf{G}} \\
&= \sum_{I \subset S/F} (-1)^{|I|} \sum_{K \subset I} (-1)^{|K|} R_{\mathbf{L}_I}^{\mathbf{G}} \circ R_{\mathbf{L}_K}^{\mathbf{L}_I} \circ {}^* R_{\mathbf{L}_K}^{\mathbf{L}_I} \circ {}^* R_{\mathbf{L}_I}^{\mathbf{G}} \\
&= \sum_{K \subset S/F} (-1)^{|K|} \left(\sum_{I \supset K} (-1)^{|I|} \right) R_{\mathbf{L}_K}^{\mathbf{G}} \circ {}^* R_{\mathbf{L}_K}^{\mathbf{G}} = \mathrm{Id};
\end{aligned}
$$

the last equality uses the fact that $\sum_{I \supset K} (-1)^{|I|} = 0$ if $K \neq S/F$. $\qquad\square$

The following result shows that the dual of $\gamma \in \mathrm{Irr}(\mathbf{G}^F)$ is irreducible up to a sign determined by the Harish-Chandra series of γ.

Corollary 7.2.9 *Let $(\mathbf{L}, \lambda)$ be a cuspidal pair and let $\gamma \in \mathrm{Irr}(\mathbf{G}^F)$ be such that $\langle \gamma, R_{\mathbf{L}}^{\mathbf{G}} \lambda \rangle_{\mathbf{G}^F} \neq 0$. Then $(-1)^{r(\mathbf{L})} D_{\mathbf{G}}(\gamma) \in \mathrm{Simp}(\mathbf{G}^F \mid (\mathbf{L}, \lambda))$.*

Proof We have $\langle \gamma, \gamma \rangle_{\mathbf{G}^F} = \langle D_{\mathbf{G}} \gamma, D_{\mathbf{G}} \gamma \rangle_{\mathbf{G}^F}$ (see 7.2.8), so $D_{\mathbf{G}}(\gamma)$ is an irreducible character up to sign. As λ is cuspidal, all summands in the formula for $D_{\mathbf{L}}(\lambda)$ are zero except one of them, thus $D_{\mathbf{L}}(\lambda) = (-1)^{r(\mathbf{L})} \lambda$. So, by 7.2.3, 7.2.4 and this last fact, we have

$$
\begin{aligned}
(-1)^{r(\mathbf{L})} \langle D_{\mathbf{G}} \gamma, R_{\mathbf{L}}^{\mathbf{G}} \lambda \rangle_{\mathbf{G}^F} &= (-1)^{r(\mathbf{L})} \langle \gamma, D_{\mathbf{G}} R_{\mathbf{L}}^{\mathbf{G}} \lambda \rangle_{\mathbf{G}^F} \\
&= (-1)^{r(\mathbf{L})} \langle \gamma, R_{\mathbf{L}}^{\mathbf{G}} D_{\mathbf{L}} \lambda \rangle_{\mathbf{G}^F} = \langle \gamma, R_{\mathbf{L}}^{\mathbf{G}} \lambda \rangle_{\mathbf{G}^F} > 0,
\end{aligned}
$$

whence the result. $\qquad\square$

We want now to describe the action of $D_{\mathbf{G}}$ on the principal series. The result will be given in 7.2.13. We start with some lemmas.

Lemma 7.2.10 *Let (W, S) be a Coxeter system. For any character $\chi \in \mathrm{Irr}(\mathbb{C}W)$ we have $\sum_{I \subset S} (-1)^{|I|} \mathrm{Ind}_{W_I}^{W} \mathrm{Res}_{W_I}^{W} \chi = \chi \otimes \mathrm{sgn}$.*

Proof Since $\mathrm{Ind}_{W_I}^{W} \mathrm{Res}_{W_I}^{W} \chi = \chi \otimes (\mathrm{Ind}_{W_I}^{W} \mathrm{Res}_{W_I}^{W} \mathbf{1})$, the lemma results from the case $\chi = \mathbf{1}$. Now for $w \in W$ we have

$$
\mathrm{Ind}_{W_I}^{W} \mathbf{1}(w) = |W_I|^{-1} |\{x \in W \mid w \in {}^x W_I\}|.
$$

We first give the proof for Weyl groups, placing ourselves in the space $E = X(\mathbf{T}) \otimes \mathbb{R}$, where $\mathbf{T}$ is a torus of a reductive group having W as Weyl group, and we apply the methods of the proof of 7.2.5. If Φ^+ is the positive root system corresponding to S, to an x in the above set we attach the facet $x(\mathcal{F}_I)$ where $\mathcal{F}_I = \{e \in E \mid \langle \alpha, e \rangle = 0 \text{ for } \alpha \in \Phi_I^\vee, \langle \alpha, e \rangle > 0 \text{ for } \alpha \in \Phi^{+\vee} - \phi_I^\vee\}$ is the facet attached to the standard parabolic subgroup subgroup $\mathbf{P}_I$. The condition that ${}^xW_I \ni w$ is equivalent to $x(\mathcal{F}_I) \subset E^w$ (the fixed space of the parabolic subgroup ${}^xW_I$ is a subspace of the fixed space of one of its elements w). Further the stabiliser in W of $\mathcal{F}_I$ is W_I, so the formula above becomes $\operatorname{Ind}_{W_I}^W \mathbf{1}(w) = |\{\mathcal{F} \sim \mathcal{F}_I \mid \mathcal{F} \subset E^w\}|$ where $\sim$ means that $\mathcal{F}$ is a facet conjugate to $\mathcal{F}_I$, and the total sum becomes $\sum_{\mathcal{F} \subset E^w} (-1)^{\operatorname{codim} \mathcal{F}}$, since every facet is conjugate to some $\mathcal{F}_I$ (every parabolic subgroup is conjugate to some $\mathbf{P}_I$) and $\operatorname{codim} \mathcal{F}_I = |I|$. By the argument at the end of 7.2.5 this sums to $\operatorname{codim} E^w$, which is equal to $(-1)^{l(w)}$ by the argument in the proof of 7.1.2 applied to $\tau = w$.

For the general case where W is not a Weyl group, the above proof still applies verbatim using for E the space of the reflection representation of W, defined for example in Bourbaki (1968, V, §4, Proposition 3).   □

Lemma 7.2.11 *Let $\mathbf{L}_J$ be an F-stable standard Levi subgroup of $\mathbf{G}^F$; for $\psi \in \operatorname{Irr} W_J^F$ and $\chi \in \operatorname{Irr} W^F$ we have $\langle {}^*R_{\mathbf{L}_J}^{\mathbf{G}} \gamma_\chi, \lambda_\psi \rangle_{\mathbf{L}_J^F} = \langle \operatorname{Res}_{W_J^F}^{W^F} \chi, \psi \rangle_{W_J^F}$, where γ_χ (resp. λ_ψ) are the characters of $\mathbf{G}^F$ (resp. $\mathbf{L}_J^F$) corresponding to χ (resp. ψ) by 6.2.14(i).*

Proof Let $H := \operatorname{End}_{\mathbb{C}\mathbf{G}^F}(\operatorname{Ind}_{\mathbf{B}^F}^{\mathbf{G}^F} \mathbf{1})$ and $H_J := \operatorname{End}_{\mathbb{C}\mathbf{L}_J^F}(\operatorname{Ind}_{\mathbf{B}^F \cap \mathbf{L}_J}^{\mathbf{L}_J^F} \mathbf{1})$. Let χ_q (resp. ψ_q) be the characters of H (resp. H_J) corresponding to χ (resp. ψ). We have $H_J = e_{\mathbf{B}^F \cap \mathbf{L}_J} \mathbb{C}\mathbf{L}_J^F e_{\mathbf{B}^F \cap \mathbf{L}_J}$. Note that if $e \in H_J$ is a primitive idempotent such that $H_J e$ is the representation with character ψ_q, then e is primitive in $\mathbb{C}\mathbf{L}_J^F$ and affords λ_ψ. Indeed e is an idempotent which, being multiple of $e_{\mathbf{B}^F \cap \mathbf{L}_J}$, acts by 0 on irreducible modules not in the principal series and acts also by 0 on principal series modules other that the one with character ψ_q since the action of H_J on the principal series is by restriction from $\mathbf{L}_J^F$ (see Exercise 6.1.3(ii)). Moreover $\lambda_\psi(e) = \psi_q(e) = 1$, hence e is primitive in $\mathbb{C}\mathbf{L}_J^F$.

Thus, if $\mathbf{U}_J$ is the unipotent radical of the parabolic subgroup $\mathbf{P}_J$, the idempotent $ee_{\mathbf{U}_J^F}$ affords the inflation of λ_ψ to $\mathbf{P}_J^F$ and $\mathbb{C}\mathbf{G}^F ee_{\mathbf{U}_J^F}$ is a representation of $\mathbf{G}^F$ with character $R_{\mathbf{L}_J}^{\mathbf{G}} \lambda_\psi$ (see Exercise 6.1.3(i)). Similarly He is a representation of H with character $\operatorname{Ind}_{H_J}^H \psi_q$. Thus $\lambda_\chi(ee_{\mathbf{U}_J^F}) = \langle \lambda_\chi, R_{\mathbf{L}_J}^{\mathbf{G}} \lambda_\psi \rangle_{\mathbf{G}^F}$ (see Exercise 6.1.3(i)), and similarly $\chi_q(e)$ is the multiplicity of χ_q in $\operatorname{Ind}_{H_J}^H \psi_q$ which is also the multiplicity of χ in $\operatorname{Ind}_{W_J}^W (\psi)$ as can be seen using the fact that the scalar product on the generic algebra specialises to the scalar product $\langle ., . \rangle_H$ and also that – by 6.3.7 and 6.3.12 – it specialises to $|W^F| \langle ., . \rangle_{W^F}$.

Since $H = e_{\mathbf{B}^F} \mathbb{C} \mathbf{G}^F e_{\mathbf{B}^F}$ and $\mathbf{U}_J \subset \mathbf{B}$ we have $He = Hee_{\mathbf{U}^F}$ and $\chi_q(e) = \chi_q(ee_{\mathbf{U}^F}) = \lambda_\chi(ee_{\mathbf{U}^F})$, since χ_q is the restriction to H of λ_χ. Thus the two multiplicities are equal, whence the result by adjunction. $\qquad\square$

Corollary 7.2.12 *With notation as in 7.2.11, if* $\mathbf{T}$ *is a quasi-split torus of* $\mathbf{L}_J$, *the map* $\psi \mapsto \gamma_\psi$ *extends to an isometry between the span of* $\mathrm{Irr}(W_J^F)$ *and the span of the irreducible constituents of* $R_{\mathbf{T}}^{\mathbf{L}_J}(\mathbf{1})$ *which, for any* $\chi \in \mathrm{Irr}\, W^F$, *maps* $\mathrm{Res}_{W_J^F}^{W^F} \chi$ *to* ${}^*R_{\mathbf{L}_J}^{\mathbf{G}}(\gamma_\chi)$. *In particular* ${}^*R_{\mathbf{L}_J}^{\mathbf{G}}(\mathbf{1}_{\mathbf{G}^F}) = \mathbf{1}_{\mathbf{L}_J^F}$.

Proof This follows from 7.2.11 and the fact that ${}^*R_{\mathbf{L}}^{\mathbf{G}}$ preserves Harish-Chandra series. $\qquad\square$

Proposition 7.2.13 *With same notation as in 7.2.11, for* $\chi \in \mathrm{Irr}\, W^F$, *we have* $D_{\mathbf{G}}\gamma_\chi = \gamma_{\chi \otimes \mathrm{sgn}}$ *where* sgn *is the sign character of the Coxeter group* W^F.

Proof We have $D_{\mathbf{G}} = \sum_{I \subset S/F} (-1)^{|I|} R_{\mathbf{L}_I}^{\mathbf{G}} {}^*R_{\mathbf{L}_I}^{\mathbf{G}}$ (see the proof of 7.2.8).

By 7.2.12, for $\chi, \chi' \in \mathrm{Irr}\, W^F$ we have

$$\langle {}^*R_{\mathbf{L}_J}^{\mathbf{G}}\gamma_\chi, {}^*R_{\mathbf{L}_J}^{\mathbf{G}}\gamma_{\chi'}\rangle_{\mathbf{L}_J^F} = \langle \mathrm{Res}_{W_J^F}^{W^F} \chi, \mathrm{Res}_{W_J^F}^{W^F} \chi'\rangle_{W_J^F},$$

whence by adjunction

$$\langle R_{\mathbf{L}_J}^{\mathbf{G}} {}^*R_{\mathbf{L}_J}^{\mathbf{G}}\gamma_\chi, \gamma_{\chi'}\rangle_{\mathbf{G}^F} = \langle \mathrm{Ind}_{W_J^F}^{W^F} \mathrm{Res}_{W_J^F}^{W^F} \chi, \chi'\rangle_{W^F}.$$

We deduce

$$\langle D_{\mathbf{G}}\gamma_\chi, \gamma_{\chi'}\rangle_{\mathbf{G}^F} = \sum_{I \subset S/F} (-1)^{|I|} \langle \mathrm{Ind}_{W_I^F}^{W^F} \mathrm{Res}_{W_I^F}^{W^F} \chi, \chi'\rangle_{W^F} = \langle \chi \otimes \mathrm{sgn}, \chi'\rangle_{W^F},$$

the last equality by 7.2.10. This gives the result using 7.2.9 and using again the isometry between the spans of $\mathrm{Irr}(W^F)$ and of the constituents of $\mathrm{Ind}_{\mathbf{B}^F}^{\mathbf{G}^F} \mathbf{1}$. $\qquad\square$

Corollary 7.2.14 *Under the bijection of 6.2.14(i) the sign character of* W^F *parametrises the Steinberg character of* $\mathbf{G}^F$ *(see 7.4.1 for its definition).*

Proof This comes from 7.2.13 and the facts that the trivial character of W^F parametrises the trivial character of $\mathbf{G}^F$ by 6.2.14(i) and that $D_{\mathbf{G}}\mathrm{St}_{\mathbf{G}^F} = \mathbf{1}_{\mathbf{G}^F}$. $\qquad\square$

7.3 Restriction to Centralisers of Semi-Simple Elements

Before stating the main result 7.3.7 of this section, we need to state some further properties of Harish-Chandra induction. Let $\mathbf{P} = \mathbf{U} \rtimes \mathbf{L}$ be an F-stable Levi decomposition of an F-stable parabolic subgroup of $\mathbf{G}$. We begin with a finite group theoretic result.

Proposition 7.3.1   *Let s be the semi-simple part of an element $l \in \mathbf{L}^F$; then the map $\mathbf{U}^F \times (C_{\mathbf{G}}(s) \cap \mathbf{U}^F) \to l\mathbf{U}^F$ defined by $(y, z) \mapsto {}^y(lz)$ is onto and all its fibres have the same cardinality equal to $|C_{\mathbf{G}}(s) \cap \mathbf{U}^F|$.*

Proof   Let v be the unipotent part of l and let $V = \langle v, \mathbf{U}^F \rangle$; the group V is the unique Sylow p-subgroup of $\langle l, \mathbf{U}^F \rangle$ because it is normal and the quotient is cyclic generated by a p'-element (the image of s). For any $u \in \mathbf{U}^F$ we must count the number of solutions $(y, z) \in \mathbf{U}^F \times (C_{\mathbf{G}}(s) \cap \mathbf{U}^F)$ to the equation ${}^y(lz) = lu$. Let $s'u'$ be the Jordan decomposition of lu; since v normalises $C_{\mathbf{G}}(s) \cap \mathbf{U}^F$, the element vz is unipotent in $C_{\mathbf{G}}(s)$ and $s.vz$ is the Jordan decomposition of lz, so the equation above is equivalent to the system of two equations ${}^y s = s'$ and ${}^y(vz) = u'$.

But $\langle s \rangle$ and $\langle s' \rangle$ are two p'-complements of the normal subgroup V of $\langle l, \mathbf{U}^F \rangle$ so, by the Schur–Zassenhaus theorem – see Gorenstein (1980, 6.2.1) – they are conjugate by an element of V. As s cannot be conjugate to one of its powers other than itself in the quotient group $\langle l, \mathbf{U}^F \rangle / V$, there exists $y \in V$ such that ${}^y s = s'$. The element y can be taken in $\mathbf{U}^F$ as v centralises s. If we prove that the element z defined by $z = v^{-1}.{}^{y^{-1}} u'$ is in $C_{\mathbf{G}}(s) \cap \mathbf{U}^F$ then we are done, as there are $|C_{\mathbf{G}}(s) \cap \mathbf{U}^F|$ solutions y to ${}^y s = s'$ and the value of z is uniquely defined by that of y.

But v and s commute, and ${}^{y^{-1}} u'$ commutes with ${}^{y^{-1}} s' = s$, so z commutes with s. It remains for us to show that z is in $\mathbf{U}^F$: we have $lz = {}^{y^{-1}}(lu) \in {}^{y^{-1}}(l\mathbf{U}^F) = l\mathbf{U}^F$ (this last equality because $y \in \mathbf{U}^F$), whence the result.   □

The following corollaries show the usefulness of 7.3.1.

Definition 7.3.2

(i) We denote by $C(H)_{p'}$ the set of class functions f on a finite group H such that $f(x) = f(x_{p'})$ for any $x \in H$, where $x_{p'}$ is the p'-part of the element x.

(ii) We denote by χ_p (or $\chi_p^{\mathbf{G}}$ to specify the group) the characteristic function of $\mathbf{G}_u^F$; note that this function is in $C(\mathbf{G}^F)_{p'}$.

Corollary 7.3.3   *If $f \in C(\mathbf{G}^F)_{p'}$, it satisfies $f(lu) = f(l)$ for any $(u, l) \in \mathbf{U}^F \times \mathbf{L}^F$.*

Proof   By 7.3.1, if $l = sv$ is the Jordan decomposition of l, there exists $y \in \mathbf{U}^F$ and $z \in C_{\mathbf{G}}(s) \cap \mathbf{U}^F$ such that ${}^y(lz) = lu$, so, in particular $f(lz) = f(lu)$. As we have seen in the proof of 7.3.1 $s.vz$ is the Jordan decomposition of lz, so $f(lz) = f(s) = f(l)$, whence the result.   □

Corollary 7.3.4   *If $f \in C(\mathbf{G}^F)_{p'}$, then ${}^* R_{\mathbf{L}}^{\mathbf{G}} f = \mathrm{Res}_{\mathbf{L}^F}^{\mathbf{G}^F} f$ and ${}^* R_{\mathbf{L}}^{\mathbf{G}}(\chi f) = ({}^* R_{\mathbf{L}}^{\mathbf{G}} \chi) . \mathrm{Res}_{\mathbf{L}^F}^{\mathbf{G}^F} f$ for any $\chi \in C(\mathbf{G}^F)$.*

Proof This result follows from 7.3.3 and 5.1.12. □

Corollary 7.3.5 *If $f \in C(\mathbf{G}^F)_{p'}$, then $R_{\mathbf{L}}^{\mathbf{G}}(\chi \operatorname{Res}_{\mathbf{L}^F}^{\mathbf{G}^F} f) = (R_{\mathbf{L}}^{\mathbf{G}} \chi).f$ for any $\chi \in C(\mathbf{L}^F)$.*

Proof This is what we get by adjunction from 7.3.4. □

The next property has been proved and used, under somewhat stronger assumptions, in Curtis (1980); we shall generalise it to Lusztig induction in Chapter 10.

Corollary 7.3.6 *Let s be the semi-simple part of $l \in \mathbf{L}^F$; then*

$$(\operatorname{Res}_{C_{\mathbf{L}}(s)^{0F}}^{\mathbf{L}^F} {}^*R_{\mathbf{L}}^{\mathbf{G}} \chi)(l) = ({}^*R_{C_{\mathbf{L}}(s)^0}^{C_{\mathbf{G}}(s)^0} \operatorname{Res}_{C_{\mathbf{G}}(s)^{0F}}^{\mathbf{G}^F} \chi)(l).$$

Proof First let us note that the above formula has a well-defined meaning because $l \in C_{\mathbf{L}}(s)^0$ by 3.5.3, and $C_{\mathbf{L}}(s)^0$ is a Levi subgroup of $C_{\mathbf{G}}(s)^0$. Indeed, since $C_{\mathbf{G}}(s)^0$ is a connected reductive subgroup of maximal rank of $\mathbf{G}$ by 3.5.1(i), by 3.4.10(ii) and (iii) $\mathbf{P} \cap C_{\mathbf{G}}(s)^0$ is a parabolic subgroup of $C_{\mathbf{G}}(s)^0$ with Levi decomposition $(\mathbf{U} \cap C_{\mathbf{G}}(s)^0) \rtimes (\mathbf{L} \cap C_{\mathbf{G}}(s)^0)$. So $\mathbf{L} \cap C_{\mathbf{G}}(s)^0$ is connected and $\mathbf{L} \cap C_{\mathbf{G}}(s)^0 = C_{\mathbf{L}}(s)^0$.

We have by 5.1.12 and 7.3.1

$$({}^*R_{\mathbf{L}}^{\mathbf{G}} \chi)(l) = |\mathbf{U}^F|^{-1} \sum_{u \in \mathbf{U}^F} \chi(lu) = |C_{\mathbf{G}}(s) \cap \mathbf{U}^F|^{-1} \sum_{z \in C_{\mathbf{G}}(s) \cap \mathbf{U}^F} \chi(lz).$$

We then get the result since $C_{\mathbf{G}}(s) \cap \mathbf{U} = C_{\mathbf{G}}(s)^0 \cap \mathbf{U}$, see 3.5.3 again. □

Proposition 7.3.7 *If s is the semi-simple part of $x \in \mathbf{G}^F$, then*

$$\varepsilon_{\mathbf{G}}(D_{\mathbf{G}} \chi)(x) = \varepsilon_{C_{\mathbf{G}}(s)^0}(D_{C_{\mathbf{G}}(s)^0} \circ \operatorname{Res}_{C_{\mathbf{G}}(s)^0}^{\mathbf{G}} \chi)(x).$$

Proof If $\mathbf{L}$ is an F-stable Levi subgroup of an F-stable parabolic subgroup $\mathbf{P}$ of $\mathbf{G}$, we have by 7.2.2 $R_{\mathbf{L}}^{\mathbf{G}} \circ {}^*R_{\mathbf{L}}^{\mathbf{G}} \chi = \operatorname{Ind}_{\mathbf{P}^F}^{\mathbf{G}^F}(\chi')$, if $\chi'(p) = {}^*R_{\mathbf{L}}^{\mathbf{G}}(\chi)(\overline{p})$ where $\overline{p}$ is the image of p in $\mathbf{L}$, whence

$$(R_{\mathbf{L}}^{\mathbf{G}} \circ {}^*R_{\mathbf{L}}^{\mathbf{G}} \chi)(x) = |\mathbf{P}^F|^{-1} \sum_{\{g \in \mathbf{G}^F | {}^g\mathbf{P} \ni x\}} ({}^*R_{\mathbf{L}}^{\mathbf{G}} \chi)(\overline{g^{-1}x}) = \sum_{\{\mathbf{P}' \underset{\mathbf{G}^F}{\sim} \mathbf{P} | \mathbf{P}' \ni x\}} ({}^*R_{\mathbf{L}'}^{\mathbf{G}} \chi)(\overline{x})$$

where $\mathbf{P}'$ in the last summation denotes an F-stable parabolic subgroup of $\mathbf{G}$, where $\mathbf{L}'$ is an F-stable Levi subgroup of $\mathbf{P}'$, and $\overline{x}$ denotes the image of x in $\mathbf{L}'$. Using this in the definition of $D_{\mathbf{G}}$ gives

$$(D_{\mathbf{G}} \chi)(x) = \sum_{\mathbf{P} \supset \mathbf{B}} (-1)^{r(\mathbf{P})} \sum_{\{\mathbf{P}' \underset{\mathbf{G}^F}{\sim} \mathbf{P} | \mathbf{P}' \ni x\}} ({}^*R_{\mathbf{L}'}^{\mathbf{G}} \chi)(\overline{x}) = \sum_{\mathbf{P}' \ni x} (-1)^{r(\mathbf{P}')} ({}^*R_{\mathbf{L}'}^{\mathbf{G}} \chi)(\overline{x}).$$

$$(7.3.1)$$

By 7.3.6 we then get

$$(D_{\mathbf{G}}\chi)(x) = \sum_{\mathbf{P}\ni x}(-1)^{r(\mathbf{P})}\left({}^*R_{\mathbf{L}\cap C_{\mathbf{G}}(s)^0}^{C_{\mathbf{G}}(s)^0}\operatorname{Res}_{C_{\mathbf{G}}(s)^{0F}}^{\mathbf{G}^F}\chi\right)(\bar{x})$$

$$= \sum_{\mathbf{P}'}\left(\sum_{\{\mathbf{P}\mid\mathbf{P}\cap C_{\mathbf{G}}(s)^0=\mathbf{P}'\}}(-1)^{r(\mathbf{P})}\right)\left({}^*R_{\mathbf{L}'}^{C_{\mathbf{G}}(s)^0}\operatorname{Res}_{C_{\mathbf{G}}(s)^{0F}}^{\mathbf{G}^F}\chi\right)(\bar{x})$$

where $\mathbf{P}'$ runs over the set of F-stable parabolic subgroups of $C_{\mathbf{G}}(s)^0$ and where $\mathbf{L}'$ is any F-stable Levi subgroup of $\mathbf{P}'$. If we now apply 7.2.5 with $\mathbf{H} = C_{\mathbf{G}}(s)^0$ and compare with the equality (7.3.1) applied in $C_{\mathbf{G}}(s)^0$, we get the result.   □

7.4 The Steinberg Character

In this section we use duality to define and study the famous "Steinberg character" which was originally defined in Steinberg (1956, 1957). Its definition below as the dual of the trivial character is due to Curtis.

Definition 7.4.1   The irreducible character $\operatorname{St}_{\mathbf{G}^F} = D_{\mathbf{G}}(\mathbf{1}_{\mathbf{G}^F})$ where $\mathbf{1}_{\mathbf{G}^F}$ is the trivial character of $\mathbf{G}^F$ is called the **Steinberg character** of $\mathbf{G}^F$.

Note that $\operatorname{St}_{\mathbf{G}^F}$ is a true character (and not the negative of a character) by 7.2.9 since – as we have seen in 6.2.14(i) – $\mathbf{1}_{\mathbf{G}^F}\in\operatorname{Simp}(\mathbf{G}^F\mid(\mathbf{T},1))$ where $\mathbf{T}$ is a quasi-split torus of $\mathbf{G}$, and $r(\mathbf{T}) = 0$.

Proposition 7.4.2 ${}^*R_{\mathbf{L}}^{\mathbf{G}}\operatorname{St}_{\mathbf{G}^F} = \operatorname{St}_{\mathbf{L}^F}$.

Proof   As the duality and the Harish-Chandra restriction commute we get

$${}^*R_{\mathbf{L}}^{\mathbf{G}}\operatorname{St}_{\mathbf{G}^F} = {}^*R_{\mathbf{L}}^{\mathbf{G}}D_{\mathbf{G}}(\mathbf{1}_{\mathbf{G}^F}) = D_{\mathbf{L}}{}^*R_{\mathbf{L}}^{\mathbf{G}}(\mathbf{1}_{\mathbf{G}^F}) = D_{\mathbf{L}}(\mathbf{1}_{\mathbf{L}^F}) = \operatorname{St}_{\mathbf{L}^F},$$

the third equality by 7.2.12, whence the result.   □

In the case of a torus we have the following more precise result.

Lemma 7.4.3   *Let* $\mathbf{T}$ *be an* F-*stable maximal torus of an* F-*stable Borel subgroup* $\mathbf{B}$ *of* $\mathbf{G}$; *then*

$$\operatorname{Res}_{\mathbf{B}^F}^{\mathbf{G}^F}\operatorname{St}_{\mathbf{G}^F} = \operatorname{Ind}_{\mathbf{T}^F}^{\mathbf{B}^F}\mathbf{1}.$$

Proof   Using the definitions of $\operatorname{St}_{\mathbf{G}^F}$ and of $D_{\mathbf{G}}$ we get

$$\operatorname{St}_{\mathbf{G}^F} = \sum_{I\subset S/F}(-1)^{|I|}\operatorname{Ind}_{\mathbf{P}_I}^{\mathbf{G}^F}\mathbf{1}$$

(the notation is the same as in the proof of 7.2.8). So we have

$$\operatorname{Res}_{\mathbf{B}^F}^{\mathbf{G}^F}(\operatorname{St}_{\mathbf{G}^F}) = \sum_{I \subseteq S/F} (-1)^{|I|} \operatorname{Res}_{\mathbf{B}^F}^{\mathbf{G}^F} \operatorname{Ind}_{\mathbf{P}_I^F}^{\mathbf{G}^F} \mathbf{1}$$

$$= \sum_{I \subseteq S/F} (-1)^{|I|} \sum_{w \in Z_I} \operatorname{Ind}_{\mathbf{B}^F \cap {}^w \mathbf{P}_I}^{\mathbf{B}^F} \operatorname{Res}_{\mathbf{B}^F \cap {}^w \mathbf{P}_I}^{\mathbf{P}_I^F} \mathbf{1},$$

the last equality following from the ordinary Mackey formula, where we have denoted by Z_I the set of F-stable reduced-I' elements of $W = W_{\mathbf{G}}(\mathbf{T})$ where $I' \subset S$ is the preimage of $I \subset S/F$ – the set Z_I is a set of representatives for the F-stable double cosets $\mathbf{B} \backslash \mathbf{G}/\mathbf{P}_I$ and is also a set of representatives for the double cosets $\mathbf{B}^F \backslash \mathbf{G}^F/\mathbf{P}_I^F$; see 3.2.3(ii), 4.4.8 and 5.2.2(ii). We have $\mathbf{B} \cap {}^w \mathbf{P}_I = \mathbf{B} \cap {}^w \mathbf{B}$. Indeed, since for any $v \in W_{I'}$ we have $l(w) + l(v) = l(wv)$, for any $v \in W_{I'}$ we have $w\mathbf{B}v\mathbf{B} \subset \mathbf{B}wv\mathbf{B}$; whence $\mathbf{B}w \cap w\mathbf{B}v\mathbf{B} \subset \mathbf{B}w \cap \mathbf{B}wv\mathbf{B}$, and this last intersection is empty unless $v = 1$. Hence we have

$$\mathbf{B}w \cap w\mathbf{P}_I = \mathbf{B}w \cap w\mathbf{B}W_{I'}\mathbf{B} = \mathbf{B}w \cap \left(\bigsqcup_{v \in W_{I'}} w\mathbf{B}v\mathbf{B} \right) = \mathbf{B}w \cap w\mathbf{B}.$$

Using this result in the formula for $\operatorname{Res}_{\mathbf{B}^F}^{\mathbf{G}^F} \operatorname{St}_{\mathbf{G}^F}$ and exchanging the summations give

$$\operatorname{Res}_{\mathbf{B}^F}^{\mathbf{G}^F} \operatorname{St}_{\mathbf{G}^F} = \sum_{w \in W} \left(\sum_{\{I \subseteq S/F \mid w \in Z_I\}} (-1)^{|I|} \right) \operatorname{Ind}_{\mathbf{B}^F \cap {}^w \mathbf{B}}^{\mathbf{B}^F} \mathbf{1}.$$

For $w \in W^F$ let $S_w := \{O \in S/F \mid \exists s \in O, l(ws) > l(w)\} = \{O \in S/F \mid \forall s \in O, l(ws) > l(w)\}$, the last equality since $w \in W^F$. We have $Z_I = \{w \in W^F \mid I \subset S_w\}$, so

$$\sum_{\{I \subseteq S/F \mid w \in Z_I\}} (-1)^{|I|} = \sum_{I \subset S_w} (-1)^{|I|},$$

which is different from zero only if $S_w = \emptyset$, in which case w is the longest element of W, see 2.1.5(i), and $\mathbf{B} \cap {}^w \mathbf{B} = \mathbf{T}$, whence the result. $\qquad \square$

Corollary 7.4.4 *For $x \in \mathbf{G}^F$, we have*

$$\operatorname{St}_{\mathbf{G}^F}(x) = \begin{cases} \varepsilon_{\mathbf{G}} \varepsilon_{(C_{\mathbf{G}}(x)^0)} |C_{\mathbf{G}}(x)^{0F}|_p & \text{if } x \text{ is semi-simple,} \\ 0 & \text{otherwise.} \end{cases}$$

Proof Let s be the semi-simple part of x. We have

$$\operatorname{St}_{\mathbf{G}^F}(x) = D_{\mathbf{G}}(\mathbf{1})(x) = \varepsilon_{\mathbf{G}} \varepsilon_{C_{\mathbf{G}}(s)^0} (D_{C_{\mathbf{G}}(s)^0}(\mathbf{1}))(x) = \varepsilon_{\mathbf{G}} \varepsilon_{C_{\mathbf{G}}(s)^0} \operatorname{St}_{C_{\mathbf{G}}(s)^{0F}}(x)$$

by 7.3.7. So we may assume that s is central in $\mathbf{G}$. But then there exists an F-stable Borel subgroup $\mathbf{B}$ which contains x. Indeed the unipotent part of x

is contained in an F-stable Borel subgroup as is any F-stable unipotent element (see 4.4.3), and s, being central, is contained in all Borel subgroups. So by Lemma 7.4.3 we have $\mathrm{St}_{\mathbf{G}^F}(x) = (\mathrm{Res}_{\mathbf{B}^F}^{\mathbf{G}^F} \mathrm{St}_{\mathbf{G}^F})(x) = (\mathrm{Ind}_{\mathbf{T}^F}^{\mathbf{B}^F} \mathbf{1})(x)$. Thus $\mathrm{St}_{\mathbf{G}^F}(x) = 0$ if x has no conjugate in $\mathbf{T}^F$ — that is, x is not semisimple; and if $x = s$ we get

$$\mathrm{St}_{\mathbf{G}^F}(x) = |\mathbf{B}^F|/|\mathbf{T}^F| = |\mathbf{G}^F|_p,$$

the second equality by 4.4.1(iv), whence the result. □

Corollary 7.4.5 (of 7.4.4)   *The dual of the regular character* $\mathrm{reg}_{\mathbf{G}^F}$ *of* $\mathbf{G}^F$ *is* $D_{\mathbf{G}}(\mathrm{reg}_{\mathbf{G}^F}) = \gamma_p$, *where* γ_p *denotes the element of* $C(\mathbf{G}^F)$ *whose value is* $|\mathbf{G}^F|_{p'}$ *on unipotent elements and 0 elsewhere.*

Proof   By 7.3.4 and 7.3.5 we have $D_{\mathbf{G}}(\chi.f) = D_{\mathbf{G}}(\chi).f$ for any $\chi \in C(\mathbf{G}^F)$ and any $f \in C(\mathbf{G}^F)_{p'}$. So, since $\gamma_p \in C(\mathbf{G}^F)_{p'}$, we have

$$D_{\mathbf{G}}(\gamma_p) = D_{\mathbf{G}}(\mathbf{1}.\gamma_p) = D_{\mathbf{G}}(\mathbf{1})\gamma_p = \mathrm{St}_{\mathbf{G}^F}.\gamma_p = \mathrm{reg}_{\mathbf{G}^F},$$

the last equality by 7.4.4. □

Corollary 7.4.6 (of 7.4.5)   *The number of unipotent elements in* $\mathbf{G}^F$ *is equal to* $(|\mathbf{G}^F|_p)^2$.

Proof   This formula is easily deduced from 7.4.5 by writing

$$\langle \mathrm{reg}_{\mathbf{G}^F}, \mathrm{reg}_{\mathbf{G}^F} \rangle_{\mathbf{G}^F} = \langle \gamma_p, \gamma_p \rangle_{\mathbf{G}^F}.$$

□

Proposition 7.4.7   *Let* $\mathbf{L}$ *be an F-stable Levi subgroup of an F-stable parabolic subgroup of* $\mathbf{G}$, *and let* $\chi \in \mathrm{Irr}(\mathbf{L}^F)$; *then*

$$\mathrm{St}_{\mathbf{G}^F}.R_{\mathbf{L}}^{\mathbf{G}}(\chi) = \mathrm{Ind}_{\mathbf{L}^F}^{\mathbf{G}^F}(\mathrm{St}_{\mathbf{L}^F}.\chi).$$

Proof   Let $C(\mathbf{G}_{p'}^F)$ be the space of class functions on $\mathbf{G}_{p'}^F$ (the set of elements of order prime to p, which is also the set of semi-simple elements of $\mathbf{G}^F$; see 1.1.5); we identify $C(\mathbf{G}_{p'}^F)$ with the subspace of $C(\mathbf{G}^F)$ consisting of functions which are zero outside $\mathbf{G}_{p'}^F$. For $f \in C(\mathbf{G}_{p'}^F)$, let us denote by $\mathrm{Br}f$ the element of $C(\mathbf{G}^F)$ defined by $(\mathrm{Br}f)(x) = f(s)$ where s is the semi-simple part of x (Br is similar to the "Brauer lifting" in block theory). Then we have $\mathrm{Br}f = D_{\mathbf{G}}(\mathrm{St}_{\mathbf{G}^F}.f)$. Indeed $\mathrm{Br}f \in C(\mathbf{G}^F)_{p'}$, so – as in the proof of 7.4.5 – we have $D_{\mathbf{G}}(\mathbf{1}.\mathrm{Br}f) = D_{\mathbf{G}}(\mathbf{1}).\mathrm{Br}f = \mathrm{St}_{\mathbf{G}^F}.\mathrm{Br}f = \mathrm{St}_{\mathbf{G}^F}.f$, whence the result as $D_{\mathbf{G}}$ is an involution. So for any $\chi \in C(\mathbf{G}^F)$ we have $D_{\mathbf{G}}(\chi.\mathrm{St}_{\mathbf{G}^F}) = \mathrm{Br}(\chi|_{\mathbf{G}_{p'}^F})$. But for any $f \in C(\mathbf{G}^F)_{p'}$, as $\mathrm{Br}f \in C(\mathbf{G}^F)_{p'}$, we have, by 7.3.4,

$${}^*R_{\mathbf{L}}^{\mathbf{G}}(\mathrm{Br}f) = \mathrm{Res}_{\mathbf{L}^F}^{\mathbf{G}^F} \mathrm{Br}f = \mathrm{Br}(\mathrm{Res}_{\mathbf{L}_{p'}^F}^{\mathbf{G}_{p'}^F} f).$$

If we apply 7.2.4 and then the two previous formulae we find

$$D_{\mathbf{L}}(^*R_{\mathbf{L}}^{\mathbf{G}}(\chi.\mathrm{St}_{\mathbf{G}^F})) = {}^*R_{\mathbf{L}}^{\mathbf{G}}(D_{\mathbf{G}}(\chi.\mathrm{St}_{\mathbf{G}^F})) = {}^*R_{\mathbf{L}}^{\mathbf{G}}(\mathrm{Br}(\chi|_{\mathbf{G}^F_{p'}}))$$

$$= \mathrm{Br}(\chi|_{\mathbf{L}^F_{p'}}) = D_{\mathbf{L}}(\mathrm{St}_{\mathbf{L}^F}.\mathrm{Res}_{\mathbf{L}^F}^{\mathbf{G}^F}\chi),$$

which gives after applying $D_{\mathbf{L}}$

$$^*R_{\mathbf{L}}^{\mathbf{G}}(\chi.\mathrm{St}_{\mathbf{G}^F}) = (\mathrm{Res}_{\mathbf{L}^F}^{\mathbf{G}^F}\chi).\mathrm{St}_{\mathbf{L}^F},$$

whence the proposition by adjunction since the multiplication by $\mathrm{St}_{\mathbf{G}^F}$ is self-adjoint. $\square$

Notes

The duality operator $D_{\mathbf{G}}$ was introduced simultaneously by Curtis (1980), Alvis (1980), Kawanaka (1982), and Lusztig. Deligne and Lusztig studied it in a geometric way in Deligne and Lusztig (1982, 1983); see also Digne and Michel (1982) which we follow here, in particular for the proof of 7.2.4.

In the case where F is a Frobenius endomorphism, the notion of F-rank and semi-simple F-rank (called $\mathbb{F}_q$-rank and $\mathbb{F}_q$-semi-simple rank) can be found in, for example, Borel and Tits (1965, 4.23 and 5.3.).

Our exposition here of the properties of the Steinberg character, in particular in the proofs of 7.4.4, 7.4.5, 7.4.6 and 7.4.7, follows the ideas of Digne and Michel (1982), making a systematic use of the properties of the duality functor.

8

ℓ-Adic Cohomology

8.1 ℓ-Adic Cohomology

Let $\mathbf{X}$ be an algebraic variety over $\overline{\mathbb{F}}_p$ and ℓ be a prime number different from p. Using the results of Grothendieck one can associate to $\mathbf{X}$ groups of ℓ-adic cohomology with compact support $H^i_c(\mathbf{X}, \overline{\mathbb{Q}}_\ell)$ – see Deligne (1977, Rapport, 2.10) – which are $\overline{\mathbb{Q}}_\ell$-vector spaces of finite dimension. This cohomology theory is a powerful tool. It will be used in the next chapter to define Deligne–Lusztig induction. We recall in this chapter those properties of ℓ-adic cohomology that we need. Throughout this chapter $\mathbf{X}$ will stand for an algebraic variety over $\overline{\mathbb{F}}_p$. The reader who wants to keep within the framework of Chapter 4 may assume that all the varieties in this chapter are quasi-projective.

Proposition 8.1.1 *We have $H^i_c(\mathbf{X}, \overline{\mathbb{Q}}_\ell) = 0$ if $i \notin \{0, \dots, 2\dim\mathbf{X}\}$. If $\mathbf{X}$ is affine we have $H^i_c(\mathbf{X}, \overline{\mathbb{Q}}_\ell) = 0$ if $i \notin \{\dim\mathbf{X}, \dots, 2\dim\mathbf{X}\}$.*

References   See Grothendieck et al. (1972–1973, XVII, 5.2.8.1) for the first assertion and Grothendieck et al. (1972–1973, XIV, 3.2) together with Grothendieck et al. (1972–1973, XVIII, 3.2.6.2) for the affine case. □

Henceforth we shall write $H^i_c(\mathbf{X})$ for $H^i_c(\mathbf{X}, \overline{\mathbb{Q}}_\ell)$, since we shall consider only cohomology with coefficients in $\overline{\mathbb{Q}}_\ell$.

Proposition 8.1.2 *Any proper morphism $\mathbf{X} \to \mathbf{Y}$ induces a linear morphism $H^i_c(\mathbf{Y}) \to H^i_c(\mathbf{X})$ for any i, and this correspondence is functorial; a Frobenius root on $\mathbf{X}$ is a finite morphism and induces an automorphism of $H^i_c(\mathbf{X})$.*

Proof   The first assertion comes from Grothendieck et al. (1972–1973, XVII, 5.1.17). Now let F be a Frobenius root such that F^n is a Frobenius endomorphism. By 4.1.10 F^n is finite. It follows that F^n is separated, which implies that, as a factor of a separated morphism, F is separated, see Hartshorne (1977, II, 4.6(e)). It also follows that F^n is proper and writing $F^n = F^{n-1} \circ F$ and using that F^{n-1} is separated, we get that F is proper, see Hartshorne (1977, II, 4.8(e)). Finally, being proper and quasi-finite, F is finite, see Grothendieck (1967, 18.12.4). The assertion that F is an automorphism of the cohomology results from the same assertion for F^n, for which see for example Deligne (1977, Rapport, 1.2). □

Definition 8.1.3   If $g \in \mathrm{Aut}(\mathbf{X})$ has finite order, we call the **Lefschetz number** of g the number $\mathcal{L}(g, \mathbf{X}) = \mathrm{Trace}(g \mid H_c^*(\mathbf{X}))$, where we put $H_c^*(\mathbf{X}) := \sum_i (-1)^i H_c^i(\mathbf{X})$.

$H_c^*(\mathbf{X})$ is a virtual $\overline{\mathbb{Q}}_\ell$-vector space of finite dimension.

The fundamental property of ℓ-adic cohomology is the "Lefschetz theorem".

Theorem 8.1.4   *Let F be the Frobenius endomorphism associated to an $\mathbb{F}_q$-structure on $\mathbf{X}$. We have $|\mathbf{X}^F| = \mathrm{Trace}(F \mid H_c^*(\mathbf{X}))$.*

Reference   See Deligne (1977, Rapport, 3.2). □

Corollary 8.1.5   *Assume that $\mathbf{X}$ is defined over $\mathbb{F}_q$, with Frobenius endomorphism F, and that $g \in \mathrm{Aut}(\mathbf{X})$ is an automorphism of finite order commuting with F; then we have $\mathcal{L}(g, \mathbf{X}) = R(t)|_{t=\infty}$, where $R(t)$ is the formal series $-\sum_{n=1}^\infty |\mathbf{X}^{gF^n}| t^n$.*

This result is an example of the power of ℓ-adic cohomology: it proves that $R(t)|_{t=\infty}$ is the value at g of some virtual character for any finite subgroup of $\mathrm{Aut}(\mathbf{X})$ containing g. No other proof of this fact is known.

Proof   As g and F commute, they can be reduced to a triangular form in the same basis of $\oplus_i H_c^i(\mathbf{X})$ (note that we do not assume that F is semi-simple). Let $\lambda_1, \ldots, \lambda_k$ be the eigenvalues of F and $x_1, \ldots, x_k$ be those of g, and let $\varepsilon_i = \pm 1$ be the sign of the cohomology space (in $H_c^*(\mathbf{X})$) in which λ_i and x_i are eigenvalues. As gF^n is also a Frobenius endomorphism on $\mathbf{X}$ for any n (see 4.1.11(ii)), we have by 8.1.4

$$R(t) = -\sum_{n=1}^\infty \sum_{i=1}^k \varepsilon_i \lambda_i^n x_i t^n = \sum_{i=1}^k \varepsilon_i x_i \frac{-\lambda_i t}{1 - \lambda_i t}$$

whence $R(t)|_{t=\infty} = \sum_{i=1}^k \varepsilon_i x_i$. □

Corollary 8.1.6   *For any $g \in \mathrm{Aut}(\mathbf{X})$ of finite order, the Lefschetz number $\mathcal{L}(g, \mathbf{X})$ is a rational integer independent of ℓ.*

Proof   The independence of ℓ is a consequence of 8.1.5 as any $g \in \mathrm{Aut}(\mathbf{X})$ is defined over some finite subfield of $\overline{\mathbb{F}}_p$, see 4.1.11(v). Moreover the proof of 8.1.5 shows that $R(t)$ is in $\overline{\mathbb{Q}}_\ell(t)$. As it is a formal series with integral coefficients, it has to be in $\mathbb{Q}(t)$. So we have $\mathcal{L}(g, \mathbf{X}) \in \mathbb{Q}$. But a Lefschetz number is an algebraic integer since it is equal to $\sum_{i=1}^{k} \varepsilon_i x_i$ where all x_i are roots of unity (of order dividing that of g), whence the result.   □

In the following propositions we shall list properties of the Lefschetz numbers. When these properties can be proved directly using 8.1.5, we shall give that proof; but we shall also give, without proof, the corresponding result on ℓ-adic cohomology, if it exists.

Proposition 8.1.7

(i) *Let* $\mathbf{X} = \mathbf{X}_1 \coprod \mathbf{X}_2$ *be a partition of* $\mathbf{X}$ *into two subvarieties with* $\mathbf{X}_1$ *open* (*thus* $\mathbf{X}_2$ *closed*). *We have a long exact sequence*

$$\cdots \to H_c^i(\mathbf{X}_1) \to H_c^i(\mathbf{X}) \to H_c^i(\mathbf{X}_2) \to H_c^{i+1}(\mathbf{X}_1) \to \cdots,$$

whence $H_c^*(\mathbf{X}) = H_c^*(\mathbf{X}_1) + H_c^*(\mathbf{X}_2)$ *as virtual vector spaces.*
(ii) *If in* (i) $\mathbf{X}_1$ *is also closed, then the exact sequence of* (i) *splits, and for any* i *we have* $H_c^i(\mathbf{X}) \simeq H_c^i(\mathbf{X}_1) \oplus H_c^i(\mathbf{X}_2)$.
(iii) *Let* $\mathbf{X} = \coprod_j \mathbf{X}_j$ *be a finite partition of* $\mathbf{X}$ *into locally closed subvarieties; if* g *is an automorphism of finite order of* $\mathbf{X}$ *which stabilises* (*globally*) *this partition, we have* $\mathcal{L}(g, \mathbf{X}) = \sum_{\{j \mid {}^g\mathbf{X}_j = \mathbf{X}_j\}} \mathcal{L}(g, \mathbf{X}_j)$.

Proof For (i) and (ii) see, for example, Grothendieck et al. (1972–1973, XVII, 5.1.16.3). For (iii), let F be a Frobenius endomorphism which commutes with g and is such that all $\mathbf{X}_j$ are F-stable, see 4.1.11(v). Assertion (iii) results from 8.1.5 and the equality $|\mathbf{X}^{gF^n}| = \sum_{\{j \mid {}^g\mathbf{X}_j = \mathbf{X}_j\}} |\mathbf{X}_j^{gF^n}|$.   □

Proposition 8.1.8 *Assume that the variety* $\mathbf{X}$ *is reduced to a finite number of points.*

(i) *We have* $H_c^i(\mathbf{X}) = 0$ *if* $i \neq 0$ *and* $H_c^0(\mathbf{X}) \simeq \overline{\mathbb{Q}}_\ell \mathbf{X}$.
(ii) *Any permutation* g *of the finite set* $\mathbf{X}$ *defines an automorphism of the variety* $\mathbf{X}$, *and* $H_c^*(\mathbf{X}) \simeq \overline{\mathbb{Q}}_\ell \mathbf{X}$ *is a permutation module under* g. *We have* $\mathcal{L}(g, \mathbf{X}) = |\mathbf{X}^g|$.

Proof These results are clear from 8.1.1 and 8.1.7(ii) and (iii).   □

Proposition 8.1.9 *Let* $\mathbf{X}$ *and* $\mathbf{X}'$ *be two varieties.*

(i) *We have* $H_c^k(\mathbf{X} \times \mathbf{X}') \simeq \bigoplus_{i+j=k} H_c^i(\mathbf{X}) \otimes_{\overline{\mathbb{Q}}_\ell} H_c^j(\mathbf{X}')$ (*the "Künneth formula"*).
(ii) *Let* $g \in \operatorname{Aut}\mathbf{X}$ (*resp.* $g' \in \operatorname{Aut}\mathbf{X}'$) *be automorphisms of finite order; then we have* $\mathcal{L}(g \times g', \mathbf{X} \times \mathbf{X}') = \mathcal{L}(g, \mathbf{X})\mathcal{L}(g', \mathbf{X}')$.

Proof For (i) see Grothendieck et al. (1972–1973, XVII, 5.4.3). Let us prove (ii). We write $f * h = \sum_{i \geq 0} a_i b_i t^i$ for the Hadamard product of two formal series $f = \sum_{i \geq 0} a_i t^i$ and $h = \sum_{i \geq 0} b_i t^i$. Let F be a Frobenius endomorphism of $\mathbf{X}$

(resp. $\mathbf{X}'$) which commutes with g (resp. g'). We have to show that the series $f = \sum_{n\geq 1} |\mathbf{X}^{gF^n}| t^n$ and $h = \sum_{n\geq 1} |\mathbf{X}'^{g'F^n}| t^n$ satisfy the relation $-(f*h)|_{t=\infty} = -f|_{t=\infty} \times -h|_{t=\infty}$. But this result follows easily from the proof of 8.1.5, because these two series are linear combinations of series of the form $t/(1 - \lambda t)$ and such series satisfy the equality. □

Proposition 8.1.10 *Let $H \subset \operatorname{Aut} \mathbf{X}$ be a finite subgroup such that the quotient variety $\mathbf{X}/H$ exists (which is always true if $\mathbf{X}$ is quasi-projective). Consider an automorphism $g \in \operatorname{Aut} \mathbf{X}$ of finite order which commutes with all elements of H. Then:*

(i) *The $\overline{\mathbb{Q}}_\ell[g]$-modules $H_c^i(\mathbf{X})^H$ and $H_c^i(\mathbf{X}/H)$ are isomorphic for any i,*
(ii) *$\mathcal{L}(g, \mathbf{X}/H) = |H|^{-1} \sum_{h \in H} \mathcal{L}(gh, \mathbf{X})$.*

Proof For (i) see, for example, Grothendieck et al. (1972–1973, XVII, 6.2.5). Let us prove (ii). Let us choose a Frobenius endomorphism F on $\mathbf{X}$ such that g and all elements of H commute with F. Then F is again a Frobenius morphism on $\mathbf{X}/H$ (for example, by 4.1.8(i)), and we have

$$|(\mathbf{X}/H)^{gF^n}| = |H|^{-1} \sum_{h \in H} |\mathbf{X}^{ghF^n}|,$$

whence the result by 8.1.5. □

Proposition 8.1.11 *Let $\mathbf{X}$ be an affine space of dimension n; then:*

(i) $\dim H_c^i(\mathbf{X}) = \begin{cases} 1 & \text{if } i = 2n, \\ 0 & \text{otherwise.} \end{cases}$

(ii) *If F is the Frobenius endomorphism associated to some $\mathbb{F}_q$-structure on $\mathbf{X}$, we have $|\mathbf{X}^F| = q^n$.*

(iii) *For any finite order $g \in \operatorname{Aut}(\mathbf{X})$ we have $\mathcal{L}(g, \mathbf{X}) = 1$.*

Proof For (i) see, for example, Srinivasan (1979, 5.7). Note that (iii) is straightforward from (ii) by 8.1.5. Let us prove (ii). Let λ be the scalar by which F acts on the one-dimensional space $H_c^{2n}(\mathbf{X})$; in view of 8.1.4, we want to prove that $\lambda = q^n$. We can invoke 8.1.12(iii), or reason as follows: for any $m > 0$ we have $|\mathbf{X}^{F^m}| = \lambda^m$ so λ is a positive integer. We now show that, if $A = A_0 \otimes_{\mathbb{F}_q} \overline{\mathbb{F}}_q$ defines the $\mathbb{F}_q$-structure on $\mathbf{X}$, there exists n_0 such that $A_0 \otimes_{\mathbb{F}_q} \mathbb{F}_{q^{n_0}} \simeq \mathbb{F}_{q^{n_0}}[T_1, \ldots, T_n]$. Indeed we have $A \simeq \overline{\mathbb{F}}_q[T_1, \ldots, T_n]$, so if A_0 is generated by $t_1, \ldots, t_k$ and if n_0 is such that all t_i are in $\mathbb{F}_{q^{n_0}}[T_1, \ldots, T_n]$ we have $A_0 \otimes_{\mathbb{F}_q} \mathbb{F}_{q^{n_0}} \subset \mathbb{F}_{q^{n_0}}[T_1, \ldots, T_n]$ and this inclusion has to be an equality because these two $\mathbb{F}_{q^{n_0}}$-spaces have equal tensor products with $\overline{\mathbb{F}}_q$. For any m multiple of n_0 we have $|\mathbf{X}^{F^m}| = q^{mn}$, so $\lambda = q^n$, which proves (ii) by 8.1.4. □

Proposition 8.1.12

(i) $H_c^{2\dim \mathbf{X}}(\mathbf{X}, \overline{\mathbb{Q}}_\ell)$ *has a basis naturally indexed by the irreducible components of* $\mathbf{X}$ *of maximal dimension.*

(ii) *An automorphism of* $\mathbf{X}$ *of finite order induces on* $H_c^{2\dim \mathbf{X}}(\mathbf{X}, \overline{\mathbb{Q}}_\ell)$ *the permutation of the natural basis reflecting its action on the irreducible components of maximal dimension.*

(iii) *If* F *is the Frobenius endomorphism attached to an* $\mathbb{F}_q$*-structure on* $\mathbf{X}$, *then* F *acts on* $H_c^{2\dim \mathbf{X}}(\mathbf{X}, \overline{\mathbb{Q}}_\ell)$ *by* $q^{\dim \mathbf{X}}$ *times the permutation of the natural basis reflecting its action on the irreducible components of maximal dimension.*

Note that in (ii) and (iii), since the cohomology is contravariant the action induced on the cohomology is by the inverse of the permutation on the irreducible components.

Proof Let $\mathbf{Y}$ be a smooth dense open subvariety of $\mathbf{X}$. Since the complement of $\mathbf{Y}$ in $\mathbf{X}$ has lesser dimension than $\mathbf{X}$, the long exact sequence 8.1.7(i) shows that $H_c^{2\dim \mathbf{X}}(\mathbf{X}, \overline{\mathbb{Q}}_\ell) \simeq H_c^{2\dim \mathbf{X}}(\mathbf{Y}, \overline{\mathbb{Q}}_\ell)$. Further we may assume in (ii) (resp. (iii)) that $\mathbf{Y}$ is stable by the automorphism (resp. by F) since the union of its images by powers of the automorphism (resp. of F) has the same properties as $\mathbf{Y}$.

Thus we are reduced to the case where $\mathbf{X}$ is smooth. Then its irreducible components are connected components $\mathbf{X}_i$, thus by 8.1.7(ii) we have canonically $H_c^{2\dim \mathbf{X}}(\mathbf{X}, \overline{\mathbb{Q}}_\ell) = \oplus_i H_c^{2\dim \mathbf{X}}(\mathbf{X}_i, \overline{\mathbb{Q}}_\ell)$ where the sum is over the connected components of maximal dimension. This reduces the question to the case $\mathbf{X}$ irreducible (in the case of F, one can replace F by the composed $F_0 = F \circ \sigma$ where σ is the automorphism of $\mathbf{X}$ which just permutes the connected components in the inverse way of F, thus F_0 is a Frobenius by 4.1.11(ii) which fixes the connected components).

Finally, the case $\mathbf{X}$ irreducible is Milne (1980, Chapter VI, Lemma 11.3).

□

Proposition 8.1.13 *Let* $\mathbf{X} \xrightarrow{\pi} \mathbf{X}'$ *be an epimorphism such that all fibres are isomorphic to affine spaces of the same dimension* n. *Let* $g \in \mathrm{Aut}\, \mathbf{X}$, $g' \in \mathrm{Aut}\, \mathbf{X}'$ *be finite order automorphisms such that* $g'\pi = \pi g$.

(i) *The* $\overline{\mathbb{Q}}_\ell[g]$*-module* $H_c^i(\mathbf{X})$ *is isomorphic to* $H_c^{i-2n}(\mathbf{X}')(-n)$, *a "Tate twist" of the module* $H_c^{i-2n}(\mathbf{X}')$ *(if* g *acts on this last module by the action of* g').

(ii) *We have* $\mathcal{L}(g, \mathbf{X}) = \mathcal{L}(g', \mathbf{X}')$.

See Srinivasan (1979, III page 47) for the definition of a "Tate twist"; note that $H^j_c(\mathbf{X}')(-n)$ and $H^j_c(\mathbf{X}')$ are isomorphic vector spaces and that for any $\mathbb{F}_q$-structure on $\mathbf{X}'$ the action of F on $H^j_c(\mathbf{X}')(-n)$ is q^n times the action of F on $H^j_c(\mathbf{X}')$.

Proof For (i) see Srinivasan (1979, 5.5, 5.7). Let us prove (ii). We choose $\mathbb{F}_q$-structures on $\mathbf{X}$ and $\mathbf{X}'$ such that π is defined over $\mathbb{F}_q$. By 8.1.11(ii) we have

$$|\mathbf{X}^{gF^m}| = \sum_{y \in \mathbf{X}'^{g'F^m}} |\pi^{-1}(y)^{gF^m}| = |\mathbf{X}'^{g'F^m}| q^{mn},$$

whence the result by 8.1.5. □

Proposition 8.1.14 *Let* $\mathbf{G}$ *be a connected algebraic group acting on the variety* $\mathbf{X}$.

(i) *Any element* $g \in \mathbf{G}$ *acts trivially on* $H^i_c(\mathbf{X})$ *for any* i.
(ii) *Let* $g \in \mathbf{G}$ *be such that the induced isomorphism has finite order; then* $\mathcal{L}(g, \mathbf{X}) = \mathcal{L}(1, \mathbf{X})$.

Proof The proof of (i) may be found in Deligne and Lusztig (1976, 6.4, 6.5). Let us prove (ii). By 4.1.11(v) there exist Frobenius endomorphisms F on $\mathbf{G}$ and $\mathbf{X}$ such that the action of $\mathbf{G}$ commutes with F, that is ${}^F(gx) = {}^Fg{}^Fx$ for all $(g, x) \in \mathbf{G} \times \mathbf{X}$. If n is a positive integer, by the Lang–Steinberg theorem 4.2.9 there exists $h \in \mathbf{G}$ such that $h.{}^{F^n}h^{-1} = g$. But then $x \mapsto h^{-1}x$ defines a bijection from $\mathbf{X}^{gF^n}$ onto $\mathbf{X}^{F^n}$, so we have $|\mathbf{X}^{gF^n}| = |\mathbf{X}^{F^n}|$, whence the result by 8.1.5. □

Proposition 8.1.15 *Let* $g = su$ *be the decomposition of the finite order automorphism* $g \in \mathrm{Aut}\,\mathbf{X}$ *into the product of a* p'-*element* s *and a* p-*element* u. *Then we have* $\mathcal{L}(g, \mathbf{X}) = \mathcal{L}(u, \mathbf{X}^s)$.

Proof There is no known proof of this result by means of a computation using only Lefschetz numbers. One has to use directly the definition of ℓ-adic cohomology, see Deligne and Lusztig (1976, 3.1). □

Proposition 8.1.16 *Let* $\mathbf{T}$ *be a torus acting on a variety* $\mathbf{X}$ *and let* $g \in \mathrm{Aut}\,\mathbf{X}$ *be a finite order automorphism commuting with the action of* $\mathbf{T}$. *Then for any* $t \in \mathbf{T}$ *we have* $\mathcal{L}(g, \mathbf{X}) = \mathcal{L}(g, \mathbf{X}^t)$. *Moreover, if* $\mathbf{X}$ *is affine, we have* $\mathcal{L}(g, \mathbf{X}) = \mathcal{L}(g, \mathbf{X}^\mathbf{T})$.

Proof Let $g = su$ be the decomposition of g into its p'-part and its p-part; by 8.1.15 we have $\mathcal{L}(g, \mathbf{X}) = \mathcal{L}(u, \mathbf{X}^s)$. The action of $\mathbf{T}$ commutes with g, so commutes also with s, thus $\mathbf{T}$ acts on $\mathbf{X}^s$. As this group is connected the action of any $t \in \mathbf{T}$ on $H^*_c(\mathbf{X}^s)$ is trivial by 8.1.14. Hence for any $t \in \mathbf{T}$ we have

$$\mathcal{L}(u, \mathbf{X}^s) = \mathcal{L}(ut, \mathbf{X}^s) = \mathcal{L}(u, (\mathbf{X}^s)^t) = \mathcal{L}(u, (\mathbf{X}^t)^s) = \mathcal{L}(g, \mathbf{X}^t),$$

the second and the last equalities coming from 8.1.15. If $\mathbf{X}$ is affine 1.2.7 shows that there exists $t \in \mathbf{T}$ such that $\mathbf{X}^{\mathbf{T}} = \mathbf{X}^t$. $\square$

Notes

The theory of ℓ-adic cohomology is expounded in Grothendieck et al. (1972–1973), Deligne (1977) and Grothendieck et al. (1977). For the properties of the Lefschetz numbers one can also refer to Lusztig (1978). A good survey, with some proofs, of properties needed in our case may be found in Srinivasan (1979, Chapter V).

9

Deligne–Lusztig Induction: The Mackey Formula

We are going now to extend the construction of the functor $R_{\mathbf{L}}^{\mathbf{G}}$ to the case where $\mathbf{L}$ is an F-stable Levi subgroup of $\mathbf{G}$ which is not the Levi subgroup of any F-stable parabolic subgroup of $\mathbf{G}$.

Example We would like to generalise the construction of 5.3.2 to the case of the unitary groups (see 7.1.8), but the construction used there does not work, since the Frobenius morphism on $\mathbf{U}_{n+m}$ maps the parabolic subgroup $\begin{pmatrix} \mathbf{U}_n & * \\ 0 & \mathbf{U}_m \end{pmatrix}$ to the parabolic subgroup $\begin{pmatrix} \mathbf{U}_n & 0 \\ * & \mathbf{U}_m \end{pmatrix}$ which is not conjugate to it if $n \neq m$. Actually, in this case the F-stable Levi subgroup $\begin{pmatrix} \mathbf{U}_n & 0 \\ 0 & \mathbf{U}_m \end{pmatrix}$ is not a Levi subgroup of any F-stable parabolic subgroup.

We are again in the setting of a connected reductive group $\mathbf{G}$ with a Frobenius root F.

9.1 Deligne–Lusztig Induction

The idea that Deligne and Lusztig had was to associate to any parabolic subgroup $\mathbf{P}$ with Levi decomposition $\mathbf{P} = \mathbf{U} \rtimes \mathbf{L}$, where $\mathbf{L}$ is F-stable, a $\mathbf{G}^F$-variety-$\mathbf{L}^F$, that is an algebraic variety $\mathbf{X}_{\mathbf{U}}$ over $\overline{\mathbb{F}}_q$ endowed with commuting left action of $\mathbf{G}^F$ and right action of $\mathbf{L}^F$, such that when $\mathbf{P}$ is F-stable we have $H_c^*(\mathbf{X}_{\mathbf{U}}) \simeq \overline{\mathbb{Q}}_\ell(\mathbf{G}^F/\mathbf{U}^F)$, and to define $R_{\mathbf{L}}^{\mathbf{G}}$ as the functor associated to the $\overline{\mathbb{Q}}_\ell \mathbf{G}^F$-module-$\overline{\mathbb{Q}}_\ell \mathbf{L}^F$ afforded by $H_c^*(\mathbf{X}_{\mathbf{U}})$. This construction of Deligne and Lusztig was first published in Lusztig (1976b), and is traditionally called the "Lusztig functor" (or the "Lusztig induction"). More precisely:

Definition 9.1.1 Let $\mathbf{P} = \mathbf{U} \rtimes \mathbf{L}$ be a Levi decomposition of a parabolic subgroup of $\mathbf{G}$, where $\mathbf{L}$ is F-stable. We define the variety

$$\mathbf{X_U} := \{g\mathbf{U} \in \mathbf{G}/\mathbf{U} \mid g\mathbf{U} \cap F(g\mathbf{U}) \neq \emptyset\} = \{g\mathbf{U} \in \mathbf{G}/\mathbf{U} \mid g^{-1\,F}g \in \mathbf{U} \cdot {}^F\mathbf{U}\}$$
$$\simeq \{g \in \mathbf{G} \mid g^{-1\,F}g \in {}^F\mathbf{U}\}/(\mathbf{U} \cap {}^F\mathbf{U}).$$

We will denote by $\mathbf{X_U^G}$ this variety when needed to specify the group involved.

The first two formulae are clearly equivalent definitions of the same set; it is a variety, as well as the third set, since they are closed subvarieties of the quotient of $\mathbf{G}$ by a closed subgroup, see 1.1.4(i). To show that these varieties are isomorphic we use the following proposition.

Proposition 9.1.2   *Let $f : \mathbf{X} \to \mathbf{Y}$ be a morphism of $\overline{\mathbb{F}}_p$-varieties bijective on points. Assume that connected components of $\mathbf{X}$ are irreducible and that $\mathbf{Y}$ is normal. Then f is an isomorphism.*

Reference   See for example Digne et al. (2007, 2.2.2).   □

By the Lang–Steinberg theorem one sees that the map $g(\mathbf{U} \cap {}^F\mathbf{U}) \mapsto g\mathbf{U}$ is bijective on points from the second variety to the first and both are smooth (see below), so this map is an isomorphism.

On each of the varieties, we have actions by translation of $\mathbf{G}^F$ on the left and of $\mathbf{L}^F$ on the right, using the fact that $\mathbf{L}^F$ normalises $\mathbf{U}$ and ${}^F\mathbf{U}$.

We record now some geometric properties of the varieties $\mathbf{X_U}$ that we will not use later in this book.

- They are smooth (for example by the same argument as in the proof of Digne et al., 2007, 2.3.5).
- It is conjectured they are affine; this is known when q is greater than the Coxeter number of W (the order of the product of the elements of S, taken in any order) at least when F is a Frobenius endomorphism, see Lusztig (1990, 6.3(b)).
- They are irreducible if and only if $\mathbf{U}$ is in no F-stable proper parabolic subgroup of $\mathbf{G}$, see Bonnafé and Rouquier (2006).

Definition 9.1.3   The **Lusztig functor** $R_{\mathbf{L} \subset \mathbf{P}}^{\mathbf{G}}$, where $\mathbf{P} = \mathbf{U} \rtimes \mathbf{L}$ is a Levi decomposition of $\mathbf{P}$, is the generalised induction functor associated to the $\overline{\mathbb{Q}}_\ell \mathbf{G}^F$-module-$\overline{\mathbb{Q}}_\ell \mathbf{L}^F$ afforded by $H_c^*(\mathbf{X_U})$ with action of $(g,l) \in \mathbf{G}^F \times \mathbf{L}^F$ induced by the action $x \mapsto g^{-1}xl^{-1}$ on $\mathbf{X_U}$ (see 5.1.1).

The adjoint functor is denoted by ${}^*R_{\mathbf{L} \subset \mathbf{P}}^{\mathbf{G}}$ (and sometimes called the "Lusztig restriction").

We have taken above the action of (g^{-1}, l^{-1}) on $\mathbf{X_U}$ to have an action of (g,l) on the cohomology, since the cohomology is contravariant. But note also Proposition 9.1.4.

Proposition 9.1.4 *The $\overline{\mathbb{Q}}_\ell \mathbf{L}^F$-module-$\overline{\mathbb{Q}}_\ell \mathbf{G}^F$ afforded by $H_c^*(\mathbf{X_U})$ with action of $(g,l) \in \mathbf{G}^F \times \mathbf{L}^F$ induced by the action $x \mapsto gxl$ on $\mathbf{X_U}$ is isomorphic in the Grothendieck group of $\overline{\mathbb{Q}}_\ell \mathbf{L}^F$-modules-$\overline{\mathbb{Q}}_\ell \mathbf{G}^F$ to the module which defines $^*R^{\mathbf{G}}_{\mathbf{L} \subset \mathbf{P}}$.*

Proof By Remark 5.1.2 $^*R^{\mathbf{G}}_{\mathbf{L} \subset \mathbf{P}}$ is given by the $\mathbf{L}^F$-module-$\mathbf{G}^F$ given by H_c^* $(\mathbf{X_U})^\vee$ with transposed action. But the trace does not change by transposition. Moreover, since Lefschetz numbers are integers (see 8.1.6), (g,l) and (g^{-1}, l^{-1}) have the same trace on the cohomology of $\mathbf{X_U}$. Thus the module defining $^*R^{\mathbf{G}}_{\mathbf{L} \subset \mathbf{P}}$ and the module of the statement afford the same character. They are thus isomorphic, since group algebras over an algebraically closed field of characteristic 0 are semi-simple so an element of the Grothendieck group is defined by its character. $\qquad\square$

In general when a group G acts on a set X we write X^{opp} for the same set where $g \in G$ acts by g^{-1}. In particular we will write $H_c^*(\mathbf{X_U})^{\mathrm{opp}}$ when we consider the action 9.1.4 of $\mathbf{G}^F \times \mathbf{L}^F$ on the cohomology.

Note that if $\mathbf{P}$ is F-stable then $\mathbf{U}$ is also, in which case if $g\mathbf{U} \cap {}^F g\mathbf{U} \neq \emptyset$ then $g\mathbf{U} = {}^F g\mathbf{U}$ and $\mathbf{X_U}$ is the discrete variety $(\mathbf{G}/\mathbf{U})^F = \mathbf{G}^F/\mathbf{U}^F$ whose cohomology is $\overline{\mathbb{Q}}_\ell(\mathbf{G}^F/\mathbf{U}^F)$ placed in degree 0; thus in this case $R^{\mathbf{G}}_{\mathbf{L} \subset \mathbf{P}}$ is Harish-Chandra induction.

Note also Lemma 9.1.5.

Lemma 9.1.5 *We have an isomorphism of $\overline{\mathbb{Q}}_\ell \mathbf{G}^F$-modules-$\overline{\mathbb{Q}}_\ell \mathbf{L}^F$ between $H_c^*(\mathbf{X_U})$ and $H_c^*(\mathcal{L}^{-1}(\mathbf{U}))$.*

Proof The quotient map $\mathcal{L}^{-1}(\mathbf{U}) \rightarrow \{g \in \mathbf{G} \mid g^{-1\,F}g \in \mathbf{U}\}/({}^{F^{-1}}\mathbf{U} \cap \mathbf{U})$ has all its fibres isomorphic to an affine space. By 8.1.13(i) we get an isomorphism of $\overline{\mathbb{Q}}_\ell \mathbf{G}^F$-modules-$\overline{\mathbb{Q}}_\ell \mathbf{L}^F$ between the H_c^*. Finally F maps the second variety to $\mathbf{X_U}$ and induces an isomorphism on the cohomology, see 8.1.2. $\qquad\square$

Proposition 9.1.6 *With the above notation, for $g \in \mathbf{G}^F$ we have*

$$(R^{\mathbf{G}}_{\mathbf{L} \subset \mathbf{P}}\chi)(g) = |\mathbf{L}^F|^{-1} \sum_{l \in \mathbf{L}^F} \mathrm{Trace}((g,l) \mid H_c^*(\mathbf{X_U}))\chi(l^{-1})$$

and for $l \in \mathbf{L}^F$ we have

$$(^*R^{\mathbf{G}}_{\mathbf{L} \subset \mathbf{P}}\psi)(l) = |\mathbf{G}^F|^{-1} \sum_{g \in \mathbf{G}^F} \mathrm{Trace}((g,l) \mid H_c^*(\mathbf{X_U}))\psi(g^{-1}).$$

Proof From 5.1.5, we immediately obtain the two formulae, using for the second formula Proposition 9.1.4 and the fact that the Lefschetz numbers are integers. $\qquad\square$

Proposition 9.1.7   *Let $\pi \in \mathrm{Irr}(\mathbf{L}^F)$ and let $\overline{\pi}$ be the contragredient representation; then the contragredient of $R^{\mathbf{G}}_{\mathbf{L} \subset \mathbf{P}}(\pi)$ is $R^{\mathbf{G}}_{\mathbf{L} \subset \mathbf{P}}(\overline{\pi})$.*

Proof   The proposition results immediately from formula 9.1.6, using that the Lefschetz numbers $\mathrm{Trace}((g,l) \mid H^*_c(\mathbf{X_U}))$ are integers, and from the fact that the trace of an element x on a given representation is equal to the trace of x^{-1} on the contragredient representation.   □

Proposition 9.1.8 (Transitivity)   *Let $\mathbf{Q} \subset \mathbf{P}$ be two parabolic subgroups of $\mathbf{G}$, and let $\mathbf{M} \subset \mathbf{L}$ be F-stable Levi subgroups of $\mathbf{Q}$ and $\mathbf{P}$ respectively. Then $R^{\mathbf{G}}_{\mathbf{L} \subset \mathbf{P}} \circ R^{\mathbf{L}}_{\mathbf{M} \subset \mathbf{L} \cap \mathbf{Q}} = R^{\mathbf{G}}_{\mathbf{M} \subset \mathbf{Q}}.$*

Proof   Let $\mathbf{P} = \mathbf{U} \rtimes \mathbf{L}$ and $\mathbf{Q} = \mathbf{V} \rtimes \mathbf{M}$ be the Levi decompositions; then by 3.4.8(ii) $\mathbf{L} \cap \mathbf{Q} = (\mathbf{V} \cap \mathbf{L}) \rtimes \mathbf{M}$ is a Levi decomposition of a parabolic subgroup of $\mathbf{L}$. Following 5.1.4 and 5.1.11 we have to show that there is an isomorphism of $\overline{\mathbb{Q}}_\ell \mathbf{G}^F$-modules-$\overline{\mathbb{Q}}_\ell \mathbf{M}^F$

$$H^*_c(\mathbf{X^G_U}) \otimes_{\overline{\mathbb{Q}}_\ell \mathbf{L}^F} H^*_c(\mathbf{X^L_{V \cap L}}) \simeq H^*_c(\mathbf{X^G_V}).$$

This results from the isomorphism of $\mathbf{G}^F$-varieties-$\mathbf{M}^F$ given by

$$\mathbf{X^G_U} \times_{\mathbf{L}^F} \mathbf{X^L_{V \cap L}} \simeq \mathbf{X^G_V}$$
$$(g\mathbf{U}, l(\mathbf{V} \cap \mathbf{L})) \mapsto gl\mathbf{V},$$

(where the notation is as in 2.3.11) and the properties 8.1.9 and 8.1.10 of the cohomology.   □

9.2 Mackey Formula for Lusztig Functors

The rest of this chapter is devoted to the study of the "Mackey formula" (5.2.1) for the Lusztig functors. At the present time, we do not know of a proof in all cases; we are going to give those reductions that we can make and deduce the formula when one of the Levi subgroups is a torus. It is conjectured that the formula always holds; when F is a Frobenius endomorphism, the best known result is that it holds unless $q = 2$ and $\mathbf{G}$ has a quasi-simple component of type $^2E_6$ simply connected, E_7 simply connected or E_8, see Bonnafé and Michel (2011) and Taylor (2018). When F is a Frobenius root, we do not know of a proof for $^2F_4$.

   In what follows, we will work with $\mathcal{L}^{-1}(\mathbf{U})$ instead of $\mathbf{X_U}$, which is allowed by 9.1.5. Let us first consider the left-hand side of the Mackey formula; that is, the composite functor $^*R^{\mathbf{G}}_{\mathbf{L} \subset \mathbf{P}} \circ R^{\mathbf{G}}_{\mathbf{L}' \subset \mathbf{P}'}$ where $\mathbf{P} = \mathbf{U} \rtimes \mathbf{L}$ and $\mathbf{P}' = \mathbf{U}' \rtimes \mathbf{L}'$ are two Levi decompositions where $\mathbf{L}$ and $\mathbf{L}'$ are F-stable. Then, by 5.1.4, this composite functor is associated to the $\overline{\mathbb{Q}}_\ell \mathbf{L}^F$-module-$\overline{\mathbb{Q}}_\ell \mathbf{L}'^F$

given by $H^*_c(\mathcal{L}^{-1}(\mathbf{U}))^{\mathrm{opp}} \otimes_{\overline{\mathbb{Q}}_\ell \mathbf{G}^F} H^*_c(\mathcal{L}^{-1}(\mathbf{U}'))$ where $(g,l') \in \mathbf{G}^F \times \mathbf{L}'^F$ acts on $\mathcal{L}^{-1}(\mathbf{U}')$ by $x' \mapsto gx'l'$ and where $H^*_c(\mathcal{L}^{-1}(\mathbf{U}))^{\mathrm{opp}}$ is as defined after 9.1.4. This tensor product of cohomology spaces is isomorphic, by the Künneth formula 8.1.9 and by 8.1.10 (see proof of 9.1.8 above), to the $\overline{\mathbb{Q}}_\ell \mathbf{L}^F$-module-$\overline{\mathbb{Q}}_\ell \mathbf{L}'^F$ given by $H^*_c(\mathcal{L}^{-1}(\mathbf{U}) \times_{\mathbf{G}^F} \mathcal{L}^{-1}(\mathbf{U}'))$ – the quotient by the action of $g \in \mathbf{G}^F$ on $\mathcal{L}^{-1}(\mathbf{U}) \times \mathcal{L}^{-1}(\mathbf{U}')$ identifies (x,x') with (gx,gx').

Lemma 9.2.1 *Let $\mathbf{Z} = \{(u,u',g) \in \mathbf{U} \times \mathbf{U}' \times \mathbf{G} \mid u.^Fg = gu'\}$. The map $\varphi : (x,x') \mapsto (\mathcal{L}(x), \mathcal{L}(x'), x^{-1}x')$ is a bijective morphism of $\mathbf{L}^F$-varieties-$\mathbf{L}'^F$ from $\mathcal{L}^{-1}(\mathbf{U}) \times_{\mathbf{G}^F} \mathcal{L}^{-1}(\mathbf{U}')$ to $\mathbf{Z}$; it maps the action of $(l,l') \in \mathbf{L}^F \times \mathbf{L}'^F$ to the action on $\mathbf{Z}$ given by $(u,u',g) \mapsto (^lu, {}^{l'^{-1}}u', lgl')$.*

By 8.1.13 the map φ induces an isomorphism on cohomology groups.

Proof Let us first consider the map on $\mathcal{L}^{-1}(\mathbf{U}) \times \mathcal{L}^{-1}(\mathbf{U}')$ defined by the same formula as φ. The image of this map is clearly in $\mathbf{Z}$. On the other hand, an easy computation shows that (x,x') and (y,y') have the same image if and only if there is some $\gamma \in \mathbf{G}^F$ such that $(x,x') = (\gamma y, \gamma y')$, so the map factors through φ, and φ is injective. It remains to show that φ is surjective; let $(u,u',g) \in \mathbf{Z}$ and let x and x' be any elements of $\mathbf{G}$ such that $u = \mathcal{L}(x)$ and $u' = \mathcal{L}(x')$. We also have $u = \mathcal{L}(\gamma x)$ for any $\gamma \in \mathbf{G}^F$, so it is enough to show that there exists $\gamma \in \mathbf{G}^F$ such that $x^{-1}\gamma^{-1}x' = g$; but indeed $xgx'^{-1} \in \mathbf{G}^F$ because $u.^Fg = gu'$. The formula for the $\mathbf{L}^F \times \mathbf{L}'^F$-action on $\mathbf{Z}$ is clear. $\square$

Since by 5.2.2 we have $\mathbf{G} = \coprod_{w \in \mathbf{L}\backslash\mathcal{S}(\mathbf{L},\mathbf{L}')/\mathbf{L}'} {}^{F^{-1}}\mathbf{P}w^{F^{-1}}\mathbf{P}'$, we have

$$\mathbf{Z} = \coprod_{w \in \mathbf{L}\backslash\mathcal{S}(\mathbf{L},\mathbf{L}')/\mathbf{L}'} \mathbf{Z}_w,$$

a union of the locally closed subvarieties

$$\mathbf{Z}_w = \{(u,u',g) \in \mathbf{U} \times \mathbf{U}' \times {}^{F^{-1}}\mathbf{P}w^{F^{-1}}\mathbf{P}' \mid u.^Fg = gu'\}.$$

Introducing the new variables $w' = w.^Fw^{-1}$, $F' = w'F$ and $(u_1,u_1',g_1) = (u, {}^wu', gw^{-1})$, we have $\mathbf{Z}_w \simeq \mathbf{Z}'_w$, where

$$\mathbf{Z}'_w = \{(u_1,u_1',g_1) \in \mathbf{U} \times {}^w\mathbf{U}' \times {}^{F^{-1}}\mathbf{P}^{F'^{-1}}(^w\mathbf{P}') \mid u_1.^{F}g_1 = g_1u_1'w'\}$$

and we have used the fact that ${}^{wF^{-1}}\mathbf{P}' = {}^{F'^{-1}}(^w\mathbf{P}')$. The action of $\mathbf{L}^F \times \mathbf{L}'^F$ on $\mathbf{Z}$ induces an action of $\mathbf{L}^F \times (^w\mathbf{L}')^{F'}$ on $\mathbf{Z}'_w$ which is given in the new coordinates by the same formula as before.

At this stage we may "forget" w; that is, express everything in terms of the new variables $\mathbf{V} = {}^w\mathbf{U}'$, $\mathbf{Q} = {}^w\mathbf{P}'$ and $\mathbf{M} = {}^w\mathbf{L}'$. With this notation the variety

$$\mathbf{Z}'_w = \{(u_1,v_1,g_1) \in \mathbf{U} \times \mathbf{V} \times {}^{F^{-1}}\mathbf{P}^{F'^{-1}}\mathbf{Q} \mid u_1.^Fg_1 = g_1v_1w'\}$$

is endowed with an action of $(l,m) \in \mathbf{L}^F \times \mathbf{M}^{F'}$ given by $(u_1, v_1, g_1) \mapsto$ $({}^l u_1, {}^{m^{-1}} v_1, l g_1 m)$.

We are now going to use the decomposition ${}^{F^{-1}}\mathbf{P}^{F'^{-1}}\mathbf{Q} = {}^{F^{-1}}\mathbf{ULM}^{F'^{-1}}\mathbf{V}$.

Lemma 9.2.2 *The cohomology of the variety*

$$\mathbf{Z}''_w = \{(u, v, u', v', n) \in \mathbf{U} \times \mathbf{V} \times {}^{F^{-1}}\mathbf{U} \times {}^{F'^{-1}}\mathbf{V} \times \mathbf{LM} \mid u.{}^F n = u' n v' v w'\}$$

is isomorphic as an $\overline{\mathbb{Q}}_\ell \mathbf{L}^F$-module-$\overline{\mathbb{Q}}_\ell \mathbf{M}^{F'}$ to that of $\mathbf{Z}'_w$; the isomorphism of cohomology spaces is induced by the fibration

$$\pi : \mathbf{Z}''_w \to \mathbf{Z}'_w : (u, v, u', v', n) \mapsto (u.{}^F u'^{-1}, v.{}^{F'} v', u' n v'),$$

and the isomorphism of $\overline{\mathbb{Q}}_\ell \mathbf{L}^F$-modules-$\overline{\mathbb{Q}}_\ell \mathbf{M}^{F'}$ is for the action induced by the action of $(l,m) \in \mathbf{L}^F \times \mathbf{M}^{F'}$ on $\mathbf{Z}''_w$ given by

$$(u, v, u', v', n) \mapsto ({}^l u, {}^{m^{-1}} v, {}^l u', {}^{m^{-1}} v', lnm).$$

Proof Once we get the isomorphism of cohomology spaces, everything else is a straightforward computation. To get that isomorphism we will use 8.1.13; for that we need to compute the fibres of π; that is, determine the quintuples (u, v, u', v', n) which are mapped by π to (u_1, v_1, g_1). Since, once u_1 and v_1 are given, u and v are determined by u' and v', it is equivalent to finding the triples (u', v', n) such that $g_1 = u' n v'$, which is in turn equivalent to finding the pairs (u', v') such that $u'^{-1} g_1 v'^{-1} \in \mathbf{LM}$. Let us choose some decomposition $g_1 = u'_1 l m v'_1$ where $u'_1 \in {}^{F^{-1}}\mathbf{U}$, $v'_1 \in {}^{F'^{-1}}\mathbf{V}$, $l \in \mathbf{L}$ and $m \in \mathbf{M}$; the condition on (u', v') can be written as $l^{-1}(u'^{-1} u'_1)^m (v'_1 v'^{-1}) m \in \mathbf{LM}$; introducing the new variables $u'' = {}^{l^{-1}}(u'^{-1} u'_1)$ and $v'' = {}^m (v'_1 v'^{-1})$ the condition on the pair $(u'', v'') \in {}^{F^{-1}}\mathbf{U} \times {}^{F'^{-1}}\mathbf{V}$ becomes $u'' v'' \in \mathbf{LM}$; that is, $u''^{-1}\mathbf{L} \cap v''\mathbf{M} \neq \emptyset$. Using 5.2.4 for the parabolic subgroups ${}^{F^{-1}}\mathbf{P}$ and ${}^{F'^{-1}}\mathbf{Q}$ it is easy to see that the solutions are all pairs of the form $({}^y x^{-1} a^{-1}, ay)$ where $x \in {}^{F^{-1}}\mathbf{U} \cap \mathbf{M}$, $y \in {}^{F'^{-1}}\mathbf{V} \cap \mathbf{L}$ and $a \in {}^{F^{-1}}\mathbf{U} \cap {}^{F'^{-1}}\mathbf{V}$. Thus all fibres of π are isomorphic to the affine space $({}^{F^{-1}}\mathbf{U} \cap \mathbf{M}) \times ({}^{F'^{-1}}\mathbf{V} \cap \mathbf{L}) \times ({}^{F^{-1}}\mathbf{U} \cap {}^{F'^{-1}}\mathbf{V})$, whence the lemma. $\square$

We do not know in general how to transform the left-hand side of the Mackey formula much further towards our goal; the idea that we will use now is to find an action on $\mathbf{Z}''_w$ of a subgroup of $\mathbf{L} \times \mathbf{M}$ whose identity component is a torus which commutes with $\mathbf{L}^F \times \mathbf{M}^{F'}$; using 8.1.16 we may then replace $\mathbf{Z}''_w$ by its fixed points under that torus. This will work if we can find a large enough torus. Note that the image of some quintuple $(u, v, u', v', n) \in \mathbf{Z}''_w$ under $(l,m) \in \mathbf{L} \times \mathbf{M}$ is in $\mathbf{Z}''_w$ if and only if ${}^{F n^{-1}}(l^{-1}.{}^F l) = {}^{w^{-1}}(m^{F'} m^{-1})$.

Let $Z(\mathbf{L})$ (resp. $Z(\mathbf{M})$) be the centre of $\mathbf{L}$ (resp. of $\mathbf{M}$) and put

$$\mathbf{H}_w = \{(l,m) \in Z(\mathbf{L}) \times Z(\mathbf{M}) \mid l^{-1}.{}^F l = {}^{w^{-1}}(m^{F'} m^{-1})\}.$$

Then the identity component $\mathbf{H}_w^0$ is a torus (a connected subgroup of $Z(\mathbf{L})^0 \times Z(\mathbf{M})^0$), and the image of $(u,v,u',v',n) \in \mathbf{Z}_w''$ under $(l,m) \in \mathbf{H}_w$ is still in $\mathbf{Z}_w''$. Indeed if we write $n = \lambda\mu$ with $(\lambda,\mu) \in \mathbf{L} \times \mathbf{M}$, the condition for the image of (u,v,u',v',n) under (l,m) to be in $\mathbf{Z}_w''$ can be written ${}^{F}\lambda^{-1}(l^{-1}{}^{F}l) = {}^{w'^{-1}F'}\mu(m^{F'}m^{-1})$ which holds when $(l,m) \in \mathbf{H}_w$ since $l^{-1}{}^{F}l \in Z(\mathbf{L})$ and $m^{F'}m^{-1} \in Z(\mathbf{M})$.

The torus $\mathbf{H}_w^0$ is large enough for our purpose when one of the Levi subgroups is included in the other one, for example when $\mathbf{M} \subset \mathbf{L}$. *From now on we will assume that hypothesis* (beware that in our initial notation this means that ${}^{w}\mathbf{L}' \subset \mathbf{L}$ for the given element w). To determine $\mathbf{H}_w^0$ in that case, we will use the "norm" on a torus.

Notation 9.2.3 *Let $\mathbf{T}$ be a torus with a Frobenius root F; given a non-zero integer $n \in \mathbb{N}$, we define the morphism* **norm** *$N_{F^n/F}:\mathbf{T} \to \mathbf{T}$ by $\tau \mapsto \tau.{}^{F}\tau \ldots {}^{F^{n-1}}\tau$.*

Lemma 9.2.4

(i) *The first projection maps $\mathbf{H}_w^0$ surjectively to $Z(\mathbf{L})^0$.*

(ii) *We have $\mathbf{H}_w^0 = \{(N_{F^n/F}(\tau), N_{F^m/F'}({}^{w'}\tau^{-1})) \mid \tau \in Z(\mathbf{L})^0\}$, where n is such that ${}^{F^n}w = w$ (for such an n we have $F'^m = F^n$ and ${}^{F^n}w' = w'$) and $\mathbf{H}_w^0 \cap (\mathbf{L}^F \times \mathbf{M}^{F'})$ consists of the pairs such that $\tau \in (Z(\mathbf{L})^0)^{F^n}$.*

(iii) *$\mathbf{H}_w^0$ has a fixed point in $\mathbf{Z}_w''$ if and only if w has a representative in $\mathcal{S}(\mathbf{L},\mathbf{L}')^F$; in that case we have $\mathbf{H}_w^0 = \{(l,l^{-1}) \mid l \in Z(\mathbf{L})^0\}$ and*

$$(\mathbf{Z}_w'')^{\mathbf{H}_w^0} \simeq \{(v,v',n) \in (\mathbf{V} \cap \mathbf{L}) \times ({}^{F^{-1}}\mathbf{V} \cap \mathbf{L}) \times \mathbf{L} \mid {}^{F}n = nv'v\},$$

on which the action of $(l,m) \in \mathbf{L}^F \times \mathbf{M}^F$ is given by $(v,v',n) \mapsto ({}^{m^{-1}}v, {}^{m^{-1}}v', lnm)$.

Proof If $l \in Z(\mathbf{L})^0$ then ${}^{w'}(l^{-1}{}^{F}l) \in {}^{w'}Z(\mathbf{L})^0 = {}^{w'F}Z(\mathbf{L})^0 \subset {}^{F'}Z(\mathbf{M})^0 = Z(\mathbf{M})^0$. Thus by the Lang–Steinberg theorem there is some $m \in Z(\mathbf{M})^0$ such that ${}^{w'}(l^{-1}{}^{F}l) = m^{F'}m^{-1}$, so the first projection maps $\mathbf{H}_w$ surjectively to $Z(\mathbf{L})^0$. As $Z(\mathbf{L})^0$ is connected this projection is still surjective when restricted to $\mathbf{H}_w^0$, whence (i).

We now prove (ii). Let $\mathbf{H}_1$ stand for the group that we want $\mathbf{H}_w^0$ to be equal to. A straightforward computation shows that $\mathbf{H}_1 \subset \mathbf{H}_w$ (note that ${}^{w'}\tau \in {}^{w'}Z(\mathbf{L})^0 \subset Z(\mathbf{M})^0$). As $\mathbf{H}_1$ is the image of the connected group $Z(\mathbf{L})^0$, it is connected; thus $\mathbf{H}_1 \subset \mathbf{H}_w^0$. On the other hand, the first projection maps $\mathbf{H}_1$ surjectively to $Z(\mathbf{L})^0$ since by the Lang–Steinberg theorem any $t \in Z(\mathbf{L})^0$ can be written $s.{}^{F^n}s^{-1}$ with $s \in Z(\mathbf{L})^0$. Thus we get $t = N_{F^n/F}(\tau)$ where $\tau = s^{F}s^{-1}$, and t is the projection of $(N_{F^n/F}(\tau), N_{F^m/F'}({}^{w'}\tau^{-1}))$. As two elements of $\mathbf{H}_w$ have the same first projection only if their second projections differ by an element of $\mathbf{M}^{F'}$, the group $\mathbf{H}_1$ has finite index in $\mathbf{H}_w^0$, thus is equal to it since $\mathbf{H}_w^0$ is connected.

We prove (iii). Suppose there exists $(u, u', v, v', n) \in \mathbf{Z}_w''$ fixed by $\mathbf{H}_w^0$. Then in particular for any $(l, m) \in \mathbf{H}_w^0$ we have $lnm = n$. Since $n \in \mathbf{LM}$ and $l \in Z(\mathbf{L})$, $m \in Z(\mathbf{M})$, we get $lm = 1$, whence the inclusion $\mathbf{H}_w^0 \subset \{(l, l^{-1}) \mid l \in Z(\mathbf{L})^0\}$. Since the first projection is surjective, this gives the second assertion of (iii).

If the element $(u, v, u', v', n) \in \mathbf{Z}_w''$ is fixed by $\mathbf{H}_w^0$ then u, u', v and v' centralise $Z(\mathbf{L})^0$. But $C_{\mathbf{G}}(Z(\mathbf{L})^0) = \mathbf{L}$ (see 3.4.6), so this gives $u \in \mathbf{U} \cap \mathbf{L} = \{1\}$, $u' \in {}^{F^{-1}}\mathbf{U} \cap \mathbf{L} = \{1\}$, $v \in \mathbf{V} \cap \mathbf{L}$ and $v' \in {}^{F'^{-1}}\mathbf{V} \cap \mathbf{L}$. So the elements of $(\mathbf{Z}_w'')^{\mathbf{H}_w^0}$ are of the form $(1, v, 1, v', n)$ with ${}^F n = nv'vw'$, which gives the last assertion of (iii) and proves that $w^{-1\,F}w = w' \in \mathbf{L}$, since $n \in \mathbf{L.M} = \mathbf{L}$. Since w is determined up to left multiplication by an element of $\mathbf{L}$, it has an F-stable representative.

Conversely, if w is F-stable we have $w' = 1$ and $F' = F$; using the general form of the elements of $\mathbf{H}_w^0$ given in (ii) we see that the elements $(1, v, 1, v', n)$ are fixed by $\mathbf{H}_w^0$. This completes the proof of (iii). $\square$

We note for future reference that, by 8.1.14(i), a representation $\pi \otimes \pi' \in \mathrm{Irr}(\mathbf{L}^F \times (\mathbf{M}^{F'})^{\mathrm{opp}})$ can occur in some $H_c^i(\mathbf{Z}_w'')$ only if its restriction to $\mathbf{H}_w^0 \cap (\mathbf{L}^F \times \mathbf{M}^{F'})$ is trivial.

As stated above the virtual $\overline{\mathbb{Q}}_\ell \mathbf{L}^F$-modules-$\overline{\mathbb{Q}}_\ell \mathbf{M}^{F'}$ given by $H_c^*(\mathbf{Z}_w'')$ and $H_c^*(\mathbf{Z}_w''^{\mathbf{H}_w^0})$ are isomorphic.

We now look at a term $R_{\mathbf{L} \cap {}^w\mathbf{L}' \subset \mathbf{L} \cap {}^w\mathbf{P}'}^{\mathbf{L}} \circ {}^*R_{\mathbf{L} \cap {}^w\mathbf{L}' \subset \mathbf{P} \cap {}^w\mathbf{L}'}^{{}^w\mathbf{L}'} \circ \mathrm{ad}\, w$ of the right-hand side of the Mackey formula indexed by some $w \in \mathcal{S}(\mathbf{L}, \mathbf{L}')^F$ (we make no particular assumption on $\mathbf{L}$ and $\mathbf{L}'$ until the end of the current paragraph). As for the left-hand side, we first write the $\overline{\mathbb{Q}}_\ell \mathbf{L}^F$-module-$\overline{\mathbb{Q}}_\ell ({}^w\mathbf{L}')^F$ to which this functor is associated as

$$H_c^*(\mathcal{L}_{\mathbf{L}}^{-1}(\mathbf{L} \cap {}^w\mathbf{U}') \times_{(\mathbf{L} \cap {}^w\mathbf{L}')^F} \mathcal{L}_{{}^w\mathbf{L}'}^{-1}(\mathbf{U} \cap {}^w\mathbf{L}')).$$

As before, using $\mathbf{V} = {}^w\mathbf{U}'$, $\mathbf{M} = {}^w\mathbf{L}'$ and $\mathbf{Q} = {}^w\mathbf{P}'$ we may "forget w"; the module we study is thus the cohomology of the variety

$$\mathbf{S}_w = \{(l, m) \in \mathbf{L} \times \mathbf{M} \mid l^{-1}.{}^F l \in \mathbf{L} \cap \mathbf{V}, m^{-1}.{}^F m \in \mathbf{U} \cap \mathbf{M}\}/(\mathbf{L}^F \cap \mathbf{M}^F)$$

on which $(\lambda, \mu) \in \mathbf{L}^F \times \mathbf{M}^F$ acts by $(l, m) \mapsto (\lambda l, \mu^{-1} m)$.

The next lemma shows that the terms corresponding to the same w on both sides of the Mackey formula are equal when ${}^w\mathbf{L}'(= \mathbf{M}) \subset \mathbf{L}$.

Lemma 9.2.5   *Under the above hypothesis* $(\mathbf{M} \subset \mathbf{L})$ *the cohomology spaces* $\oplus_i H_c^i(\mathbf{S}_w)$ *and* $\oplus_i H_c^i(\mathbf{Z}_w''^{\mathbf{H}_w^0})$ *are isomorphic as* $\overline{\mathbb{Q}}_\ell \mathbf{L}^F$-*modules*-$\overline{\mathbb{Q}}_\ell \mathbf{M}^F$.

Proof We show first that the map φ from

$$(\mathbf{Z}_w'')^{\mathbf{H}_w^0} \simeq \{(v,v',n) \in (\mathbf{V} \cap \mathbf{L}) \times ({}^{F^{-1}}\mathbf{V} \cap \mathbf{L}) \times \mathbf{L} \mid {}^F n = nv'v\}$$

to $\mathcal{L}_{\mathbf{L}}^{-1}(\mathbf{V} \cap \mathbf{L})$ given by $(v,v',n) \mapsto nv'$ is surjective, with all its fibres isomorphic to the same affine space. Indeed, the image of φ is where we claim it is since $nv' \in \mathbf{L}$ and $\mathcal{L}(nv') = v'^{-1}n^{-1F}n^F v' = v^F v' \in \mathbf{V} \cap \mathbf{L}$; and if $n_1 \in \mathcal{L}_{\mathbf{L}}^{-1}(\mathbf{V} \cap \mathbf{L})$, then

$$\varphi^{-1}(n_1) = \{(n_1^{-1F}n_1{}^F v'^{-1}, v', n_1 v'^{-1}) \mid v' \in {}^{F^{-1}}\mathbf{V} \cap \mathbf{L}\},$$

which is isomorphic to the affine space ${}^{F^{-1}}\mathbf{V} \cap \mathbf{L}$.

The $\mathbf{L}^F$-action-$\mathbf{M}^F$ on $\mathbf{Z}_w''$ is clearly mapped by φ to the natural $\mathbf{L}^F$-action-$\mathbf{M}^F$ on $\mathcal{L}_{\mathbf{L}}^{-1}(\mathbf{L} \cap \mathbf{V})$ (remember that $\mathbf{M} \subset \mathbf{L}$), and thus for that action the cohomology spaces of $\mathbf{Z}_w''$ and $\mathcal{L}_{\mathbf{L}}^{-1}(\mathbf{L} \cap \mathbf{V})$ are isomorphic as $\overline{\mathbb{Q}}_\ell \mathbf{L}^F$-modules-$\overline{\mathbb{Q}}_\ell \mathbf{M}^F$.

But $\mathbf{M} \subset \mathbf{L}$ also gives $\mathbf{M} \cap \mathbf{U} = \{1\}$ whence

$$\mathbf{S}_w \simeq \{(l,m) \in \mathbf{L} \times \mathbf{M}^F \mid l^{-1F}l \in \mathbf{L} \cap \mathbf{V}\}/\mathbf{M}^F \simeq \{l \in \mathbf{L} \mid l^{-1F}l \in \mathbf{L} \cap \mathbf{V}\},$$

which gives the result. $\qquad\square$

From Lemmas 9.2.1 to 9.2.5 we get in particular the following theorem.

Theorem 9.2.6 *The Mackey formula*

$${}^*R_{\mathbf{L} \subset \mathbf{P}}^{\mathbf{G}} \circ R_{\mathbf{L}' \subset \mathbf{P}'}^{\mathbf{G}} = \sum_{w \in \mathbf{L}^F \backslash \mathcal{S}(\mathbf{L},\mathbf{L}')^F / \mathbf{L}'^F} R_{\mathbf{L} \cap {}^w\mathbf{L}' \subset \mathbf{L} \cap {}^w\mathbf{P}'}^{\mathbf{L}} \circ {}^*R_{\mathbf{L} \cap {}^w\mathbf{L}' \subset \mathbf{P} \cap {}^w\mathbf{L}'}^{{}^w\mathbf{L}'} \circ \operatorname{ad} w$$

holds when either $\mathbf{L}$ or $\mathbf{L}'$ is a maximal torus.

Proof If $\mathbf{L}'$ is a torus, then ${}^w\mathbf{L}' \subset \mathbf{L}$ for any $w \in \mathcal{S}(\mathbf{L},\mathbf{L}')$ and all the preceding lemmas hold unconditionally. Since the adjoint of the Mackey formula is the same formula with $\mathbf{L}$ and $\mathbf{L}'$ interchanged, the Mackey formula also holds when $\mathbf{L}$ is a maximal torus. $\qquad\square$

As we remarked in Chapter 6, when the Mackey formula holds, it implies that $R_{\mathbf{L} \subset \mathbf{P}}^{\mathbf{G}}$ does not depend on $\mathbf{P}$. In what follows we will usually use the notations $R_{\mathbf{L}}^{\mathbf{G}}$ and ${}^*R_{\mathbf{L}}^{\mathbf{G}}$; we leave it to the reader to check that we do that in cases where either the choice of a parabolic subgroup is irrelevant or the Mackey formula holds.

Definition 9.2.7 When $\mathbf{L} = \mathbf{T}$ is an F-stable maximal torus, $R_{\mathbf{T}}^{\mathbf{G}}(\theta)$ for $\theta \in \operatorname{Irr}(\mathbf{T}^F)$ is called a **Deligne–Lusztig character**.

These characters were introduced in Deligne and Lusztig (1976).

9.3 Consequences: Scalar Products

The Mackey formula gives the "scalar product formula for Deligne–Lusztig characters".

Corollary 9.3.1   *Let* $\mathbf{T}$ *and* $\mathbf{T}'$ *be two F-stable maximal tori of* $\mathbf{G}$. *Then:*

(i) *For* $\theta \in \mathrm{Irr}(\mathbf{T}^F)$ *and* $\theta' \in \mathrm{Irr}(\mathbf{T}'^F)$ *we have*

$$\langle R_{\mathbf{T}}^{\mathbf{G}}(\theta), R_{\mathbf{T}'}^{\mathbf{G}}(\theta') \rangle_{\mathbf{G}^F} = |\mathbf{T}^F|^{-1} \#\{n \in \mathbf{G}^F \mid {}^n\mathbf{T} = \mathbf{T}' \text{ and } {}^n\theta = \theta'\}.$$

(ii) *The functor* $R_{\mathbf{T} \subset \mathbf{B}}^{\mathbf{G}}$ *does not depend on the Borel subgroup* $\mathbf{B}$ *used in its construction.*

(iii) *The* $R_{\mathbf{T}}^{\mathbf{G}}(\theta)$ *where the pairs* $(\mathbf{T}, \theta)$ *are taken up to* $\mathbf{G}^F$-*conjugacy form an orthogonal basis of the space they span.*

Proof   (i) is just a way of writing the Mackey formula, and (ii) is clear from the remarks above 9.2.7. (iii) results from (i) and the fact that $R_{\mathbf{T}}^{\mathbf{G}}(\theta) = R_{\mathbf{T}'}^{\mathbf{G}}(\theta')$ when ${}^g(\mathbf{T}, \theta) = (\mathbf{T}', \theta')$ for some $g \in \mathbf{G}^F$.   □

In 11.1.3, in connection with Lusztig series, we will give a statement somewhat stronger than 9.3.1 which deals directly with the cohomology groups of $\mathcal{L}^{-1}(\mathbf{U})$, where $\mathbf{B} = \mathbf{U} \rtimes \mathbf{T}$. The next corollary gives 9.3.1(i) in terms of the Weyl group when θ and θ' are the trivial characters.

Corollary 9.3.2   *Let* $\mathbf{T}_w$, $\mathbf{T}_{w'}$ *be F-stable maximal tori of* $\mathbf{G}$ *of type* w *(resp. w') (see 4.2.22); then*

$$\langle R_{\mathbf{T}_w}^{\mathbf{G}}(1), R_{\mathbf{T}_{w'}}^{\mathbf{G}}(1) \rangle_{\mathbf{G}^F} = \begin{cases} |W^{wF}| & \text{if } w \text{ and } w' \text{ are } F\text{-conjugate in } W, \\ 0 & \text{otherwise.} \end{cases}$$

Proof   The scalar product is 0 unless $\mathbf{T}_w$ and $\mathbf{T}_{w'}$ are $\mathbf{G}^F$-conjugate; that is, unless w and w' are F-conjugate (see 4.2.22). In this last case we may assume $w = w'$ and $\mathbf{T}_w = \mathbf{T}_{w'}$. Then 9.3.1 gives

$$\langle R_{\mathbf{T}_w}^{\mathbf{G}}(1), R_{\mathbf{T}_w}^{\mathbf{G}}(1) \rangle_{\mathbf{G}^F} = |\mathbf{T}_w^F|^{-1} |N_{\mathbf{G}^F}(\mathbf{T}_w)| = |W(\mathbf{T}_w)^F|.$$

As the action of F on the Weyl group of $\mathbf{T}_w$ can be identified with the action of wF on W, we get the result.   □

The next proposition shows that Deligne–Lusztig induction constructs many cuspidal representations, since it applies in particular for every $\theta \in \mathrm{Irr}(\mathbf{T}_w)$ such that $N_{\mathbf{G}^F}(\mathbf{T}_w, \theta) = \mathbf{T}_w^F$.

Proposition 9.3.3 *Let W be the Weyl group of a quasi-split torus $\mathbf{T}$ of $\mathbf{G}$, and let $\mathbf{T}_w$ be an F-stable torus of type w with respect to $\mathbf{T}$ where no F-conjugate of w lies in an F-stable proper parabolic subgroup of W. Assume that $\theta \in \mathrm{Irr}(\mathbf{T}_w^F)$ is such that $R^{\mathbf{G}}_{\mathbf{T}_w}(\theta)$ is up to sign a true character. Then any irreducible constituent of $R^{\mathbf{G}}_{\mathbf{T}_w}(\theta)$ is cuspidal.*

Proof It is sufficient to show that for any F-stable Levi subgroup $\mathbf{L}$ containing $\mathbf{T}$ of an F-stable proper parabolic subgroup of $\mathbf{G}$, we have $^*R^{\mathbf{G}}_{\mathbf{L}}(R^{\mathbf{G}}_{\mathbf{T}_w}(\theta)) = 0$ since, by the assumption that $R^{\mathbf{G}}_{\mathbf{T}_w}(\theta)$ is up to sign a true character, the same property will hold for the constituents of $R^{\mathbf{G}}_{\mathbf{T}_w}(\theta)$. The Mackey formula implies that this property holds if $\mathcal{S}(\mathbf{L}, \mathbf{T}_w)^F$ is empty; that is, if no $\mathbf{G}^F$-conjugate of $\mathbf{T}_w$ is conjugate to a torus of $\mathbf{L}$. But this is a consequence of the assumption on w, since the type of any F-stable maximal torus of $\mathbf{L}$ is (up to F-conjugacy) in $W_{\mathbf{L}}$, a proper F-stable parabolic subgroup of W. □

Notes

The construction of $R^{\mathbf{G}}_{\mathbf{T}}$ is one of the fundamental ideas of Deligne and Lusztig (1976). The construction of $R^{\mathbf{G}}_{\mathbf{L}}$ was first published in Lusztig (1976b) as a natural extension of that construction. The Mackey formula when one of the Levi subgroups is a torus is given in Deligne and Lusztig (1983, Theorem 7), but the proof in that paper has an error. The proof we give here corrects this error, using an argument indicated to us by Lusztig. The case of two tori (9.3.1) was already in Deligne and Lusztig (1976).

10

The Character Formula and Other Results on Deligne–Lusztig Induction

In this chapter we give the character formulae for Deligne–Lusztig induction and restriction. We then use them to generalise the results of Chapter 5, as well as some results of Chapter 7, and we express the identity, Steinberg and regular characters, and the characteristic function of a semi-simple conjugacy class, as linear combinations of Deligne–Lusztig characters. We are in the context of a connected reductive group $\mathbf{G}$ with a Frobenius root F.

10.1 The Character Formula

Definition 10.1.1   Given $\mathbf{L}$, an F-stable Levi subgroup of $\mathbf{G}$ and a Levi decomposition $\mathbf{P} = \mathbf{U} \rtimes \mathbf{L}$ of a (possibly not F-stable) parabolic subgroup containing $\mathbf{L}$, the two-variable **Green function** $Q_{\mathbf{L}}^{\mathbf{G}} \colon \mathbf{G}_u^F \times \mathbf{L}_u^F \to \mathbb{Q}$ is defined by

$$(u, v) \mapsto |\mathbf{L}^F|^{-1} \operatorname{Trace}((u, v) \mid H_c^*(\mathbf{X}_{\mathbf{U}})).$$

We recall that $\mathbf{G}_u$ denotes the set of unipotent elements of $\mathbf{G}$. The Green function should be properly denoted by $Q_{\mathbf{L} \subseteq \mathbf{P}}^{\mathbf{G}}$ to take in account the parabolic subgroup used in the definition, but as for $R_{\mathbf{L}}^{\mathbf{G}}$ we omit $\mathbf{P}$ from the notation with the same caveats.

Proposition 10.1.2 (character formula for $R_{\mathbf{L}}^{\mathbf{G}}$ and ${}^*R_{\mathbf{L}}^{\mathbf{G}}$)   *Let $\mathbf{L}$ be an F-stable Levi subgroup of $\mathbf{G}$ and let $\psi \in \operatorname{Irr}(\mathbf{G}^F)$ and $\chi \in \operatorname{Irr}(\mathbf{L}^F)$.*

(i) *If $g = su$ is the Jordan decomposition of $g \in \mathbf{G}^F$*

$(R_{\mathbf{L}}^{\mathbf{G}} \chi)(g)$

$= |\mathbf{L}^F|^{-1} |C_{\mathbf{G}}(s)^{0F}|^{-1} \sum_{\{h \in \mathbf{G}^F \mid s \in {}^h\mathbf{L}\}} |C_{{}^h\mathbf{L}}(s)^{0F}| \sum_{v \in C_{{}^h\mathbf{L}}(s)_u^{0F}} Q_{C_{{}^h\mathbf{L}}(s)^0}^{C_{\mathbf{G}}(s)^0}(u, v^{-1}) \, {}^h\chi(sv).$

(ii) *If $l = tv$ is the Jordan decomposition of $l \in \mathbf{L}^F$*

$$({}^*R_{\mathbf{L}}^{\mathbf{G}}\psi)(l) = |C_{\mathbf{L}}(t)^{0F}||C_{\mathbf{G}}(t)^{0F}|^{-1} \sum_{u \in C_{\mathbf{G}}(t)_u^{0F}} Q_{C_{\mathbf{L}}(t)^0}^{C_{\mathbf{G}}(t)^0}(u, v^{-1})\psi(tu).$$

Proof The main step is the following lemma.

Lemma 10.1.3 *With the above notation, we have*

$$\text{Trace}((g, l) \mid H_c^*(\mathbf{X}_{\mathbf{U}}))$$

$$= |C_{\mathbf{L}}(t)^{0F}||C_{\mathbf{G}}(t)^{0F}|^{-1} \sum_{\{h \in \mathbf{G}^F \mid {}^h t = s^{-1}\}} Q_{C_{\mathbf{L}}(t)^0}^{C_{\mathbf{G}}(t)^0}({}^{h^{-1}}u, v) \qquad (*)$$

$$= |C_{\mathbf{G}}(s)^{0F}|^{-1} \sum_{\{h \in \mathbf{G}^F \mid {}^{h^{-1}}s = t^{-1}\}} |C_{{}^h\mathbf{L}}(s)^{0F}|Q_{C_{{}^h\mathbf{L}}(s)^0}^{C_{\mathbf{G}}(s)^0}(u, {}^h v). \qquad (**)$$

Proof These two equalities are clearly equivalent; we shall prove the first one. We use again the variety $\mathcal{L}^{-1}(\mathbf{U})$. From 8.1.15 we get

$$\text{Trace}((g, l) \mid H_c^*(\mathcal{L}^{-1}(\mathbf{U}))) = \text{Trace}((u, v) \mid H_c^*(\mathcal{L}^{-1}(\mathbf{U})^{(s,t)})).$$

We first show that the morphism

$$\varphi : \{h \in \mathbf{G}^F \mid {}^h t = s^{-1}\} \times \{z \in C_{\mathbf{G}}(t)^0 \mid z^{-1F}z \in \mathbf{U}\} \to \mathcal{L}^{-1}(\mathbf{U})^{(s,t)}$$

given by $(h, z) \mapsto hz$ is surjective – it is easy to check that the image of φ is in $\mathcal{L}^{-1}(\mathbf{U})^{(s,t)}$ since $\mathcal{L}^{-1}(\mathbf{U})^{(s,t)} = \{x \in \mathbf{G} \mid x^{-1F}x \in \mathbf{U} \text{ and } sxt = x\}$. If $x \in \mathcal{L}^{-1}(\mathbf{U})^{(s,t)}$ then $sxt = x$ implies $s^Fxt = {}^Fx = x(x^{-1F}x)$ whence $s^Fxt = sxt(x^{-1F}x)$; which can be written $(x^{-1F}x)t = t(x^{-1F}x)$; that is, $x^{-1F}x \in C_{\mathbf{G}}(t)$. As $x^{-1F}x$ is unipotent, we even get $x^{-1F}x \in C_{\mathbf{G}}(t)^0$ (see 3.5.3). Applying the Lang–Steinberg theorem in the group $C_{\mathbf{G}}(t)^0$ we may write $x^{-1F}x = z^{-1F}z$, where $z \in C_{\mathbf{G}}(t)^0$. If we put then $h = xz^{-1}$ we have $h \in \mathbf{G}^F$, ${}^h t = s^{-1}$ and $\varphi(h, z) = x$, whence the surjectivity of φ.

The map φ is not injective, but $\varphi(h, z) = \varphi(h', z')$ if and only if $h^{-1}h' = zz'^{-1} \in C_{\mathbf{G}}(t)^{0F}$, thus φ induces an isomorphism

$$\{h \in \mathbf{G}^F \mid {}^h t = s^{-1}\} \times_{C_{\mathbf{G}}(t)^{0F}} \{z \in C_{\mathbf{G}}(t)^0 \mid z^{-1F}z \in \mathbf{U}\} \xrightarrow{\sim} \mathcal{L}^{-1}(\mathbf{U})^{(s,t)},$$

which may be written

$$\mathcal{L}^{-1}(\mathbf{U})^{(s,t)} \simeq \coprod_{\{h \in \mathbf{G}^F/C_{\mathbf{G}}(t)^{0F} \mid {}^h t = s^{-1}\}} \{z \in C_{\mathbf{G}}(t)^0 \mid z^{-1F}z \in \mathbf{U}\}_h$$

$$= \coprod_{\{h \in \mathbf{G}^F/C_{\mathbf{G}}(t)^{0F} \mid {}^h t = s^{-1}\}} \mathcal{L}_{C_{\mathbf{G}}(t)^0}^{-1}(\mathbf{U} \cap C_{\mathbf{G}}(t)^0)_h,$$

where $(u, v) \in C_{\mathbf{G}}(s)_u^{0F} \times C_{\mathbf{L}}(t)_u^{0F}$ acts on the piece indexed by h by $z \mapsto {}^{h^{-1}}uzv$.

We thus get

$$\text{Trace}((g,l) \mid H_c^*(\mathcal{L}^{-1}(\mathbf{U})))$$
$$= |C_\mathbf{G}(t)^{0F}|^{-1} \sum_{\{h \in \mathbf{G}^F \mid {}^h t = s^{-1}\}} \text{Trace}(({}^{h^{-1}}u, v) \mid H_c^*(\mathcal{L}^{-1}_{C_\mathbf{G}(t)^0}(\mathbf{U} \cap C_\mathbf{G}(t)^0))),$$

whence the lemma. □

We now prove Proposition 10.1.2. From Proposition 9.1.6 we have

$$(R_\mathbf{L}^\mathbf{G}\chi)(g) = |\mathbf{L}^F|^{-1} \sum_{l \in \mathbf{L}^F} \text{Trace}((g,l) \mid H_c^*(\mathcal{L}^{-1}(\mathbf{U})))\chi(l^{-1})$$

and

$$({}^*R_\mathbf{L}^\mathbf{G}\psi)(l) = |\mathbf{G}^F|^{-1} \sum_{g \in \mathbf{G}^F} \text{Trace}((g,l) \mid H_c^*(\mathcal{L}^{-1}(\mathbf{U})))\psi(g^{-1}).$$

Applying Lemma 10.1.3 in both sums, and interchanging the sums, we get from the first formula and $(**)$

$$(R_\mathbf{L}^\mathbf{G}\chi)(g)$$
$$= |\mathbf{L}^F|^{-1}|C_\mathbf{G}(s)^{0F}|^{-1} \sum_{\{h \in \mathbf{G}^F \mid {}^{h^{-1}}s \in \mathbf{L}^F\}} |C_{{}^h\mathbf{L}}(s)^{0F}| \sum_{v \in C_\mathbf{L}({}^{h^{-1}}s)_u^{0F}} Q_{C_{{}^h\mathbf{L}}(s)^0}^{C_\mathbf{G}(s)^0}(u, {}^h v)\chi({}^{h^{-1}}sv^{-1})$$

and from the second formula and $(*)$

$$({}^*R_\mathbf{L}^\mathbf{G}\psi)(l) = \frac{|C_\mathbf{L}(t)^{0F}|}{|\mathbf{G}^F||C_\mathbf{G}(t)^{0F}|} \sum_{h \in \mathbf{G}^F} \sum_{u \in C_\mathbf{G}({}^h t)_u^{0F}} Q_{C_\mathbf{L}(t)^0}^{C_\mathbf{G}(t)^0}({}^{h^{-1}}u, v)\psi({}^h tu^{-1}).$$

We then get the result by changing the variable over which we sum to ${}^h v^{-1}$ in the first formula and to ${}^{h^{-1}}u^{-1}$ in the second. □

Corollary 10.1.4 *If as in 10.1.2 $g = su$ is a Jordan decomposition, and $\mathbf{M}$ is an F-stable connected reductive subgroup of $\mathbf{G}$ such that $\mathbf{M} \supset C_\mathbf{G}(s)^0$, we have*

$$(R_\mathbf{L}^\mathbf{G}\chi)(g) = \frac{1}{|C_{\mathbf{M}^F}(s)||\mathbf{L}^F|} \sum_{\{h \in \mathbf{G}^F \mid s \in {}^h\mathbf{L}\}} \frac{|({}^h\mathbf{L} \cap \mathbf{M})^F|}{|\text{class}_{\mathbf{M}^F}(s) \cap {}^h\mathbf{L}|} R_{{}^h\mathbf{L}\cap\mathbf{M}}^\mathbf{M}({}^h\chi)(g).$$

When $\mathbf{M} = C_\mathbf{G}(s)^0$ we may rewrite this as

$$(R_\mathbf{L}^\mathbf{G}\chi)(g) = \sum_{h \in \mathbf{M}^F \backslash S(\mathbf{M},\mathbf{L})^F / \mathbf{L}^F} R_{{}^h\mathbf{L}\cap\mathbf{M}}^\mathbf{M}({}^h\chi)(g)$$

where $S(\mathbf{M},\mathbf{L}) = \{h \in \mathbf{G} \mid \mathbf{M} \cap {}^h\mathbf{L} \text{ contains a maximal torus of } \mathbf{G}\}$.

Proof Note that any maximal torus containing s lies in $\mathbf{M}$, so that $\mathbf{M}$ is a subgroup of maximal rank; thus by 3.4.10(iii) $^h\mathbf{L} \cap \mathbf{M}$ is a Levi subgroup of a parabolic subgroup of $\mathbf{M}$ so that $R^{\mathbf{M}}_{^h\mathbf{L}\cap\mathbf{M}}$ makes sense. Let us show the first statement when $\mathbf{M} = C_{\mathbf{G}}(s)^0$. In the case where $s \in Z(\mathbf{G})$, Lemma 10.1.3 $(*)$ reduces to

$$|\mathbf{L}^F|^{-1} \operatorname{Trace}((g,l) \mid H^*_c(\mathcal{L}^{-1}(\mathbf{U})))$$
$$= \begin{cases} |\mathbf{G}^F|^{-1} \sum_{h\in\mathbf{G}^F} Q^{\mathbf{G}}_{\mathbf{L}}(^{h^{-1}}u,v) = Q^{\mathbf{G}}_{\mathbf{L}}(u,v) & \text{if } s = t^{-1}, \\ 0 & \text{otherwise,} \end{cases}$$

where the last equality in the first line comes from the fact that $Q^{\mathbf{G}}_{\mathbf{L}}$ is the restriction to unipotent elements of a central function on $\mathbf{G}^F \times \mathbf{L}^F$. So in that case 9.1.6 gives

$$(R^{\mathbf{G}}_{\mathbf{L}}\chi)(g) = \sum_{v\in\mathbf{L}^F_u} Q^{\mathbf{G}}_{\mathbf{L}}(u,v^{-1})\chi(sv). \tag{$*$}$$

Applying this formula to $R^{C_{\mathbf{G}}(s)^0}_{C_{^h\mathbf{L}}(s)^0}(^h\chi)(g)$ in the right-hand side of the equality we want to prove, we see that it is equivalent to 10.1.2(i) thus we have proved the first formula of the statement when $\mathbf{M} = C_{\mathbf{G}}(s)^0$.

For a more general $\mathbf{M}$, we use the formula just obtained replacing $\mathbf{G}$ by $\mathbf{M}$ and $\mathbf{L}$ by $^h\mathbf{L} \cap \mathbf{M}$ to express each term on the right-hand side of the sought formula as:

$$R^{\mathbf{M}}_{^h\mathbf{L}\cap\mathbf{M}}(^h\chi)(g)$$
$$= |C_{\mathbf{G}}(s)^{0F}|^{-1}|(^h\mathbf{L}\cap\mathbf{M})^F|^{-1} \sum_{\{h'\in\mathbf{M}^F | s\in^{h'h}\mathbf{L}\}} |(C_{\mathbf{G}}(s)^0 \cap {}^{h'h}\mathbf{L})^F| R^{C_{\mathbf{G}}(s)^0}_{C_{\mathbf{G}}(s)^0\cap^{h'h}\mathbf{L}} {}^{h'h}\chi(g).$$

Hence the sum over h in the right-hand side of the formula to prove becomes

$$\sum \frac{|(C_{\mathbf{G}}(s)^0 \cap {}^{h'h}\mathbf{L})^F|}{|\operatorname{class}_{\mathbf{M}^F}(s) \cap {}^h\mathbf{L}||C_{\mathbf{G}}(s)^{0F}|} R^{C_{\mathbf{G}}(s)^0}_{C_{\mathbf{G}}(s)^0\cap^{h'h}\mathbf{L}} {}^{h'h}\chi(g),$$

where the sum is over $\{(h,h') \in \mathbf{G}^F \times \mathbf{M}^F \mid s \in {}^h\mathbf{L} \cap {}^{h'h}\mathbf{L}\}$.

Since $|\operatorname{class}_{\mathbf{M}^F}(s) \cap {}^h\mathbf{L}| = |\operatorname{class}_{\mathbf{M}^F}(s) \cap {}^{h'h}\mathbf{L}|$ we can sum over $h'' = h'h$ with the condition $s \in {}^{h''}\mathbf{L}$, each term occurring with cardinality the number of $(h,h') \in \mathbf{G}^F \times \mathbf{M}^F$ such that $s \in {}^h\mathbf{L}$ and $h'h = h''$, or equivalently the number of h' such that $^{h'}s \in {}^{h''}\mathbf{L}$. For each $^{h'}s$ in $\operatorname{class}_{\mathbf{M}^F}(s) \cap {}^{h''}\mathbf{L}$ there are $|C_{\mathbf{M}^F}(^{h'}s)| = |C_{\mathbf{M}^F}(s)|$ such h'. Putting things together, the right-hand side of the equality we want to prove becomes its particular case for $\mathbf{M} = C_{\mathbf{G}}(s)^0$, whence the result.

We finally show the last statement of the corollary. Any h such that $s \in {}^h\mathbf{L}$ is in $\mathcal{S}(\mathbf{M},\mathbf{L})$ since $C_{^h\mathbf{L}}(s)^0$ contains a maximal torus of $\mathbf{G}$ which is in

$C_{\mathbf{G}}(s)^0 = \mathbf{M}$. Conversely, s is in all maximal tori of $\mathbf{M}$ so for any element of $\mathcal{S}(\mathbf{M},\mathbf{L})$ we have $^h\mathbf{L} \ni s$. We conclude by observing that the coefficients in the first statement count the number of $h \in \mathbf{G}^F$ which have a given image in $\mathbf{M}^F\backslash\mathcal{S}(\mathbf{M},\mathbf{L})^F/\mathbf{L}^F$. □

We now use the character formula to extend 7.3.6 to the Deligne–Lusztig induction.

Proposition 10.1.5 *Let $\mathbf{L}$ be an F-stable Levi subgroup of $\mathbf{G}$, let $\psi \in \mathrm{Irr}(\mathbf{G}^F)$ and let $l = su$ be the Jordan decomposition of some element $l \in \mathbf{L}^F$. Then*

$$((\mathrm{Res}^{\mathbf{L}^F}_{C_{\mathbf{L}}(s)^{0F}} \circ {}^*R^{\mathbf{G}}_{\mathbf{L}})\psi)(l) = (({}^*R^{C_{\mathbf{G}}(s)^0}_{C_{\mathbf{L}}(s)^0} \circ \mathrm{Res}^{\mathbf{G}^F}_{C_{\mathbf{G}}(s)^{0F}})\psi)(l).$$

Proof This results from the fact that in formula 10.1.2(ii) the right-hand side does not change if we replace $\mathbf{G}$ by $C_{\mathbf{G}}(s)^0$ and $\mathbf{L}$ by $C_{\mathbf{L}}(s)^0$. □

The next proposition and its corollary extend 7.3.4 and 7.3.5.

Proposition 10.1.6 *Let $f \in C(\mathbf{G}^F)_{p'}$; then, for any F-stable Levi subgroup $\mathbf{L}$ of $\mathbf{G}$ and any $\lambda \in C(\mathbf{L}^F)$ (resp. $\gamma \in C(\mathbf{G}^F)$), we have*

$$R^{\mathbf{G}}_{\mathbf{L}}(\lambda.\mathrm{Res}^{\mathbf{G}^F}_{\mathbf{L}^F}f) = (R^{\mathbf{G}}_{\mathbf{L}}\lambda).f, \tag{i}$$

$$({}^*R^{\mathbf{G}}_{\mathbf{L}}\gamma).\mathrm{Res}^{\mathbf{G}^F}_{\mathbf{L}^F}f = {}^*R^{\mathbf{G}}_{\mathbf{L}}(\gamma.f). \tag{ii}$$

Proof Proposition 10.1.2 gives

$$R^{\mathbf{G}}_{\mathbf{L}}(\lambda.\mathrm{Res}^{\mathbf{G}^F}_{\mathbf{L}^F}f)(g)$$
$$= |\mathbf{L}^F|^{-1}|C_{\mathbf{G}}(s)^{0F}|^{-1}\sum_{\{h\in\mathbf{G}^F|s\in{}^h\mathbf{L}\}}|C_{{}^h\mathbf{L}}(s)^{0F}|\sum_{v\in C_{{}^h\mathbf{L}}(s)^{0F}_u}Q^{C_{\mathbf{G}}(s)^0}_{C_{{}^h\mathbf{L}}(s)^0}(u,v^{-1})\,{}^h\lambda(sv)\,{}^hf(sv),$$

which gives (i) using ${}^hf(sv) = f(sv) = f(s) = f(g)$; equality (ii) is proved similarly (it can also be obtained from (i) by adjunction). □

Corollary 10.1.7 *For $f \in C(\mathbf{G}^F)_{p'}$, we have ${}^*R^{\mathbf{G}}_{\mathbf{L}}f = \mathrm{Res}^{\mathbf{G}^F}_{\mathbf{L}^F}f$.*

Proof This results from the special case of 10.1.6(ii) where $\gamma = \mathbf{1}$, and the remark that ${}^*R^{\mathbf{G}}_{\mathbf{L}}(\mathbf{1}_{\mathbf{G}^F}) = \mathbf{1}_{\mathbf{L}^F}$. Let us prove this last fact: by definition ${}^*R^{\mathbf{G}}_{\mathbf{L}}(\mathbf{1}_{\mathbf{G}^F})$ is the character afforded by the $\overline{\mathbb{Q}}_\ell\mathbf{L}^F$-module $H^*_c(\mathcal{L}^{-1}(\mathbf{U}))^{\mathbf{G}^F}$; by 8.1.10 this module is isomorphic to $H^*_c(\mathcal{L}^{-1}(\mathbf{U})/\mathbf{G}^F)$, and the Lang map induces an isomorphism from $\mathcal{L}^{-1}(\mathbf{U})/\mathbf{G}^F$ to $\mathbf{U}$, whence the result by 8.1.11. □

10.2 Uniform Functions

As discussed in the proof of 7.2.4 the validity of the Mackey formula when one of the Levi subgroups is a torus (see 9.2.6) allows us to state the following analogue of 7.2.4, using the fact that $D_{\mathbf{T}} = \mathrm{Id}$ for a torus $\mathbf{T}$.

Theorem 10.2.1 *For any F-stable maximal torus $\mathbf{T}$ of $\mathbf{G}$ we have*

$$\varepsilon_{\mathbf{G}} D_{\mathbf{G}} \circ R_{\mathbf{T}}^{\mathbf{G}} = \varepsilon_{\mathbf{T}} R_{\mathbf{T}}^{\mathbf{G}}.$$

As in 7.4.2 we deduce, using the value of $^*R_{\mathbf{T}}^{\mathbf{G}}(\mathbf{1})$ given in the proof of 10.1.7 instead of 7.2.12 that

$$^*R_{\mathbf{T}}^{\mathbf{G}} \mathrm{St}_{\mathbf{G}^F} = \varepsilon_{\mathbf{G}} \varepsilon_{\mathbf{T}} \, \mathrm{St}_{\mathbf{T}^F} = \varepsilon_{\mathbf{G}} \varepsilon_{\mathbf{T}} \mathbf{1}_{\mathbf{T}^F}.$$

We can now give the dimension of the (virtual) characters $R_{\mathbf{T}}^{\mathbf{G}}(\theta)$.

Proposition 10.2.2 *For any F-stable maximal torus $\mathbf{T}$ of $\mathbf{G}$ and any $\theta \in \mathrm{Irr}(\mathbf{T}^F)$, we have* $\dim R_{\mathbf{T}}^{\mathbf{G}}(\theta) = \varepsilon_{\mathbf{G}} \varepsilon_{\mathbf{T}} |\mathbf{G}^F|_{p'} |\mathbf{T}^F|^{-1}$.

Proof Taking the scalar product with θ of the equality $^*R_{\mathbf{T}}^{\mathbf{G}} \mathrm{St}_{\mathbf{G}^F} = \varepsilon_{\mathbf{G}} \varepsilon_{\mathbf{T}} \mathbf{1}_{\mathbf{T}^F}$, we get $\langle R_{\mathbf{T}}^{\mathbf{G}}(\theta), \mathrm{St}_{\mathbf{G}^F} \rangle_{\mathbf{G}^F} = \varepsilon_{\mathbf{T}} \varepsilon_{\mathbf{G}} \delta_{1,\theta}$, whence $\langle \sum_\theta R_{\mathbf{T}}^{\mathbf{G}}(\theta), \mathrm{St}_{\mathbf{G}^F} \rangle_{\mathbf{G}^F} = \varepsilon_{\mathbf{T}} \varepsilon_{\mathbf{G}}$. But $\sum_\theta R_{\mathbf{T}}^{\mathbf{G}}(\theta)$ is the character afforded by the module $H_c^*(\mathcal{L}^{-1}(\mathbf{U}))$, where $\mathbf{U}$ is the unipotent radical of some Borel subgroup containing $\mathbf{T}$. By 8.1.15 this character vanishes on all non-unipotent elements, as a non-trivial semi-simple element has no fixed points on $\mathcal{L}^{-1}(\mathbf{U})$. Since $\mathrm{St}_{\mathbf{G}^F}$ vanishes outside semi-simple elements, the scalar product above reduces to

$$|\mathbf{G}^F|^{-1} \mathrm{St}_{\mathbf{G}^F}(1) \sum_\theta \dim(R_{\mathbf{T}}^{\mathbf{G}}(\theta)).$$

But by 10.1.2 $\dim(R_{\mathbf{T}}^{\mathbf{G}}(\theta))$ does not depend on θ. This gives the result after replacing $\mathrm{St}_{\mathbf{G}^F}(1)$ by its value $|\mathbf{G}^F|_p$. $\qquad\qquad \square$

To get further properties of the Lusztig functor, in particular the dimension of $R_{\mathbf{L}}^{\mathbf{G}}(\chi)$ and the analogue of 7.4.7, we first need to prove that the identity and the regular representations are both linear combinations of Deligne–Lusztig characters. We will use the following terminology.

Definition 10.2.3   We call **uniform functions** the class functions on $\mathbf{G}^F$ that are linear combinations of Deligne–Lusztig characters.

Note that by 9.3.1 the $R_{\mathbf{T}}^{\mathbf{G}}(\theta)$ where $(\mathbf{T}, \theta)$ runs over $\mathbf{G}^F$-conjugacy classes of pairs of an F-stable maximal torus $\mathbf{T}$ and of $\theta \in \mathrm{Irr}(\mathbf{T}^F)$ form an orthogonal (but not orthonormal) basis of the uniform functions.

In the formulae up to the end of this chapter, we will let $\mathcal{T}$ denote the set of all F-stable maximal tori of $\mathbf{G}$, and $[\mathcal{T}/\mathbf{G}^F]$ denote a set of representatives of $\mathbf{G}^F$-conjugacy classes of F-stable maximal tori. We fix a quasi-split $\mathbf{T}_1 \in \mathcal{T}$ and we put $W = W(\mathbf{T}_1)$. We recall that $\mathcal{T}/\mathbf{G}^F$ is in one-to-one correspondence with the F-classes of W (see 4.2.22). Finally for each $w \in W$ we choose some F-stable maximal torus $\mathbf{T}_w$ of type w.

Proposition 10.2.4 *Denote by p (or $p_\mathbf{G}$ if needed to specify the group) the orthogonal projection of class functions onto the subspace of uniform functions. Then:*

$$p = |W|^{-1} \sum_{w \in W} R_{\mathbf{T}_w}^{\mathbf{G}} \circ {}^* R_{\mathbf{T}_w}^{\mathbf{G}} = \sum_{\mathbf{T} \in [\mathcal{T}/\mathbf{G}^F]} |W(\mathbf{T})^F|^{-1} R_{\mathbf{T}}^{\mathbf{G}} \circ {}^* R_{\mathbf{T}}^{\mathbf{G}}$$

$$= |\mathbf{G}^F|^{-1} \sum_{\mathbf{T} \in \mathcal{T}} |\mathbf{T}^F| R_{\mathbf{T}}^{\mathbf{G}} \circ {}^* R_{\mathbf{T}}^{\mathbf{G}}.$$

Proof The equality of the three expressions for p results from a straightforward computation. Let us check that the middle one is a projector on uniform functions. Since $p(\chi)$ is clearly uniform for any $\chi \in \mathrm{Irr}(\mathbf{G}^F)$, it is enough to check that for any $\mathbf{T} \in [\mathcal{T}/\mathbf{G}^F]$ and any $\theta \in \mathrm{Irr}(\mathbf{T}^F)$, we have $\langle \chi, R_{\mathbf{T}}^{\mathbf{G}}(\theta) \rangle_{\mathbf{G}^F} = \langle p(\chi), R_{\mathbf{T}}^{\mathbf{G}}(\theta) \rangle_{\mathbf{G}^F}$. We have

$$\langle p(\chi), R_{\mathbf{T}}^{\mathbf{G}}(\theta) \rangle_{\mathbf{G}^F} = \langle \sum_{\mathbf{T}' \in [\mathcal{T}/\mathbf{G}^F]} |W(\mathbf{T}')^F|^{-1} R_{\mathbf{T}'}^{\mathbf{G}} {}^* R_{\mathbf{T}'}^{\mathbf{G}} \chi, R_{\mathbf{T}}^{\mathbf{G}}(\theta) \rangle_{\mathbf{G}^F}$$

$$= \sum_{\mathbf{T}' \in [\mathcal{T}/\mathbf{G}^F]} \langle |W(\mathbf{T}')^F|^{-1} {}^* R_{\mathbf{T}'}^{\mathbf{G}} \chi, {}^* R_{\mathbf{T}'}^{\mathbf{G}} R_{\mathbf{T}}^{\mathbf{G}}(\theta) \rangle_{\mathbf{G}^F}$$

but, by 9.3.1 we have:

$$ {}^* R_{\mathbf{T}'}^{\mathbf{G}} R_{\mathbf{T}}^{\mathbf{G}}(\theta) = \begin{cases} \sum_{w \in W(\mathbf{T})^F} {}^w \theta & \text{if } \mathbf{T} = \mathbf{T}' \\ 0 & \text{if } \mathbf{T} \text{ and } \mathbf{T}' \text{ are not } \mathbf{G}^F\text{-conjugate} \end{cases} $$

so

$$\langle p(\chi), R_{\mathbf{T}}^{\mathbf{G}}(\theta) \rangle_{\mathbf{G}^F} = \langle {}^* R_{\mathbf{T}}^{\mathbf{G}}(\chi), |W(\mathbf{T})^F|^{-1} \sum_{w \in W(\mathbf{T})^F} {}^w \theta \rangle_{\mathbf{G}^F} = \langle \chi, R_{\mathbf{T}}^{\mathbf{G}}(\theta) \rangle_{\mathbf{G}^F}$$

the rightmost equality since for any θ we have $R_{\mathbf{T}}^{\mathbf{G}}({}^w \theta) = R_{\mathbf{T}}^{\mathbf{G}}(\theta)$. □

Proposition 10.2.5

(i) $\mathbf{1}_{\mathbf{G}^F}$ *is a uniform function; we have*

$$\mathbf{1}_{\mathbf{G}^F} = |W|^{-1} \sum_{w \in W} R_{\mathbf{T}_w}^{\mathbf{G}}(\mathbf{1}) = \sum_{\mathbf{T} \in [\mathcal{T}/\mathbf{G}^F]} |W(\mathbf{T})^F|^{-1} R_{\mathbf{T}}^{\mathbf{G}}(\mathbf{1})$$

$$= |\mathbf{G}^F|^{-1} \sum_{\mathbf{T} \in \mathcal{T}} |\mathbf{T}^F| R_{\mathbf{T}}^{\mathbf{G}}(\mathbf{1}),$$

(ii) $\mathrm{St}_{\mathbf{G}^F}$ *is a uniform function; we have*

$$\mathrm{St}_{\mathbf{G}^F} = |W|^{-1} \sum_{w \in W} \varepsilon_{\mathbf{G}} \varepsilon_{\mathbf{T}_w} R_{\mathbf{T}_w}^{\mathbf{G}}(1) = \sum_{\mathbf{T} \in [\mathcal{T}/\mathbf{G}^F]} |W(\mathbf{T})^F|^{-1} \varepsilon_{\mathbf{G}} \varepsilon_{\mathbf{T}} R_{\mathbf{T}}^{\mathbf{G}}(1)$$

$$= |\mathbf{G}^F|^{-1} \sum_{\mathbf{T} \in \mathcal{T}} |\mathbf{T}^F| \varepsilon_{\mathbf{G}} \varepsilon_{\mathbf{T}} R_{\mathbf{T}}^{\mathbf{G}}(1).$$

Proof By 10.2.1, (ii) is equivalent to (i). We prove (i). Since by 10.1.7 we have $^*R_{\mathbf{T}}^{\mathbf{G}}(\mathbf{1}_{\mathbf{G}^F}) = \mathbf{1}_{\mathbf{T}^F}$, the three expressions in (i) above all represent $p(\mathbf{1}_{\mathbf{G}^F})$. It is enough to check that $\mathbf{1}_{\mathbf{G}^F}$ has the same scalar product with one of these expressions as with itself. But indeed we have

$$\langle \mathbf{1}, |W|^{-1} \sum_{w \in W} R_{\mathbf{T}_w}^{\mathbf{G}}(\mathbf{1}) \rangle_{\mathbf{G}^F} = |W|^{-1} \sum_{w \in W} \langle ^*R_{\mathbf{T}_w}^{\mathbf{G}}(\mathbf{1}), \mathbf{1} \rangle_{\mathbf{T}_w^F} = 1$$

using again that $^*R_{\mathbf{T}_w}^{\mathbf{G}}(\mathbf{1}_{\mathbf{G}^F}) = \mathbf{1}_{\mathbf{T}_w^F}$. $\qquad\square$

Corollary 10.2.6 *The character* $\mathrm{reg}_{\mathbf{G}^F}$ *of the regular representation of* $\mathbf{G}^F$ *is a uniform function; we have*

$$\mathrm{reg}_{\mathbf{G}^F} = |W|^{-1} \sum_{w \in W} \dim(R_{\mathbf{T}_w}^{\mathbf{G}}(\mathbf{1})) R_{\mathbf{T}_w}^{\mathbf{G}}(\mathrm{reg}_{\mathbf{T}_w^F})$$

$$= |\mathbf{G}^F|_p^{-1} \sum_{\mathbf{T} \in \mathcal{T}} \varepsilon_{\mathbf{G}} \varepsilon_{\mathbf{T}} R_{\mathbf{T}}^{\mathbf{G}}(\mathrm{reg}_{\mathbf{T}^F}) = |\mathbf{G}^F|_p^{-1} \sum_{\substack{\mathbf{T} \in \mathcal{T} \\ \theta \in \mathrm{Irr}(\mathbf{T}^F)}} \varepsilon_{\mathbf{G}} \varepsilon_{\mathbf{T}} R_{\mathbf{T}}^{\mathbf{G}}(\theta).$$

Proof Again, the equality of the three expressions is straightforward. Let us get the first one. As seen in the proof of 7.4.5 $\mathrm{reg}_{\mathbf{G}^F} = \mathrm{St}_{\mathbf{G}^F} \gamma_p$. Using the first formula for $\mathrm{St}_{\mathbf{G}^F}$ in 10.2.5(ii), is enough to see that $\varepsilon_{\mathbf{G}} \varepsilon_{\mathbf{T}_w} R_{\mathbf{T}_w}^{\mathbf{G}}(\mathbf{1}) \gamma_p = \dim R_{\mathbf{T}_w}^{\mathbf{G}}(\mathbf{1}) R_{\mathbf{T}_w}^{\mathbf{G}}(\mathrm{reg}_{\mathbf{T}_w^F})$. This comes from the equality $R_{\mathbf{T}_w}^{\mathbf{G}}(\mathbf{1}) \gamma_p = R_{\mathbf{T}_w}^{\mathbf{G}}(\mathrm{Res}_{\mathbf{T}_w^F}^{\mathbf{G}^F}(\gamma_p))$ given by 10.1.6, from the fact that $\mathrm{Res}_{\mathbf{T}_w^F}^{\mathbf{G}^F}(\gamma_p)$ has value $|\mathbf{G}^F|_{p'}$ at 1 and 0 elsewhere, so is equal to $|\mathbf{G}^F|_{p'} |\mathbf{T}_w^F|^{-1} \mathrm{reg}_{\mathbf{T}_w^F}$, and from 10.2.2. $\qquad\square$

The next corollary is in Steinberg (1968, 14.14) with a completely different proof. We follow here the arguments of Digne and Michel (1987).

Corollary 10.2.7 *The number of F-stable maximal tori of* $\mathbf{G}$ *is equal to* $|\mathbf{G}^F|_p^2$.

Proof For any F-stable maximal torus $\mathbf{T}$, we have $\mathrm{Ind}_{\mathbf{T}^F}^{\mathbf{G}^F}(\mathrm{reg}_{\mathbf{T}^F}) = \mathrm{reg}_{\mathbf{G}^F} = |\mathcal{T}|^{-1} \sum_{\mathbf{T} \in \mathcal{T}} \mathrm{Ind}_{\mathbf{T}^F}^{\mathbf{G}^F} \mathrm{reg}_{\mathbf{T}^F}$. We now use Lemma 10.2.8.

Lemma 10.2.8 *For any F-stable maximal torus* $\mathbf{T}$ *we have*

$$\mathrm{Ind}_{\mathbf{T}^F}^{\mathbf{G}^F}(\mathrm{reg}_{\mathbf{T}^F}) = \varepsilon_{\mathbf{T}} \varepsilon_{\mathbf{G}} R_{\mathbf{T}}^{\mathbf{G}}(\mathrm{reg}_{\mathbf{T}^F}) \mathrm{St}_{\mathbf{G}^F}.$$

Proof   This is just the special case of 10.2.10 below for $R_{\mathbf{T}}^{\mathbf{G}}$. We can give this lemma because the proof of 10.2.10 in this special case needs only 10.2.2 above and not 10.2.9.   □

Applying this lemma, we get $\mathrm{reg}_{\mathbf{G}^F} = |\mathcal{T}|^{-1}\mathrm{St}_{\mathbf{G}^F}\sum_{\mathbf{T}\in\mathcal{T}}\varepsilon_{\mathbf{T}}\varepsilon_{\mathbf{G}}R_{\mathbf{T}}^{\mathbf{G}}(\mathrm{reg}_{\mathbf{T}^F})$ and the right-hand side of this equality is equal to $|\mathcal{T}|^{-1}\mathrm{St}_{\mathbf{G}^F}|\mathbf{G}^F|_p\,\mathrm{reg}_{\mathbf{G}^F}$ by 10.2.6. Taking the value at 1 of both sides, we get the result, as $\mathrm{St}_{\mathbf{G}^F}(1) = |\mathbf{G}^F|_p$.   □

We will now give, using 10.2.2, the analogue for $R_{\mathbf{L}}^{\mathbf{G}}$ of 10.2.2 and 10.2.8. These properties could be proved directly, using the same arguments as for 10.2.2 and 10.2.8, if we knew the validity of the Mackey formula in general, and thus the truth of 7.2.4 for a general Lusztig functor.

Proposition 10.2.9   *Let* $\mathbf{L}$ *be a Levi subgroup of* $\mathbf{G}$, *and let* $\varphi \in \mathrm{Irr}(\mathbf{L}^F)$; *then*

$$\dim(R_{\mathbf{L}}^{\mathbf{G}}\varphi) = \varepsilon_{\mathbf{G}}\varepsilon_{\mathbf{L}}|\mathbf{G}^F/\mathbf{L}^F|_{p'}\dim(\varphi).$$

Proof   We have $\varphi(1) = \langle\varphi,\mathrm{reg}_{\mathbf{L}^F}\rangle_{\mathbf{L}^F}$ and similarly

$$(R_{\mathbf{L}}^{\mathbf{G}}\varphi)(1) = \langle R_{\mathbf{L}}^{\mathbf{G}}\varphi,\mathrm{reg}_{\mathbf{G}^F}\rangle_{\mathbf{G}^F} = \langle R_{\mathbf{L}}^{\mathbf{G}}\varphi,|\mathbf{G}^F|_p^{-1}\sum_{\mathbf{T}\in\mathcal{T}}\varepsilon_{\mathbf{G}}\varepsilon_{\mathbf{T}}R_{\mathbf{T}}^{\mathbf{G}}(\mathrm{reg}_{\mathbf{T}^F})\rangle_{\mathbf{G}^F},$$

the last equality by 10.2.6. We now use adjunction to transform the last term above, then apply to it the Mackey formula 9.2.6, and then again take adjoints; we get

$$(R_{\mathbf{L}}^{\mathbf{G}}\varphi)(1) = |\mathbf{G}^F|_p^{-1}\sum_{\mathbf{T}\in\mathcal{T}}\varepsilon_{\mathbf{T}}\varepsilon_{\mathbf{G}}\langle {}^*R_{\mathbf{T}}^{\mathbf{G}}\circ R_{\mathbf{L}}^{\mathbf{G}}\varphi,\mathrm{reg}_{\mathbf{T}^F}\rangle_{\mathbf{T}^F}$$

$$= |\mathbf{G}^F|_p^{-1}\sum_{\mathbf{T}\in\mathcal{T}}\varepsilon_{\mathbf{T}}\varepsilon_{\mathbf{G}}\langle\sum_{\mathbf{L}^F\backslash\{x\in\mathbf{G}^F|{}^x\mathbf{T}\subset\mathbf{L}\}}\mathrm{ad}\,x^{-1}\circ{}^*R_{{}^x\mathbf{T}}^{\mathbf{L}}\varphi,\mathrm{reg}_{\mathbf{T}^F}\rangle_{\mathbf{T}^F}$$

$$= |\mathbf{G}^F|_p^{-1}\sum_{\mathbf{T}\in\mathcal{T}}\varepsilon_{\mathbf{T}}\varepsilon_{\mathbf{G}}\sum_{\mathbf{L}^F\backslash\{x\in\mathbf{G}^F|{}^x\mathbf{T}\subset\mathbf{L}\}}\langle\varphi,R_{{}^x\mathbf{T}}^{\mathbf{L}}\,{}^x\mathrm{reg}_{\mathbf{T}^F}\rangle_{\mathbf{L}^F}.$$

In the last expression we may take as a new variable ${}^x\mathbf{T}$ which is equivalent to summing over all F-stable maximal tori of $\mathbf{L}$, if we multiply the expression by $|\mathbf{G}^F|/|\mathbf{L}^F|$. We get $|\mathbf{G}^F|_{p'}|\mathbf{L}^F|^{-1}\varepsilon_{\mathbf{G}}\sum_{\mathbf{T}\subset\mathbf{L}}\varepsilon_{\mathbf{T}}\langle\varphi,R_{\mathbf{T}}^{\mathbf{L}}\,\mathrm{reg}_{\mathbf{T}^F}\rangle_{\mathbf{L}^F}$, which is equal by 10.2.6 to $\varepsilon_{\mathbf{G}}\varepsilon_{\mathbf{L}}|\mathbf{G}^F|_{p'}/|\mathbf{L}^F|_{p'}\langle\varphi,\mathrm{reg}_{\mathbf{L}^F}\rangle_{\mathbf{L}^F}$, whence the result.   □

Corollary 10.2.10   *For any* $\varphi \in \mathrm{Irr}(\mathbf{L}^F)$ *(resp.* $\psi \in \mathrm{Irr}(\mathbf{G}^F)$) *we have*

(i) $(R_{\mathbf{L}}^{\mathbf{G}}\varphi).\varepsilon_{\mathbf{G}}\mathrm{St}_{\mathbf{G}^F} = \mathrm{Ind}_{\mathbf{L}^F}^{\mathbf{G}^F}(\varphi.\varepsilon_{\mathbf{L}}\,\mathrm{St}_{\mathbf{L}^F}).$

(ii) ${}^*R_{\mathbf{L}}^{\mathbf{G}}(\psi.\varepsilon_{\mathbf{G}}\mathrm{St}_{\mathbf{G}^F}) = \varepsilon_{\mathbf{L}}\,\mathrm{St}_{\mathbf{L}^F}.\,\mathrm{Res}_{\mathbf{L}^F}^{\mathbf{G}^F}\psi.$

Proof It is enough to prove (ii) as (i) is its adjoint. Applying 10.1.2, and using the value of $\mathrm{St}_{\mathbf{L}^F}$, the truth of (ii) is equivalent to the fact that for any $\psi \in \mathrm{Irr}(\mathbf{G}^F)$ and any semi-simple element $s \in \mathbf{L}^F$ we have

$$\varepsilon_{C_{\mathbf{G}}(s)^0}|C_{\mathbf{G}}(s)^{0F}|_{p'}^{-1}|C_{\mathbf{L}}(s)^{0F}|\psi(s)Q_{C_{\mathbf{L}}(s)^0}^{C_{\mathbf{G}}(s)^0}(1, v^{-1})$$

$$= \begin{cases} 0 & \text{if } v \neq 1, \\ \varepsilon_{C_{\mathbf{L}}(s)^0}|C_{\mathbf{L}}(s)^{0F}|_p\psi(s) & \text{if } v = 1 \end{cases} \quad (*)$$

(we have used the fact that the Steinberg character vanishes outside semi-simple elements).

On the other hand, 10.2.9 gives that for any $\varphi \in \mathrm{Irr}(\mathbf{L}^F)$ we have

$$\langle R_{\mathbf{L}}^{\mathbf{G}}\varphi, \mathrm{reg}_{\mathbf{G}^F}\rangle_{\mathbf{G}^F} = \varepsilon_{\mathbf{G}}\varepsilon_{\mathbf{L}}|\mathbf{G}^F/\mathbf{L}^F|_{p'}\langle\varphi, \mathrm{reg}_{\mathbf{L}^F}\rangle_{\mathbf{L}^F}$$

whence

$${}^*R_{\mathbf{L}}^{\mathbf{G}}\,\mathrm{reg}_{\mathbf{G}^F} = \varepsilon_{\mathbf{G}}\varepsilon_{\mathbf{L}}|\mathbf{G}^F/\mathbf{L}^F|_{p'}\,\mathrm{reg}_{\mathbf{L}^F}.$$

If in the equality above we apply the character formula 10.1.2 to ${}^*R_{\mathbf{L}}^{\mathbf{G}}\,\mathrm{reg}_{\mathbf{G}^F}$, and compute both sides at some given unipotent element v of $\mathbf{L}^F$, we get

$$Q_{\mathbf{L}}^{\mathbf{G}}(1, v^{-1}) = \begin{cases} 0 & \text{if } v \neq 1, \\ \varepsilon_{\mathbf{G}}\varepsilon_{\mathbf{L}}|\mathbf{G}^F/\mathbf{L}^F|_{p'} & \text{if } v = 1. \end{cases} \quad (**)$$

The equality $(*)$ we want to prove results immediately from $(**)$ applied with $\mathbf{G}$ replaced by $C_{\mathbf{G}}(s)^0$ and $\mathbf{L}$ replaced by $C_{\mathbf{L}}(s)^0$. $\square$

10.3 The Characteristic Function of a Semi-Simple Class

We will show in 13.3.4 that the characteristic function of a geometric conjugacy class is uniform. We end this chapter with a characterisation of uniform functions which proves in particular that the characteristic function of a semi-simple class is uniform.

Definition 10.3.1 For $s \in \mathbf{G}^F$ semi-simple, we define a map $d_s^{\mathbf{G}} : C(\mathbf{G}^F) \to C((C_{\mathbf{G}}(s)^0)^F)$ by

$$(d_s^{\mathbf{G}}f)(x) = \begin{cases} f(sx) & \text{if } x \text{ is unipotent}, \\ 0 & \text{otherwise}. \end{cases}$$

Proposition 10.3.2 *We have*

(i) $d_s^{\mathbf{L}} \circ {}^*R_{\mathbf{L}}^{\mathbf{G}} = {}^*R_{C_{\mathbf{L}}(s)^0}^{C_{\mathbf{G}}(s)^0} \circ d_s^{\mathbf{G}}$,

(ii) $d_s^{\mathbf{G}} \circ R_{\mathbf{L}}^{\mathbf{G}} = R_{\mathbf{L}}^{\mathbf{G}} \circ d_s^{\mathbf{L}}$ *for* $s \in Z(\mathbf{G}^F)$.

Proof We have $d_s^{\mathbf{G}}(f) = t_s(\chi_p^{C_{\mathbf{G}}(s)^0} \cdot \mathrm{Res}_{C_{\mathbf{G}}(s)^{0F}}^{\mathbf{G}^F} f)$, where $\chi_p^{C_{\mathbf{G}}(s)^0}$ is as in 7.3.2 and t_s denotes the translation by the central element s. By the character formula 10.1.2(ii) $^*R_{C_{\mathbf{L}}(s)^0}^{C_{\mathbf{G}}(s)^0}$ commutes with t_s. By 10.1.6(ii), it commutes with the multiplication by $\chi_p^{C_{\mathbf{G}}(s)^0}$. We then get (i) by 10.1.5.

For $s \in Z(\mathbf{G}^F)$ we have $d_s^{\mathbf{G}}(f) = t_s(\chi_p^{\mathbf{G}}f)$. Since t_s commutes with $R_{\mathbf{L}}^{\mathbf{G}}$ by the character formula 10.1.2(i) and the multiplication by $\chi_p^{\mathbf{G}}$ also commutes with $R_{\mathbf{L}}^{\mathbf{G}}$ by 10.1.6(i), we get (ii). $\qquad\square$

Proposition 10.3.3 *For $s \in \mathbf{G}^F$ semi-simple we have $d_s^{\mathbf{G}} \circ p_{\mathbf{G}} = p_{C_{\mathbf{G}}(s)^0} \circ d_s^{\mathbf{G}}$.*

Proof If $x \in (C_{\mathbf{G}}(s)^0)^F$ is not unipotent we have $(d_s^{\mathbf{G}} \circ p_{\mathbf{G}}f)(x) = (p_{C_{\mathbf{G}}(s)^0} \circ d_s^{\mathbf{G}}f)(x) = 0$. For $u \in (C_{\mathbf{G}}(s)^0)_u^F$ we have

$$(d_s^{\mathbf{G}} \circ p_{\mathbf{G}}f)(u) = p_{\mathbf{G}}f(su) = \sum_{\mathbf{T} \in \mathcal{T}} \frac{|\mathbf{T}^F|}{|\mathbf{G}^F|}(R_{\mathbf{T}}^{\mathbf{G}}\,{}^*R_{\mathbf{T}}^{\mathbf{G}}f)(su)$$

$$= \sum_{\mathbf{T} \in \mathcal{T}} \frac{|\mathbf{T}^F|}{|C_{\mathbf{G}}(s)^{0F}||\mathbf{G}^F|} \sum_{h \in \mathbf{G}^F | {}^h\mathbf{T} \ni s} R_{{}^h\mathbf{T}}^{C_{\mathbf{G}}(s)^0}({}^{h*}R_{\mathbf{T}}^{\mathbf{G}}f)(su)$$

$$= \sum_{\mathbf{T} \in \mathcal{T}} \frac{|\mathbf{T}^F||N_{\mathbf{G}^F}(\mathbf{T})|}{|C_{\mathbf{G}}(s)^{0F}||\mathbf{G}^F|} \sum_{\mathbf{T}' \ni s} R_{\mathbf{T}'}^{C_{\mathbf{G}}(s)^0}(^*R_{\mathbf{T}'}^{\mathbf{G}}f)(su)$$

$$= \sum_{\mathbf{T} \in \mathcal{T}, \mathbf{T} \ni s} \frac{|\mathbf{T}^F|}{|C_{\mathbf{G}}(s)^{0F}|}R_{\mathbf{T}}^{C_{\mathbf{G}}(s)^0}(^*R_{\mathbf{T}}^{\mathbf{G}}f)(su)$$

where the first line is by the definitions, the second line is using the first formula in 10.1.4 for $\mathbf{M} = C_{\mathbf{G}}(s)^0$, the third line is taking $\mathbf{T}' = {}^h\mathbf{T}$ as a variable and the last line results from counting tori. Now we have

$$R_{\mathbf{T}}^{C_{\mathbf{G}}(s)^0}(^*R_{\mathbf{T}}^{\mathbf{G}}f)(su) = d_s^{C_{\mathbf{G}}(s)^0}(R_{\mathbf{T}}^{C_{\mathbf{G}}(s)^0}(^*R_{\mathbf{T}}^{\mathbf{G}}f))(u)$$

$$= R_{\mathbf{T}}^{C_{\mathbf{G}}(s)^0}(d_s^{\mathbf{T}}\,{}^*R_{\mathbf{T}}^{\mathbf{G}}f)(u)$$

$$= R_{\mathbf{T}}^{C_{\mathbf{G}}(s)^0}(^*R_{\mathbf{T}}^{C_{\mathbf{G}}(s)^0}d_s^{\mathbf{G}}f)(u)$$

where the second equality is 10.3.2(ii) and the third equality is 10.3.2(i). Applying this transformation to the terms above, one recognises $(p_{C_{\mathbf{G}}(s)^0} \circ d_s^{\mathbf{G}}f)(u)$. $\qquad\square$

From the above proposition we immediately get the following corollary.

Corollary 10.3.4 $f \in C(\mathbf{G}^F)$ *is uniform if and only if $d_s f$ is uniform for all $s \in \mathbf{G}^F$ semi-simple.*

It results that if f is the characteristic function of a $\mathbf{G}^F$ semi-simple class, it is uniform since $d_s^{\mathbf{G}}f$ is 0 or a multiple of $\mathrm{reg}_{C_{\mathbf{G}}(s)^{0F}}$, which is uniform. We can get actually a nice formula for the "normalised" characteristic function of the class of s.

Notation 10.3.5 *Given a finite group H and some $x \in H$, we denote by π_x^H the function whose value is $|C_H(x)|$ on the H-conjugacy class of x and 0 on other elements of H.*

With this notation, we can write the following proposition.

Proposition 10.3.6 *Let s be a semi-simple element of $\mathbf{G}^F$; then*

$$\pi_s^{\mathbf{G}^F} = |W^0(s)|^{-1} \sum_{w \in W^0(s)} \dim(R_{\mathbf{T}_w}^{C_{\mathbf{G}}(s)^0}(1)) R_{\mathbf{T}_w}^{\mathbf{G}}(\pi_w^{\mathbf{T}_w^F})$$

$$= \varepsilon_{C_{\mathbf{G}}(s)^0} |C_{\mathbf{G}}(s)^{0F}|_p^{-1} \sum_{\substack{\mathbf{T} \in \mathcal{T} \\ \mathbf{T} \ni s}} \varepsilon_{\mathbf{T}} R_{\mathbf{T}}^{\mathbf{G}}(\pi_s^{\mathbf{T}^F}),$$

where in the first sum $W^0(s)$ is as in 3.5.2, and $\mathbf{T}_w$ has type w with respect to some quasi-split torus of $C_{\mathbf{G}}(s)^0$.

Note that if $\mathbf{T}_0$ is a quasi-split torus of $C_{\mathbf{G}}(s)^0$, of type v with respect to a quasi-split torus $\mathbf{T}_1$ of $\mathbf{G}$, then $(\mathbf{T}_w, F)$ is $C_{\mathbf{G}}(s)^0$-conjugate to $(\mathbf{T}_0, wF)$, and there is $g \in \mathbf{G}$ such that $^g(\mathbf{T}_0, wF) = (\mathbf{T}_1, v^g wF)$, thus $\mathbf{T}_w$ is of type $v^g w$ in $\mathbf{G}$.

Proof By 7.4.4 we have $\mathrm{St}_{\mathbf{G}^F} \pi_s^{\mathbf{G}^F} = \varepsilon_{\mathbf{G}} \varepsilon_{C_{\mathbf{G}}(s)^0} |C_{\mathbf{G}}(s)^{0F}|_p \pi_s^{\mathbf{G}^F}$, thus it is enough to give a formula as a uniform function for $\mathrm{St}_{\mathbf{G}^F} \pi_s^{\mathbf{G}^F}$, and for that, since we know that $\pi_s^{\mathbf{G}^F}$ is uniform, to compute

$$p_{\mathbf{G}}(\mathrm{St}_{\mathbf{G}^F} \pi_s^{\mathbf{G}^F}) = \sum_{\mathbf{T} \in \mathcal{T}} \frac{|\mathbf{T}^F|}{|\mathbf{G}^F|} R_{\mathbf{T}}^{\mathbf{G}} {}^* R_{\mathbf{T}}^{\mathbf{G}}(\mathrm{St}_{\mathbf{G}^F} \pi_s^{\mathbf{G}^F})$$

$$= \sum_{\mathbf{T} \in \mathcal{T}} \varepsilon_{\mathbf{G}} \varepsilon_{\mathbf{T}} \frac{|\mathbf{T}^F|}{|\mathbf{G}^F|} R_{\mathbf{T}}^{\mathbf{G}} \mathrm{Res}_{\mathbf{T}^F}^{\mathbf{G}^F} \pi_s^{\mathbf{G}^F},$$

where the second equality is 10.2.10(ii) and the fact that $\mathrm{St}_{\mathbf{T}^F} = \mathbf{1}_{\mathbf{T}^F}$. For $t \in \mathbf{T}^F$ we have $\mathrm{Res}_{\mathbf{T}^F}^{\mathbf{G}^F} \pi_s^{\mathbf{G}^F}(t) = \begin{cases} |C_{\mathbf{G}^F}(t)| & \text{if } t \text{ is conjugate to } s \\ 0 & \text{otherwise} \end{cases}$ which is equal to $|\{g \in \mathbf{G}^F \mid {}^g s = t\}| = |\mathbf{T}^F|^{-1} \sum_{\{g \in \mathbf{G}^F \mid {}^g s \in \mathbf{T}^F\}} \pi_{{}^g s}^{\mathbf{T}^F}(t)$, hence $\mathrm{Res}_{\mathbf{T}^F}^{\mathbf{G}^F} \pi_s^{\mathbf{G}^F} = |\mathbf{T}^F|^{-1} \sum_{\{g \in \mathbf{G}^F \mid {}^g s \in \mathbf{T}^F\}} \pi_{{}^g s}^{\mathbf{T}^F}$. Plugging everything back together we get

$$p_{\mathbf{G}}(\pi_s^{\mathbf{G}^F}) = \frac{\varepsilon_{C_{\mathbf{G}}(s)^0}}{|C_{\mathbf{G}}(s)^{0F}|_p |\mathbf{G}^F|} \sum_{\mathbf{T} \in \mathcal{T}} \varepsilon_{\mathbf{T}} \sum_{\{g \in \mathbf{G}^F \mid {}^g s \in \mathbf{T}^F\}} R_{\mathbf{T}}^{\mathbf{G}} \pi_{{}^g s}^{\mathbf{T}^F}$$

which, taking $^{g^{-1}}\mathbf{T}$ as a new variable and counting tori, gives the second formula of the statement. The first one results from a straightforward computation. □

Remark 10.3.7 Another corollary of 10.3.4 is that any function in $C(\mathbf{G}^F)_{p'}$ is uniform.

Exercise 10.3.8 Let $x = su$ be the Jordan decomposition of some element $x \in \mathbf{G}^F$, and let $\mathbf{L}$ be an F-stable Levi subgroup of $\mathbf{G}$ containing $C_\mathbf{G}(s)^0$; prove that

$$R_\mathbf{L}^\mathbf{G}(\pi_x^{\mathbf{L}^F}) = \pi_x^{\mathbf{G}^F}.$$

Hint Use 10.1.5 to compute $(^*R_\mathbf{L}^\mathbf{G}\psi)(x)$, for any $\psi \in \operatorname{Irr}\mathbf{G}^F$. □

Notes

The character formula 10.1.2 for the functor $R_\mathbf{T}^\mathbf{G}$ is in Deligne and Lusztig (1976, 4.2). The generalisation to $R_\mathbf{L}^\mathbf{G}$ was first published in Digne and Michel (1983); we have since learned that it was already known by Deligne at the time of Deligne and Lusztig (1976).

11

Geometric Conjugacy and the Lusztig Series

We now introduce Lusztig's classification of irreducible characters of $\mathbf{G}^F$. In this classification, a character is parametrised by a pair: an F^*-stable semisimple class (s) of the dual group $\mathbf{G}^*$ of $\mathbf{G}$, and a "unipotent" irreducible character χ of the centraliser $C_{\mathbf{G}^{*F^*}}(s)$.

11.1 Geometric Conjugacy

In this chapter, we have all the tools to deal with the ingredient (s) of the classification; see Chapter 14 for the ingredient χ. We start with an immediate corollary of 10.2.6.

Proposition 11.1.1 *For any $\chi \in \mathrm{Irr}(\mathbf{G}^F)$, there exists an F-stable maximal torus $\mathbf{T}$ and $\theta \in \mathrm{Irr}(\mathbf{T}^F)$ such that $\langle \chi, R_{\mathbf{T}}^{\mathbf{G}}(\theta) \rangle_{\mathbf{G}^F} \neq 0$.*

We note also that 9.3.1(i) implies that $R_{\mathbf{T}}^{\mathbf{G}}(\theta)$ is irreducible up to sign if and only if no non-trivial element of $W(\mathbf{T})^F$ stabilises θ. Furthermore, by 10.2.2, in that case $\varepsilon_{\mathbf{T}}\varepsilon_{\mathbf{G}}R_{\mathbf{T}}^{\mathbf{G}}(\theta)$ is a true character.

We now give a condition for two Deligne–Lusztig characters to have a common irreducible constituent, using the norm $N_{F^n/F}$ on an F-stable torus $\mathbf{T}$, see 9.2.3. Note that $N_{F^n/F}$ maps $\mathbf{T}^{F^n}$ to $\mathbf{T}^F$. By 9.3.1(i), if the pairs $(\mathbf{T}, \theta)$ and $(\mathbf{T}', \theta')$ are not $\mathbf{G}^F$-conjugate, then $R_{\mathbf{T}}^{\mathbf{G}}(\theta)$ and $R_{\mathbf{T}'}^{\mathbf{G}}(\theta')$ are orthogonal to each other; but they may have a common constituent as they are virtual characters.

Definition 11.1.2 Let $\mathbf{T}$ and $\mathbf{T}'$ be two F-stable maximal tori, and let θ and θ' be characters respectively of $\mathbf{T}^F$ and $\mathbf{T}'^F$. We say that the pairs $(\mathbf{T}, \theta)$ and $(\mathbf{T}', \theta')$ are **geometrically conjugate** by $g \in \mathbf{G}$ if $\mathbf{T} = {}^g\mathbf{T}'$ and for any n such that $g \in \mathbf{G}^{F^n}$ the two characters of $\mathbf{T}'^{F^n}$ given by $\theta \circ N_{F^n/F} \circ \mathrm{ad}\,g$ and by $\theta' \circ N_{F^n/F}$ coincide.

Proposition 11.1.3 *Let* $\mathbf{T}$ *(resp.* $\mathbf{T}'$*) be an F-stable maximal torus, and let* $\mathbf{U}$ *(resp.* $\mathbf{U}'$*) be the unipotent radical of some Borel subgroup containing* $\mathbf{T}$ *(resp.* $\mathbf{T}'$*). Let* $\theta \in \mathrm{Irr}(\mathbf{T}^F)$ *(resp.* $\theta' \in \mathrm{Irr}(\mathbf{T}'^F)$*). Assume that there exist i and j such that the* $\overline{\mathbb{Q}}_\ell \mathbf{G}^F$*-modules* $H_c^i(\mathcal{L}^{-1}(\mathbf{U})) \otimes_{\overline{\mathbb{Q}}_\ell \mathbf{T}^F} \theta$ *and* $H_c^j(\mathcal{L}^{-1}(\mathbf{U}')) \otimes_{\overline{\mathbb{Q}}_\ell \mathbf{T}'^F} \theta'$ *have a common irreducible constituent. Then the pairs* $(\mathbf{T},\theta)$ *and* $(\mathbf{T}',\theta')$ *are geometrically conjugate.*

Proof Let χ be the common irreducible constituent of the statement. We remark first that, by 8.1.6, χ occurring in $H_c^i(\mathcal{L}^{-1}(\mathbf{U})) \otimes_{\overline{\mathbb{Q}}_\ell \mathbf{T}^F} \theta$ is equivalent to $\overline{\chi}$ occurring in $H_c^i(\mathcal{L}^{-1}(\mathbf{U})) \otimes_{\overline{\mathbb{Q}}_\ell \mathbf{T}^F} \overline{\theta}$, which implies that $\overline{\chi}^{\mathrm{opp}}$ occurs in $\overline{\theta}^{\mathrm{opp}} \otimes_{\overline{\mathbb{Q}}_\ell \mathbf{T}^F} H_c^i(\mathcal{L}^{-1}(\mathbf{U}))^{\mathrm{opp}}$. As χ occurs in $H_c^j(\mathcal{L}^{-1}(\mathbf{U}')) \otimes_{\overline{\mathbb{Q}}_\ell \mathbf{T}'^F} \theta'$, the representation $\overline{\theta} \otimes \theta'^{\mathrm{opp}}$ of $\mathbf{T}^F \times \mathbf{T}'^F$ occurs in the module $H_c^i(\mathcal{L}^{-1}(\mathbf{U}))^{\mathrm{opp}} \otimes_{\overline{\mathbb{Q}}_\ell \mathbf{G}^F} H_c^j(\mathcal{L}^{-1}(\mathbf{U}'))$, which with the notation of 9.2.1 is a submodule of $H_c^{i+j}(\mathbf{Z})$. However, we have the following lemma.

Lemma 11.1.4 *If the* $\overline{\mathbb{Q}}_\ell \mathbf{T}^F$*-module-*$\overline{\mathbb{Q}}_\ell{}^w(\mathbf{T}'^F)$ *given by* $\overline{\theta} \otimes {}^w\theta'^{\mathrm{opp}}$ *occurs in some cohomology group of* $\mathbf{Z}_w''$ *for some* $w \in \mathcal{S}(\mathbf{T},\mathbf{T}')$ *(see 9.2.2) and if* $n > 0$ *is such that* $F^n w = w$*, then* $\theta \circ N_{F^n/F} = \theta' \circ N_{F^n/F} \circ \mathrm{ad}\, {}^F w^{-1}$.

Proof Using the remark which follows the proof of 9.2.4, we get the result from 9.2.4(ii) applied with $\mathbf{L} = \mathbf{T}$ and $\mathbf{M} = {}^w\mathbf{T}'$, using the fact that $N_{F^m/F'}({}^{w'}\tau^{-1}) = {}^w N_{F^n/F}({}^{F_w^{-1}}\tau^{-1})$.   □

Under the hypothesis of 11.1.4 the element ${}^F w$ is the required element g of 11.1.2, and the proposition is proved. As the cohomology of the $\mathbf{T}^F$-variety-${}^w\mathbf{T}'^F$ given by $\mathbf{Z}_w''$ is isomorphic to that of the $\mathbf{T}^F$-variety-$\mathbf{T}'^F$ given by $\mathbf{Z}_w$ (via ad w), the following lemma thus completes the proof of the proposition.

Lemma 11.1.5 *If the character* $\theta \otimes \theta'^{\mathrm{opp}}$ *does not occur in* $H_c^k(\mathbf{Z}_w)$ *for any* k *and* w*, then* $\theta \otimes \theta'^{\mathrm{opp}}$ *does not occur in* $H_c^i(\mathbf{Z})$ *for any* i.

Proof We use the following lemma.

Lemma 11.1.6

(i) *With the notation of 9.2.1, if* $\mathbf{L} = \mathbf{T}$ *and* $\mathbf{L}' = \mathbf{T}'$ *are tori, and if we let* $\mathbf{B}$ *and* $\mathbf{B}'$ *denote the Borel subgroups* ${}^{F^{-1}}\mathbf{P}$ *and* ${}^{F^{-1}}\mathbf{P}'$*, then for any* $v \in W(\mathbf{T})$ *the union* $\bigcup_{v' \leq v} \mathbf{Z}_{v'w_1}$ *is closed in* $\mathbf{Z}$ *where* $w_1 \in \mathcal{S}(\mathbf{T},\mathbf{T}')$ *is such that* ${}^{w_1}\mathbf{B}' = \mathbf{B}$ *and where* $v' \leq v$ *is the Bruhat–Chevalley order of 3.2.8.*

(ii) *The connected components of the union* $\bigcup_{l(v)=n} \mathbf{Z}_{vw_1}$ *are the* $\mathbf{Z}_{vw_1}$.

Proof Property (i) results from the fact that in the present case $\mathbf{Z}_{vw_1}$ is the inverse image in $\mathbf{Z}$ by the third projection of the subset $\mathbf{B}vw_1\mathbf{B}' = \mathbf{B}v\mathbf{B}w_1$ of $\mathbf{G}$, and from 3.2.8. Furthermore, (i) shows that the closure of $\mathbf{Z}_{vw_1}$ in $\mathbf{Z}$ does not meet $\mathbf{Z}_{v'w_1}$ if $l(v) = l(v')$ and $v \neq v'$, whence (ii). □

This lemma will allow us to apply repeatedly the long exact sequence 8.1.7(i) to $\mathbf{Z} = \bigcup_v \mathbf{Z}_{vw_1}$. We now prove 11.1.5 by showing by induction on n that the hypothesis of 11.1.5 implies that $H_c^i(\bigcup_{l(v) \leq n} \mathbf{Z}_{vw_1})_{\theta \otimes \theta'^{\text{opp}}} = 0$, where the subscript $\theta \otimes \theta'^{\text{opp}}$ denotes the subspace of the cohomology where $\mathbf{T}^F \times \mathbf{T}'^F$ acts through $\theta \otimes \theta'^{\text{opp}}$. This is true for $n = 0$ by hypothesis. Suppose it for $n - 1$. We have

$$\bigcup_{l(v) \leq n} \mathbf{Z}_{vw_1} = (\bigcup_{l(v)=n} \mathbf{Z}_{vw_1}) \bigcup (\bigcup_{l(v) \leq n-1} \mathbf{Z}_{vw_1}),$$

and this last union is closed since by 11.1.6(i) it is a finite union of closed subsets. There is thus a cohomology long exact sequence relating these three unions; this sequence remains exact restricted to the subspaces where $\mathbf{T}^F \times \mathbf{T}'^F$ acts through $\theta \otimes \theta'^{\text{opp}}$. But 11.1.6(ii) implies, according to 8.1.7(ii), that for any k we have $H_c^k(\bigcup_{l(v)=n} \mathbf{Z}_{vw_1}) = \bigoplus_{l(v)=n} H_c^k(\mathbf{Z}_{vw_1})$, so by assumption $H_c^k(\bigcup_{l(v)=n} \mathbf{Z}_{vw_1})_{\theta \otimes \theta'^{\text{opp}}} = 0$. Since, by the induction hypothesis, for any k we have $H_c^k(\bigcup_{l(v) \leq n-1} \mathbf{Z}_{vw_1})_{\theta \otimes \theta'^{\text{opp}}} = 0$, at each step two out of three terms of the long exact sequence are 0, so the third one $H_c^k(\bigcup_{l(v) \leq n} \mathbf{Z}_{vw_1})_{\theta \otimes \theta'^{\text{opp}}}$ is also 0. □

□

We will now introduce the dual of $\mathbf{G}$, which will allow us to give a nice interpretation of geometric conjugacy (and a justification of this term), using the duality between the groups $X(\mathbf{T})$ and $Y(\mathbf{T})$ for a torus $\mathbf{T}$.

We shall assume we have chosen once and for all an isomorphism $\overline{\mathbb{F}}_p^\times \xrightarrow{\sim} (\mathbb{Q}/\mathbb{Z})_{p'}$ and an embedding $\overline{\mathbb{F}}_p^\times \hookrightarrow \overline{\mathbb{Q}}_\ell^\times$.

Proposition 11.1.7 *Let $\mathbf{T}$ be a torus with a Frobenius root F; as in 4.2.2 the action of F on $X := X(\mathbf{T})$ (resp. $Y := Y(\mathbf{T})$) is $\alpha \mapsto \alpha \circ F$ (resp. $\beta \mapsto F \circ \beta$). Then:*

(i) *The sequence*

$$0 \to X \xrightarrow{F-1} X \to \mathrm{Irr}(\mathbf{T}^F) \to 1$$

is exact, where the right map is the restriction to $\mathbf{T}^F$ of characters (for this to make sense we identify $\mathrm{Hom}(\mathbf{T}^F, \mathbb{G}_m)$ with $\mathrm{Irr}(\mathbf{T}^F)$, using the chosen embedding $\overline{\mathbb{F}}_p^\times \hookrightarrow \overline{\mathbb{Q}}_\ell^\times$).

(ii) *The sequence*

$$0 \to Y \xrightarrow{F-1} Y \xrightarrow{\pi} \mathbf{T}^F \to 1$$

is exact, where π is defined by $y \mapsto N_{F^n/F}(\pi_n(y))$ for any n such that F^n is a split Frobenius on $\mathbf{T}$, where $\pi_n(y) = y(1/(q^n - 1))$. Here q is the real number associated with F, and we have used the chosen isomorphism to identify $1/(q^n - 1) \in (\mathbb{Q}/\mathbb{Z})_{p'}$ with an element of $\overline{\mathbb{F}}_p^{\times}$.

Proof We have seen (i) in the proof of 4.4.9. Let us prove (ii). Let n be the smallest integer such that F^n is a split Frobenius on $\mathbf{T}$ – with the notation of 4.2.5, this is the lcm of the order of τ and of the smallest m such that q^m is an integer, and the set of m such that F^m is a split Frobenius is $\{kn \mid k \in \mathbb{N} - \{0\}\}$. For any positive integer k, we have

$$N_{F^n/F}(\pi_n(y)) = N_{F^n/F}(N_{F^{kn}/F^n}(\pi_{kn}(y))) = N_{F^{kn}/F}(\pi_{kn}(y)),$$

so the map π is well-defined.

To show the exactness we use the commutative diagram

$$
\begin{array}{ccccccccc}
0 & \to & Y & \xrightarrow{F^n-1} & Y & \xrightarrow{\pi_n} & \mathbf{T}^{F^n} & \to & 1 \\
  &   & \downarrow{\scriptstyle N_{F^n/F}} & & \| & & \downarrow{\scriptstyle N_{F^n/F}} & & \\
0 & \to & Y & \xrightarrow{F-1} & Y & \xrightarrow{\pi} & \mathbf{T}^F & \to & 1
\end{array}
$$

where $N_{F^n/F} := \mathrm{Id} + F + \cdots + F^{n-1}$ on Y. We show first the exactness of the top sequence: the injectivity of $F^n - 1$ is clear as it is the multiplication by $q^n - 1$, since $\mathbf{T}$ is split over $\mathbb{F}_{q^n}$; the surjectivity of π_n is straightforward by 1.2.10 applied with $k = \mathbb{F}_{q^n}$. Moreover if y is in the kernel of π_n then it is trivial on all $(q^n - 1)$-th roots of unity, so is constant on all fibres of the map $x \mapsto x^{q^n-1}$ from $\mathbb{G}_m$ to itself; so it factors through that endomorphism of $\mathbb{G}_m$; that is, there exists $y_1 \in Y$ such that $y = (q^n - 1)y_1$; that is, y is in the image of $F^n - 1$. The exactness of the bottom sequence is deduced from that of the top sequence: the surjectivity of π follows by the surjectivity of $N_{F^n/F} : \mathbf{T}^{F^n} \to \mathbf{T}^F$. The injectivity of $F - 1$ is clear as F is the transpose of the endomorphism $q\tau$ of $X(\mathbf{T})$, see 4.2.5, so has no eigenvalue 1. The image of $F - 1$ is obviously in the kernel of π; Take now y in the kernel of π; that is, such that $N_{F^n/F}(y(1/(q^n - 1))) = 1$. This means that $N_{F^n/F}(y)$ is in the kernel of π_n, so $N_{F^n/F}(y) \in (F^n - 1)Y = N_{F^n/F} \circ (F - 1)(Y)$. As $N_{F^n/F} : Y \to Y$ is injective, since the composite $(F - 1) \circ N_{F^n/F} = F^n - 1$ is injective, we get $y \in (F - 1)Y$. □

We note that by 11.1.7(ii) any character of $\mathbf{T}^F$ gives a character of $Y(\mathbf{T})$ with values in the p'-roots of unity in $\overline{\mathbb{Q}}_\ell$; using our chosen isomorphism, we will consider it as an element of $\mathrm{Hom}(Y(\mathbf{T}), (\mathbb{Q}/\mathbb{Z})_{p'})$. We may now reinterpret the notion of geometric conjugacy.

Proposition 11.1.8 *Let $\mathbf{T}$ and $\mathbf{T}'$ be two F-stable maximal tori of $\mathbf{G}$ and let $\theta \in \mathrm{Irr}(\mathbf{T}^F)$ and $\theta' \in \mathrm{Irr}(\mathbf{T}'^F)$; then $(\mathbf{T}, \theta)$ and $(\mathbf{T}', \theta')$ are geometrically conjugate by $g \in \mathbf{G}$ if and only if $\mathbf{T} = {}^g\mathbf{T}'$ and g conjugates θ', considered* *as an element of* $\mathrm{Hom}(Y(\mathbf{T}'),(\mathbb{Q}/\mathbb{Z})_{p'})$ *via 11.1.7(ii), to* θ, *considered as an element of* $\mathrm{Hom}(Y(\mathbf{T}),(\mathbb{Q}/\mathbb{Z})_{p'})$.

Proof The characters θ and $\theta \circ N_{F^n/F}$ are identified with the same element of $\mathrm{Hom}(Y(\mathbf{T}),(\mathbb{Q}/\mathbb{Z})_{p'})$ by the construction of 11.1.7(ii), as seen in the proof of that proposition. This gives the result. □

Corollary 11.1.9 *Let us fix an F-stable maximal torus* $\mathbf{T}$; *then geometric conjugacy classes of pairs* $(\mathbf{T}',\theta')$ *(where* $\theta' \in \mathrm{Irr}(\mathbf{T}'^F)$*) are in one-to-one correspondence with F-stable* $W(\mathbf{T})$*-orbits in* $X(\mathbf{T}) \otimes (\mathbb{Q}/\mathbb{Z})_{p'}$.

Proof Since all tori are $\mathbf{G}$-conjugate we may assume that $\mathbf{T}' = \mathbf{T}$, so $^g\mathbf{T}' = \mathbf{T}$ becomes $g \in W(\mathbf{T})$. The result then comes from the remark that $\mathrm{Hom}(Y(\mathbf{T}),$ $(\mathbb{Q}/\mathbb{Z})_{p'})$ is isomorphic to $X(\mathbf{T}) \otimes (\mathbb{Q}/\mathbb{Z})_{p'}$. □

We now introduce the dual of a reductive group.

Definition 11.1.10 Two connected reductive algebraic groups $\mathbf{G}$ and $\mathbf{G}^*$ are said to be **dual** to each other if there exists maximal tori $\mathbf{T}$ of $\mathbf{G}$ and $\mathbf{T}^*$ of $\mathbf{G}^*$ such that their root data are dual; that is, there is an isomorphism $X(\mathbf{T}) \xrightarrow{\sim} Y(\mathbf{T}^*)$ inducing an isomorphism

$$(X(\mathbf{T}),Y(\mathbf{T}),\Phi,\Phi^\vee) \xrightarrow{\sim} (Y(\mathbf{T}^*),X(\mathbf{T}^*),\Phi^{*\vee},\Phi^*).$$

In particular, two tori $\mathbf{T}$ and $\mathbf{T}^*$ are said to be dual to each other if we have been given an isomorphism $X(\mathbf{T}) \xrightarrow{\sim} Y(\mathbf{T}^*)$. When we talk of groups dual to each other, we will always assume that we have chosen corresponding dual tori; we will say that $(\mathbf{G},\mathbf{T})$ is dual to $(\mathbf{G}^*,\mathbf{T}^*)$ when we need to specify the tori used. Note that by 2.4.2 every connected reductive group has a dual and, given $\mathbf{G}$, the isomorphism class of $\mathbf{G}^*$ is well defined. Using the exact pairing between $X(\mathbf{T})$ and $Y(\mathbf{T})$, being given a duality $X(\mathbf{T}) \xrightarrow{\sim} Y(\mathbf{T}^*)$ is equivalent to being given a duality $X(\mathbf{T}^*) \xrightarrow{\sim} Y(\mathbf{T})$. Given $\varphi \in \mathrm{Hom}(X(\mathbf{T}),X(\mathbf{T}'))$ we will denote by $\varphi^* \in \mathrm{Hom}(X(\mathbf{T}'^*),X(\mathbf{T}^*))$ the transposed of the image of φ by the duality isomorphism.

Proposition 11.1.11 *Given an isogeny* $\mathbf{G} \xrightarrow{f} \mathbf{G}'$ *inducing the p-morphism* $X(\mathbf{T}') \xrightarrow{\varphi} X(\mathbf{T})$, *then* φ^* *is a p-morphism induced by an isogeny* $\mathbf{G}'^* \xrightarrow{f^*} \mathbf{G}^*$, *that we call the* **dual isogeny** *to f.*

Proof We just need to check that the dual of a p-morphism is a p-morphism. This results from the definition and from the fact that a map φ between lattices has a finite cokernel K if and only if its transposed is injective. Indeed, by the left exactness of $\mathrm{Hom}(.,\mathbb{Z})$ the exact sequence $X \xrightarrow{\varphi} X' \to K \to 0$

gives the exact sequence $0 \rightarrow \mathrm{Hom}(K,\mathbb{Z}) \rightarrow \mathrm{Hom}(X',\mathbb{Z}) \xrightarrow{\varphi^{\vee}} \mathrm{Hom}(X,\mathbb{Z})$ and $\mathrm{Hom}(K,\mathbb{Z}) = 0$ if and only if K is finite. □

Definition 11.1.12 If F is a Frobenius root on $\mathbf{G}$ which stabilises $\mathbf{T}$ and F^* is the dual isogeny, we say then that the pair $(\mathbf{G},F)$ is dual to the pair $(\mathbf{G}^*,F^*)$.

Note that F^* is a Frobenius root by the sentence just after 4.3.2.

Note that $w \mapsto w^*$ is an anti-isomorphism $W(\mathbf{T}) \rightarrow W(\mathbf{T}^*)$, and that through that anti-isomorphism the action of F on $W(\mathbf{T})$ is identified with the inverse of the action of F^* on $W(\mathbf{T}^*)$.

Examples 11.1.13 The group $\mathbf{GL}_n$ is its own dual, and the same is true for the unitary group which is $\mathbf{GL}_n$ with the Frobenius endomorphism which sends a matrix $(a_{i,j})$ to ${}^t(a_{i,j}^q)^{-1}$, see 4.3.3. The groups $\mathbf{SL}_n$ and $\mathbf{PGL}_n$ are dual to each other. The group $\mathbf{SO}_{2l}$ is its own dual, with either the split Frobenius endomorphism or the Frobenius endomorphism of the non-split group. The symplectic group $\mathbf{Sp}_{2n}$ is dual to $\mathbf{SO}_{2n+1}$.

It follows from the definitions that the dual of a semi-simple group is semi-simple, and that the dual of an adjoint group is simply connected, and conversely.

We get as an immediate corollary of 11.1.7:

Proposition 11.1.14 *If $(\mathbf{T}^*,F^*)$ is dual to $(\mathbf{T},F)$ then $\mathrm{Irr}(\mathbf{T}^F) \simeq \mathbf{T}^{*F^*}$.*

The next proposition interprets geometric conjugacy using $\mathbf{G}^*$.

Proposition 11.1.15 *Assume that $(\mathbf{G},F)$ and $(\mathbf{G}^*,F^*)$ are dual to each other with corresponding dual tori $\mathbf{T}$ and $\mathbf{T}^*$. Geometric conjugacy classes of pairs $(\mathbf{T}',\theta')$ in $\mathbf{G}$ are in one-to-one correspondence with F^*-stable conjugacy classes of semi-simple elements of $\mathbf{G}^*$.*

F^*-stable conjugacy classes of semi-simple elements of $\mathbf{G}^*$ are in bijection with their intersection with $\mathbf{G}^{*F^*}$, the semi-simple geometric conjugacy classes of $\mathbf{G}^{*F^*}$.

Proof We apply 11.1.9. Using the fact that the isomorphism from $X(\mathbf{T})$ to $Y(\mathbf{T}^*)$, is compatible with the action of the Weyl group, we get a bijection between geometric conjugacy classes of pairs $(\mathbf{T}',\theta')$ and F^*-stable $W(\mathbf{T}^*)$-orbits in $Y(\mathbf{T}^*) \otimes (\mathbb{Q}/\mathbb{Z})_{p'}$. But by 1.2.10 there is an isomorphism from $Y(\mathbf{T}^*) \otimes (\mathbb{Q}/\mathbb{Z})_{p'}$ to $\mathbf{T}^*$ (which depends on the fixed isomorphism $\overline{\mathbb{F}}_q^{\times} \simeq (\mathbb{Q}/\mathbb{Z})_{p'}$), so geometric conjugacy classes of pairs $(\mathbf{T}',\theta')$ are in one-to-one correspondence with F^*-stable $W(\mathbf{T}^*)$-orbits in $\mathbf{T}^*$. As any semi-simple element is in a maximal torus and all maximal tori are conjugate, any semi-simple class meets $\mathbf{T}^*$. Furthermore – by 1.3.6(iv) – $W(\mathbf{T}^*)$-orbits in $\mathbf{T}^*$ are the intersections of

$\mathbf{G}^*$-conjugacy classes with $\mathbf{T}^*$, hence F^*-stable $W(\mathbf{T}^*)$-orbits are the intersections of F^*-stable conjugacy classes with $\mathbf{T}^*$, whence the result. $\qquad\square$

After 9.3.1 we remarked that the $R_{\mathbf{T}}^{\mathbf{G}}(\theta)$ are parametrised by $\mathbf{G}^F$-conjugacy classes of pairs $(\mathbf{T}, \theta)$. Using the dual group, we may give another parametrisation

Proposition 11.1.16 *The $\mathbf{G}^F$-conjugacy classes of pairs $(\mathbf{T}, \theta)$ where $\mathbf{T}$ is an F-stable maximal torus of $\mathbf{G}$ and $\theta \in \mathrm{Irr}(\mathbf{T}^F)$ are in one-to-one correspondence with the $\mathbf{G}^{*F^*}$-conjugacy classes of pairs $(\mathbf{T}^*, s)$ where s is a semi-simple element of $\mathbf{G}^{*F^*}$ and $\mathbf{T}^*$ is an F^*-stable maximal torus containing s.*

Proof Let $\mathbf{T}$ be a fixed F-stable maximal torus. By the remark after 4.2.22 the $\mathbf{G}^F$-conjugacy classes of pairs $(\mathbf{T}', \theta')$ correspond one-to-one to the conjugacy classes under $W(\mathbf{T})$ of pairs (wF, θ) with $\theta \in \mathrm{Irr}(\mathbf{T}^{wF})$. If $(\mathbf{G}^*, \mathbf{T}^*)$ is dual to $(\mathbf{G}, \mathbf{T})$, these pairs are by 11.1.14 in one-to-one correspondence with the conjugacy classes under $W(\mathbf{T}^*)$ of pairs (F^*w^*, s) with $s \in \mathbf{T}^{*F^*w^*}$ which are in turn, by the remark after 4.2.22, in one-to-one correspondence with the $\mathbf{G}^{*F^*}$-conjugacy classes of F^*-stable pairs $(\mathbf{T}'^*, s')$. $\qquad\square$

Using this proposition, we will, in what follows, sometimes use the notation $R_{\mathbf{T}^*}^{\mathbf{G}}(s)$ for $R_{\mathbf{T}}^{\mathbf{G}}(\theta)$.

11.2 More on Centralisers of Semi-Simple Elements

Lusztig's classification of characters is considerably simpler when centralisers of semi-simple elements in $\mathbf{G}^*$ are connected. We will sometimes need this hypothesis in this chapter and the next. We now show that it holds if the centre of $\mathbf{G}$ is connected, and give some other consequences of the connectedness of $Z(\mathbf{G})$ we will need in the next chapter.

Lemma 11.2.1 *Let $\mathbf{T}$ be a maximal torus of $\mathbf{G}$ and Φ be the set of roots of $\mathbf{G}$ relative to $\mathbf{T}$; then:*

(i) *The group $\mathrm{Irr}(Z(\mathbf{G})/Z(\mathbf{G})^0)$ is canonically isomorphic to the torsion group of $X(\mathbf{T})/((\langle \Phi \rangle_{\mathbf{T}}^{\perp})_{X(\mathbf{T})}^{\perp}$.*
(ii) *If the centre of $\mathbf{G}$ is connected, then the centre of any Levi subgroup of $\mathbf{G}$ is also connected.*
(iii) *If $(\mathbf{G}^*, \mathbf{T}^*)$ is dual to $(\mathbf{G}, \mathbf{T})$, then for any $s \in \mathbf{T}^*$ the group $W(s)/W^0(s)$ (isomorphic to $C_{\mathbf{G}^*}(s)/C_{\mathbf{G}^*}(s)^0$, see 3.5.2) is isomorphic to a subgroup of $\mathrm{Irr}(Z(\mathbf{G})/Z(\mathbf{G})^0)$.*

Proof By 1.2.3(i) and 1.2.4 the torsion of $X(\mathbf{D})$ is isomorphic to $\mathrm{Irr}(\mathbf{D}/\mathbf{D}^0)$ for any diagonalisable group $\mathbf{D}$. As we have $Z(\mathbf{G}) = \langle\Phi\rangle^\perp_{\mathbf{T}}$, it follows using 1.2.4 that $X(Z(\mathbf{G})) = X(\mathbf{T})/(\langle\Phi\rangle^\perp_{\mathbf{T}})^\perp_{X(\mathbf{T})}$, whence (i).

It follows that $Z(\mathbf{G})$ is connected if and only if $X(\mathbf{T})/(\langle\Phi\rangle^\perp_{\mathbf{T}})^\perp_{X(\mathbf{T})}$ has no torsion, which is equivalent to $X(\mathbf{T})/\langle\Phi\rangle$ having no p'-torsion by 1.2.13; but the subgroup $\langle\Phi_\mathbf{L}\rangle$ of $\langle\Phi\rangle$ spanned by the roots of a Levi subgroup $\mathbf{L}$ containing $\mathbf{T}$ is a direct summand of $\langle\Phi\rangle$, so if $X(\mathbf{T})/\langle\Phi\rangle$ has no p'-torsion then $X(\mathbf{T})/\langle\Phi_\mathbf{L}\rangle$ has none either, so $Z(\mathbf{L})$ is connected, whence (ii).

By 1.2.10 an element $s \in \mathbf{T}^*$ can be identified with an element of $Y(\mathbf{T}^*) \otimes (\mathbb{Q}/\mathbb{Z})_{p'}$; that is, of $X(\mathbf{T}) \otimes (\mathbb{Q}/\mathbb{Z})_{p'}$. Let us write such an element x/m with $x \in X(\mathbf{T})$ and $m \in \mathbb{Z}$ relatively prime to p. We have

$$W(s) = \{w \in W \mid w(x) - x \in mX(\mathbf{T})\}.$$

Consider the map

$$W(s) \to X(\mathbf{T})/\langle\Phi\rangle$$

$$w \mapsto \frac{w(x) - x}{m} \quad (\mathrm{mod} \ \langle\Phi\rangle);$$

it is a group morphism because, as W acts trivially on $X(\mathbf{T})/\langle\Phi\rangle$ (this is true for a reflection, whence also for an arbitrary product of reflections), we have

$$\frac{ww'(x) - w(x)}{m} \equiv \frac{w'(x) - x}{m} \quad (\mathrm{mod} \ \langle\Phi\rangle)$$

which is equivalent to

$$\frac{ww'(x) - x}{m} \equiv \frac{w'(x) - x}{m} + \frac{w(x) - x}{m} \quad (\mathrm{mod} \ \langle\Phi\rangle).$$

The kernel of this morphism is $\{w \in W \mid w(x) - x \in m\langle\Phi\rangle\}$ which is a reflection group by Bourbaki (1968, VI, Ex. 1 of §2), so is equal to $W^0(s)$ by 3.5.2. Thus we get an embedding of $W(s)/W^0(s)$ into $X(\mathbf{T})/\langle\Phi\rangle$. The image is a p'-torsion group since for any w we have $w(x) - x \in \langle\Phi\rangle$, so $m((w(x) - x)/m) \in \langle\Phi\rangle$, which means that the exponent of the image divides m. We then get (iii) by (i) and 1.2.13. $\square$

Remark 11.2.2

(i) Note that the above proof, using 3.5.2, shows that, for any semi-simple element s of a connected algebraic group $\mathbf{G}$, the exponent of $C_\mathbf{G}(s)/C_\mathbf{G}(s)^0$ divides the order of s.

(ii) A consequence of 11.2.1(iii) is that, if the centre of $\mathbf{G}$ is connected, the centraliser of any semi-simple element of $\mathbf{G}^*$ is connected.

We will now give further information on $C_{\mathbf{G}}(s)/C_{\mathbf{G}}(s)^0$. We first show that we can reduce to adjoint groups.

Proposition 11.2.3   *Let $s \in \mathbf{G}$ be semi-simple and let $\bar{s}$ be the image of s in $\overline{\mathbf{G}} := \mathbf{G}/Z(\mathbf{G})$; then $C_{\mathbf{G}}(s)/C_{\mathbf{G}}(s)^0$ identifies with a subgroup of $C_{\overline{\mathbf{G}}}(\bar{s})/C_{\overline{\mathbf{G}}}(\bar{s})^0$.*

Proof   Let $\mathbf{T}$ be a maximal torus containing s and let $\overline{\mathbf{T}}$ be its image in $\overline{\mathbf{G}}$; the roots of $\mathbf{G}$ with respect to $\mathbf{T}$ identify with those of $\overline{\mathbf{G}}$ with respect to $\overline{\mathbf{T}}$ and the $\mathbf{U}_\alpha$ are isomorphic. By 3.5.1 $C_{\mathbf{G}}(s)^0 = \langle \mathbf{T}, \mathbf{U}_\alpha \mid \alpha(s) = 1 \rangle$ and $C_{\overline{\mathbf{G}}}(\bar{s})^0 = \langle \overline{\mathbf{T}}, \mathbf{U}_\alpha \mid \alpha(\bar{s}) = 1 \rangle$ but $\alpha(s) = \alpha(\bar{s})$, hence $C_{\overline{\mathbf{G}}}(\bar{s})^0 = C_{\mathbf{G}}(s)^0/Z(\mathbf{G})$. The proposition follows. $\square$

Now, if $\mathbf{G}$ is adjoint it is a direct product of quasi-simple adjoint groups, so we may reduce the study of $C_{\mathbf{G}}(s)/C_{\mathbf{G}}(s)^0$ to the quasi-simple and adjoint case. Then $\mathbf{G}^*$ is quasi-simple simply connected. We will describe $Z(\mathbf{G}^*)$ in this case, from which the description of the groups $C_{\mathbf{G}}(s)/C_{\mathbf{G}}(s)^0$ follows by 11.2.1(iii).

We first need:

Proposition 11.2.4   *Let Φ be an irreducible crystallographic root system, and let Φ^+ be a positive subsystem and Π the corresponding basis of Φ. Then there is a unique root (in Φ^+), called the **highest root**, such that the sum of its coefficients on Π is maximal.*

Proof   Denote by λ the linear form which takes value 1 on Π. Let α be a root with maximal value of λ. Note that $\langle \alpha, \beta \rangle \geq 0$ for any $\beta \in \Phi^+$ otherwise by 3.3.3(i) $\alpha + \beta$ would be a root with greater value of λ. Now write $\alpha = \sum_{\gamma \in \Pi} n_\gamma \gamma$. We show that no n_γ is 0; otherwise let $I = \{\gamma \in \Pi \mid n_\gamma = 0\}$ and let $J = \Pi - I$. Then for $\delta \in I$ we have $0 \leq \langle \alpha, \delta \rangle = \langle \sum_{\gamma \in J} n_\gamma \gamma, \delta \rangle \leq 0$, the left inequality by what we just showed and the right one by 2.2.5. It follows that we must have $\langle \gamma, \delta \rangle = 0$ for any $\gamma \in J, \delta \in I$ which contradicts the irreducibility of Φ if $I \neq \emptyset$. We deduce the uniqueness of α: if β is another root with maximal value of λ, in $0 \leq \langle \alpha, \beta \rangle = \langle \sum_{\gamma \in \Pi} n_\gamma \gamma, \beta \rangle$ we must have some term $\langle \gamma, \beta \rangle \neq 0$ otherwise $\beta = 0$. It follows that $\langle \alpha, \beta \rangle > 0$ thus by 3.3.3(i) both $\alpha - \beta$ and $\beta - \alpha$ are roots. One of them must be positive, which contradicts the maximality of $\lambda(\alpha)$ or that of $\lambda(\beta)$. $\square$

Definition 11.2.5   Let Φ be an irreducible crystallographic root system, let Φ^+ be a positive subsystem and Π be the corresponding basis. We call **extended basis** of Φ the set $\tilde{\Pi} = \Pi \cup \{\alpha_0\}$ where α_0 is the negative of the highest root.

Proposition 11.2.6   *Let* **G** *be a quasi-simple simply connected reductive group, and let* $\tilde{\Pi}$ *be an extended root basis of the root system of* **G**. *Then* $Z(\mathbf{G})$ *is isomorphic to the group* Ω *of permutations of* $\tilde{\Pi}$ *induced by the elements of* W *which stabilise* $\tilde{\Pi}$.

Reference   See for example Bonnafé (2006, Proposition 5.3).   □

Table 11.1 illustrates such groups Ω. The extended root basis gives rise to an **extended Coxeter diagram** which records the scalar products between elements of $\tilde{\Pi}$ (or the order of st, where s and t are the corresponding reflections). The extended Coxeter diagram is also the Coxeter diagram describing a Coxeter group, the **affine Weyl group**, which is the affine reflection group in $Y(\mathbf{T}) \otimes \mathbb{R}$ generated by W and the reflection with respect to the affine hyperplane of equation $\langle \alpha_0, y \rangle = 1$, see for example Bourbaki (1968, Chap. VI, §2.3). We describe the extended diagrams and generators of the group Ω, described by their cycle decomposition, in each case. In Table 11.1, we just write $0, 1, \ldots, n$ for the elements $\alpha_0, \alpha_1, \ldots, \alpha_n$ of an extended root basis, where $\Pi = \{\alpha_1, \ldots, \alpha_n\}$.

We give another application of the extended basis of an irreducible root system Φ, due to Borel and De Siebenthal:

Proposition 11.2.7   *The sets of the form* $\tilde{\Pi} - \{\alpha\}$, *for* $\alpha \in \tilde{\Pi}$, *are bases of closed root subsystems of* Φ. *Every maximal proper closed root subsystem of* Φ *has a basis conjugate to one of these sets.*

Reference   See for example Bourbaki (1968, VI, Ex. 4 of §4).   □

11.3 The Lusztig Series

The next definition is a first step towards the classification of characters of $\mathbf{G}^F$.

Definition 11.3.1   The **Lusztig series** $\mathcal{E}(\mathbf{G}^F, (s))$ associated to a semi-simple geometric conjugacy class $(s) \subset \mathbf{G}^{*F^*}$ is the set of irreducible characters of $\mathbf{G}^F$ which occur in some $R_{\mathbf{T}^*}^{\mathbf{G}}(s)$, for some F^*-stable maximal torus $\mathbf{T}^*$ of $\mathbf{G}^*$ containing an element of (s).

The series $\mathcal{E}(\mathbf{G}^F, (s))$ are sometimes called "geometric Lusztig series"; we will also (see 12.4.4) consider "$\mathbf{G}^{*F^*}$-Lusztig series" where the $\mathbf{G}^{*F^*}$-class of s is fixed. The two notions coincide, by 4.2.15(ii), when $C_{\mathbf{G}^*}(s)$ is connected.

Proposition 11.3.2   *We have* $\mathrm{Irr}(\mathbf{G}^F) = \coprod_{(s)} \mathcal{E}(\mathbf{G}^F, (s))$ *where* (s) *runs over the semi-simple geometric conjugacy classes of* $\mathbf{G}^{*F^*}$.

Table 11.1. *Groups Ω for extended Coxeter diagrams*

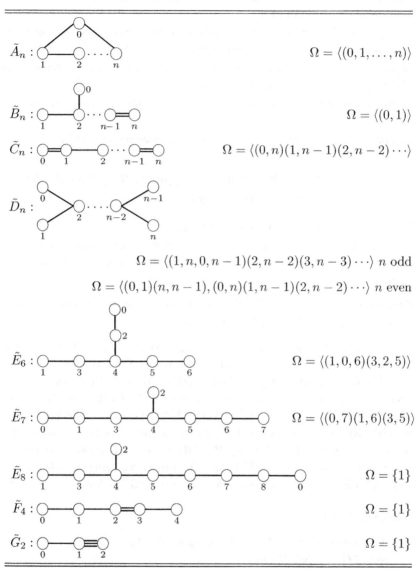

$$\tilde{A}_n : \qquad \Omega = \langle (0, 1, \ldots, n) \rangle$$

$$\tilde{B}_n : \qquad \Omega = \langle (0, 1) \rangle$$

$$\tilde{C}_n : \qquad \Omega = \langle (0, n)(1, n - 1)(2, n - 2) \cdots \rangle$$

$$\tilde{D}_n :$$

$$\Omega = \langle (1, n, 0, n - 1)(2, n - 2)(3, n - 3) \cdots \rangle \ n \text{ odd}$$

$$\Omega = \langle (0, 1)(n, n - 1), (0, n)(1, n - 1)(2, n - 2) \cdots \rangle \ n \text{ even}$$

$$\tilde{E}_6 : \qquad \Omega = \langle (1, 0, 6)(3, 2, 5) \rangle$$

$$\tilde{E}_7 : \qquad \Omega = \langle (0, 7)(1, 6)(3, 5) \rangle$$

$$\tilde{E}_8 : \qquad \Omega = \{1\}$$

$$\tilde{F}_4 : \qquad \Omega = \{1\}$$

$$\tilde{G}_2 : \qquad \Omega = \{1\}$$

Proof By 11.1.3 and 11.1.15 two Deligne–Lusztig characters in different series have no common constituent, and by 11.1.1 any irreducible character is in some series. □

Remark 11.3.3 If $\mathbf{U} \rtimes \mathbf{T}$ is a Levi decomposition of a Borel subgroup of $\mathbf{G}$, by 11.1.3 any $\gamma \in \mathrm{Irr}(\mathbf{G}^F)$ occurring in some $H_c^i(\mathcal{L}^{-1}(\mathbf{U})) \otimes_{\overline{\mathbb{Q}}_\ell \mathbf{T}^F} \theta$ is in the series

$\mathcal{E}(\mathbf{G}^F, (s))$, where (s) is the semi-simple conjugacy class of $\mathbf{G}^*$ corresponding to the geometric class of $(\mathbf{T}, \theta)$.

A particularly important series, which is a kind of "prototype" for the other ones, is the series associated to the identity element of $\mathbf{G}^*$.

Definition 11.3.4   The elements of $\mathcal{E}(\mathbf{G}^F, 1)$ (that is, the irreducible constituents of the $R_{\mathbf{T}}^{\mathbf{G}}(1)$) are called **unipotent characters**.

We introduce a variety whose cohomology gives directly the unipotent characters.

Definition 11.3.5   If $\mathbf{B}$ is a Borel subgroup containing an F-stable maximal torus, we define the variety

$$\mathbf{X_B} := \{g\mathbf{B} \in \mathbf{G}/\mathbf{B} \mid g\mathbf{B} \cap F(g\mathbf{B}) \neq \emptyset\} = \{g\mathbf{B} \in \mathbf{G}/\mathbf{B} \mid g^{-1\,F}g \in \mathbf{B} \cdot {}^F\mathbf{B}\}$$
$$\simeq \{g \in \mathbf{G} \mid g^{-1\,F}g \in {}^F\mathbf{B}\}/(\mathbf{B} \cap {}^F\mathbf{B}).$$

The group $\mathbf{G}^F$ acts by left translations on $\mathbf{X_B}$.

The same arguments as after 9.1.1 show that the three formulas define isomorphic varieties. They are smooth by, for example, the proof of Digne et al. (2007, 2.3.5).

Proposition 11.3.6   *If $\mathbf{B} = \mathbf{U} \rtimes \mathbf{T}$ is a Levi decomposition of a Borel subgroup where $\mathbf{T}$ is F-stable then $\mathbf{X_U}/\mathbf{T}^F \simeq \mathbf{X_B}$ as $\mathbf{G}^F$-varieties.*

Proof   We work with the models $\mathbf{X_U} = \{g \in \mathbf{G} \mid g^{-1\,F}g \in {}^F\mathbf{U}\}/(\mathbf{U} \cap {}^F\mathbf{U})$ and $\mathbf{X_B} = \{g \in \mathbf{G} \mid g^{-1\,F}g \in {}^F\mathbf{B}\}/(\mathbf{B} \cap {}^F\mathbf{B})$. By 3.4.8(ii), we have $\mathbf{B} \cap {}^F\mathbf{B} = (\mathbf{U} \cap {}^F\mathbf{U})\mathbf{T}$. The morphism $g(\mathbf{U} \cap {}^F\mathbf{U}) \mapsto g(\mathbf{B} \cap {}^F\mathbf{B})$ maps $\mathbf{X_U}$ to $\mathbf{X_B}$. If $g(\mathbf{B} \cap {}^F\mathbf{B}) \in \mathbf{X_B}$, we have $g^{-1\,F}g = tu$ with $t \in \mathbf{T}$ and $u \in {}^F\mathbf{U}$. By the Lang–Steinberg theorem in $\mathbf{T}$, we write $t = \tau.{}^F\tau^{-1}$ so that $g\tau(\mathbf{U} \cap {}^F\mathbf{U})$ is in $\mathbf{X_U}$ and is a preimage of $g(\mathbf{B} \cap {}^F\mathbf{B})$. Moreover τ is well defined up to multiplication by $\mathbf{T}^F$. This shows that the morphism $g(\mathbf{U} \cap {}^F\mathbf{U}) \mapsto g(\mathbf{B} \cap {}^F\mathbf{B})$ is surjective and that it defines a bijective morphism from $\mathbf{X_U}/\mathbf{T}^F$ to $\mathbf{X_B}$. The variety $\mathbf{X_B}$ being smooth, it is an isomorphism by 9.1.2, which is compatible with the $\mathbf{G}^F$-actions.   □

Corollary 11.3.7

(i)   *With notation as in 11.3.6, the character $R_{\mathbf{T}}^{\mathbf{G}}(\mathbf{1})$ is the character of the virtual $\mathbf{G}^F$-module $H_c^*(\mathbf{X_B})$.*

(ii)   *The unipotent characters of $\mathbf{G}^F$ are the irreducible constituents of the $H_c^i(\mathbf{X_B})$ when $\mathbf{B}$ runs over Borel subgroups containing an F-stable maximal torus.*

Proof   Both statements are consequences of the proposition and of 8.1.10(i), using Remark 11.3.3 for (ii).   □

The next statement shows that the set of unipotent characters depends only on the isomorphism type of the root system of **G**.

Proposition 11.3.8 *Let* $(\mathbf{G}, F)$ *and* $(\mathbf{G}_1, F_1)$ *be two connected reductive groups with Frobenius roots, and let* $f : \mathbf{G} \to \mathbf{G}_1$ *be a morphism of algebraic groups with a central kernel, such that* $f \circ F = F_1 \circ f$ *and such that* $f(\mathbf{G})$ *contains* Der $\mathbf{G}_1$; *then the unipotent characters of* $\mathbf{G}^F$ *are the* $\chi \circ f$, *where* χ *runs over the unipotent characters of* $\mathbf{G}_1^{F_1}$.

Proof We first remark that in the situation of the proposition we have $\mathbf{G}_1 = Z(\mathbf{G}_1)^0.f(\mathbf{G})$ by 2.3.8. The Borel subgroups, the maximal tori and the parabolic subgroups of $f(\mathbf{G})$ are thus the intersections of those of $\mathbf{G}_1$ with $f(\mathbf{G})$. On the other hand, f induces a bijection from the Borel subgroups, the maximal tori and the parabolic subgroups of $\mathbf{G}$ to those of $f(\mathbf{G})$, so there is a bijection from the sets of Borel subgroups, maximal tori and parabolic subgroups of $\mathbf{G}$ to the corresponding sets for $\mathbf{G}_1$.

The unipotent characters of $\mathbf{G}^F$ are those which occur in some cohomology space of some variety $\mathbf{X_B}$ where $\mathbf{B}$ runs over Borel subgroups of $\mathbf{G}$ containing an F-stable maximal torus. Let $\mathbf{B}_1 = f(\mathbf{B})Z(\mathbf{G}_1)^0$ be the Borel subgroup of $\mathbf{G}_1$ corresponding to $\mathbf{B}$ by the above bijection. The morphism $g\mathbf{B} \mapsto f(g\mathbf{B})$: $\mathbf{X_B} \to \mathbf{X_{B_1}}$ is compatible with the actions of $\mathbf{G}^F$ and $\mathbf{G}_1^{F_1}$. Since Ker$f \subset \mathbf{B}$ this morphism is injective. Let us show that it is an isomorphism, using the second model of the varieties. Let $x_1 \in \mathbf{G}_1$ be such that $x_1^{-1}{}^{F_1}x_1 \in \mathbf{B}_1{}^{F_1}\mathbf{B}_1$; multiplying by an element of $Z(\mathbf{G}_1)^0$ does not change $x_1\mathbf{B}_1$, hence we can assume $x_1 \in f(\mathbf{G})$. A preimage x of x_1 in $\mathbf{G}$ satisfies $x^{-1}{}^Fx \in {}^F\mathbf{B}$ whence the surjectivity. We conclude using that $\mathbf{X_{B_1}}$ is smooth and 9.1.2.

We have thus proved that $R_\mathbf{T}^\mathbf{G}(1) = R_{\mathbf{T}_1}^{\mathbf{G}_1}(1) \circ f$, where $\mathbf{T}$ is an F-stable maximal torus of $\mathbf{B}$, and $\mathbf{T}_1 = f(\mathbf{T})Z(\mathbf{G}_1)^0$ is the corresponding F_1-stable maximal torus of $\mathbf{G}_1$. Moreover as $\mathbf{G}^F$-classes of F-stable maximal tori are in one-to-one correspondence with $\mathbf{G}_1^{F_1}$-classes (they are parametrised by the F-classes of the Weyl group), the "restriction through f" defines a bijection from the set of Deligne–Lusztig characters $R_\mathbf{T}^\mathbf{G}(1)$ of $\mathbf{G}^F$ onto the similar set for $\mathbf{G}_1^{F_1}$. It remains to see that the restriction through f maps the irreducible constituents of one to those of the other.

We have $\mathbf{G}_1^{F_1} = \mathbf{T}_1^{F_1}.f(\mathbf{G}^F)$. Indeed, any $y \in \mathbf{G}_1$ can be written $y = zf(x)$ with $z \in Z(\mathbf{G}_1)^0$ and $x \in \mathbf{G}$, and y is F_1-stable if and only if $f(x^Fx^{-1}) = z^{-1}{}^{F_1}z$. This element is in $Z(\mathbf{G}_1)^0 \cap f(\mathbf{G}) \subset Z(f(\mathbf{G}))$ and this last group is in $f(\mathbf{T})$ which is a maximal torus of the reductive group $f(\mathbf{G})$ (see the beginning of the present proof). So $x^Fx^{-1} \in \mathbf{T}$, since the kernel of f is in $\mathbf{T}$, and by the

Lang–Steinberg theorem applied in the group $\mathbf{T}$ we can find $t \in \mathbf{T}$ such that $tx \in \mathbf{G}^F$. So $y = (zf(t^{-1}))f(tx) \in \mathbf{T}_1^{F_1} \cdot f(\mathbf{G}^F)$.

It follows that $\mathbf{G}_1^{F_1}/f(\mathbf{G}^F)$ is a quotient of $\mathbf{T}_1^{F_1}$ thus commutative, so we can use the following result from Clifford's theory, see Curtis and Reiner (1981, 11.4, 11.5).

Lemma 11.3.9 *Let G be a finite group and H be a normal subgroup of G such that G/H is abelian. Then for $\chi, \chi' \in \mathrm{Irr}(G)$, we have*

$$\langle \mathrm{Res}_H^G \chi, \mathrm{Res}_H^G \chi' \rangle_H = \#\{\zeta \in \mathrm{Irr}(G/H) \mid \chi\zeta = \chi'\}.$$

From this lemma we see that the restriction of a unipotent character χ of $\mathbf{G}_1^{F_1}$ to $f(\mathbf{G}^F)$ is irreducible if and only if for any irreducible non-trivial character ζ of $\mathbf{G}_1^{F_1}/f(\mathbf{G}^F)$ we have $\chi \neq \chi\zeta$. But $\chi\zeta$ is not unipotent if ζ is not trivial, because if χ is a constituent of $R_{\mathbf{T}_1}^{\mathbf{G}_1}(1)$ then $\chi\zeta$ is a constituent of $R_{\mathbf{T}_1}^{\mathbf{G}_1}(\mathrm{Res}_{\mathbf{T}_1^{F_1}}^{\mathbf{G}_1^{F_1}} \zeta)$ and the restriction of ζ to $\mathbf{T}_1^{F_1}$ is not trivial, since $\mathbf{G}_1^{F_1} = \mathbf{T}_1^{F_1} \cdot f(\mathbf{G}^F)$, so cannot be geometrically conjugate to the trivial character, whence the proposition. $\square$

There is a similar – though weaker – result for a general Lusztig functor.

Proposition 11.3.10 *We assume the same hypotheses as in 11.3.8, and assume, in addition, that the kernel of f is connected. Then, if $\mathbf{L}_1$ is an F_1-stable Levi subgroup of some parabolic subgroup $\mathbf{P}_1$ of $\mathbf{G}_1$ and if $\lambda_1 \in \mathrm{Irr}(\mathbf{L}_1^{F_1})$, $\gamma_1 \in \mathrm{Irr}(\mathbf{G}_1^{F_1})$, we have*

$$R_{\mathbf{L}_1 \subset \mathbf{P}_1}^{\mathbf{G}_1}(\lambda_1) \circ f = R_{f^{-1}(\mathbf{L}_1) \subset f^{-1}(\mathbf{P}_1)}^{\mathbf{G}}(\lambda_1 \circ f)$$

and

$${}^*R_{\mathbf{L}_1 \subset \mathbf{P}_1}^{\mathbf{G}_1}(\gamma_1) \circ f = {}^*R_{f^{-1}(\mathbf{L}_1) \subset f^{-1}(\mathbf{P}_1)}^{\mathbf{G}}(\gamma_1 \circ f).$$

Proof By the remarks we made at the beginning of the proof of 11.3.8, the group $\mathbf{P} := f^{-1}(\mathbf{P}_1) = f^{-1}(\mathbf{P}_1 \cap f(\mathbf{G}))$ is a parabolic subgroup of $\mathbf{G}$ which has $\mathbf{L} := f^{-1}(\mathbf{L}_1) = f^{-1}(\mathbf{L}_1 \cap f(\mathbf{G}))$ as Levi subgroup, so the statement makes sense. If $\mathbf{P} = \mathbf{U} \rtimes \mathbf{L}$ is the Levi decomposition of $\mathbf{P}$, and $\mathbf{P}_1 = \mathbf{U}_1 \rtimes \mathbf{L}_1$ that of $\mathbf{P}_1$, then we have $\mathbf{U}_1 = f(\mathbf{U})$ since $\mathbf{U}_1 \subset \mathrm{Der}\,\mathbf{G}_1 \subset f(\mathbf{G})$.

To prove the proposition, we will use the following.

Lemma 11.3.11

(i) *If f is surjective, it induces isomorphisms $\mathcal{L}_{\mathbf{G}}^{-1}(\mathbf{U})/(\mathrm{Ker}\,f)^F \simeq \mathcal{L}_{\mathbf{G}_1}^{-1}(\mathbf{U}_1)$ and $H_c^*(\mathcal{L}_{\mathbf{G}}^{-1}(\mathbf{U}))^{(\mathrm{Ker}\,f)^F} \simeq H_c^*(\mathcal{L}_{\mathbf{G}_1}^{-1}(\mathbf{U}_1))$.*

(ii) *If f is an embedding of $\mathbf{G}$ as a closed subgroup of $\mathbf{G}_1$ containing $\mathrm{Der}\,\mathbf{G}_1$, we have $\mathbf{U} = \mathbf{U}_1$ and $\mathcal{L}_{\mathbf{G}_1}^{-1}(\mathbf{U}) = \coprod_{g \in \mathbf{G}_1^{F_1}/\mathbf{G}^F} g \cdot \mathcal{L}_{\mathbf{G}}^{-1}(\mathbf{U}) = \coprod_{l \in \mathbf{L}_1^{F_1}/\mathbf{L}^F} \mathcal{L}_{\mathbf{G}}^{-1}(\mathbf{U}) \cdot l.$*

Proof For (i) we first prove the surjectivity of the morphism induced by f. If $\mathcal{L}_{\mathbf{G}_1}(g_1) = u_1 \in \mathbf{U}_1$ then any $g \in f^{-1}(g_1)$ satisfies $\mathcal{L}_{\mathbf{G}}(g) = uz$ with $z \in \mathrm{Ker} f$ and $u \in \mathbf{U}$. Since $\mathrm{Ker} f$ is connected we may find $z' \in \mathrm{Ker} f$ such that $\mathcal{L}_{\mathbf{G}}(z') = z^{-1}$. Then $\mathcal{L}_{\mathbf{G}}(gz') = u$ (using that z' is central), thus $gz' \in \mathcal{L}_{\mathbf{G}}^{-1}(\mathbf{U})$.

We now show that the fibres of f on $\mathcal{L}^{-1}(\mathbf{U})$ are isomorphic to $(\mathrm{Ker}\, z)^F$. If g and gz are in $\mathcal{L}^{-1}(\mathbf{U})$, where $z \in \mathrm{Ker} f$, then $\mathcal{L}_{\mathbf{G}}(gz) = \mathcal{L}_{\mathbf{G}}(z)\mathcal{L}_{\mathbf{G}}(g) \in \mathcal{L}_{\mathbf{G}}(z)\mathbf{U}$, and $\mathcal{L}_{\mathbf{G}}(z)\mathbf{U}$ is disjoint from $\mathbf{U}$ unless $\mathcal{L}_{\mathbf{G}}(z) = 1$, that is $z \in (\mathrm{Ker} f)^F$. Using the smoothness of $\mathcal{L}^{-1}(\mathbf{U})$ and 9.1.2, we get the first assertion of (i). The second assertion of (i) follows from 8.1.10(i).

For the first equality of (ii) it is clear that $\mathcal{L}_{\mathbf{G}}^{-1}(\mathbf{U}) \subset \mathcal{L}_{\mathbf{G}_1}^{-1}(\mathbf{U})$ and since the right variety is stable by left translation by $\mathbf{G}_1^{F_1}$, it is sufficient to prove that it is covered by $\mathbf{G}_1^{F_1}$-translates of $\mathcal{L}_{\mathbf{G}}^{-1}(\mathbf{U})$. If $g_1 \in \mathcal{L}_{\mathbf{G}_1}^{-1}(\mathbf{U})$ then by the Lang–Steinberg theorem there exists $g \in \mathcal{L}_{\mathbf{G}}^{-1}(\mathbf{U})$ such that $\mathcal{L}_{\mathbf{G}}(g) = \mathcal{L}_{\mathbf{G}_1}(g_1)$, which implies that $g_1 g^{-1} \in \mathbf{G}_1^{F_1}$, whence the result. We deduce the last equality using $\mathbf{G}_1^{F_1}/\mathbf{G}^F \simeq \mathbf{L}_1^{F_1}/\mathbf{L}^F$. $\qquad\square$

The restriction from $\mathbf{L}_1^{F_1}$ to $\mathbf{L}^F$ through f is $\overline{\mathbb{Q}}_\ell \mathbf{L}_1^{F_1} \otimes_{\overline{\mathbb{Q}}_\ell \mathbf{L}_1^{F_1}} \cdot$. Decomposing f into a surjective map followed by an embedding, we thus get $R_{\mathbf{L}}^{\mathbf{G}} \circ \mathrm{Res}_{\mathbf{L}^F}^{\mathbf{L}_1^{F_1}} = H_c^*(\mathcal{L}_{\mathbf{G}}^{-1}(\mathbf{U})) \otimes_{\overline{\mathbb{Q}}_\ell \mathbf{L}^F} \overline{\mathbb{Q}}_\ell \mathbf{L}_1^{F_1} \otimes_{\overline{\mathbb{Q}}_\ell \mathbf{L}_1^{F_1}} \cdot$. Since both cases of the lemma imply that $H_c^*(\mathcal{L}_{\mathbf{G}}^{-1}(\mathbf{U})) \otimes_{\overline{\mathbb{Q}}_\ell \mathbf{L}^F} \overline{\mathbb{Q}}_\ell \mathbf{L}_1^{F_1} \simeq H_c^*(\mathcal{L}_{\mathbf{G}_1}^{-1}(\mathbf{U}_1))$ as $\mathbf{G}^F$-modules-$\mathbf{L}_1^{F_1}$, the action of $\mathbf{G}^F$ on the right-hand side being by restriction through f we get the first equality of the proposition.

The proof of the second equality is similar, using 9.1.4, interchanging $\mathbf{G}$ and $\mathbf{L}$ and simultaneously $\mathbf{G}_1$ and $\mathbf{L}_1$ and using that both cases of the lemma imply $H_c^*(\mathcal{L}_{\mathbf{G}}^{-1}(\mathbf{U}))^{\mathrm{opp}} \otimes_{\overline{\mathbb{Q}}_\ell \mathbf{G}^F} \overline{\mathbb{Q}}_\ell \mathbf{G}_1^{F_1} \simeq H_c^*(\mathcal{L}_{\mathbf{G}_1}^{-1}(\mathbf{U}_1))^{\mathrm{opp}}$ as $\mathbf{L}^F$-modules-$\mathbf{G}_1^{F_1}$, the action of $\mathbf{L}^F$ on the right-hand side being by restriction through f. $\qquad\square$

11.4 Lusztig's Jordan Decomposition of Characters: The Levi Case

We explain now Lusztig's "Jordan decomposition of characters" which gives a parametrisation of $\mathcal{E}(\mathbf{G}^F, (s))$ by the unipotent characters of $C_{\mathbf{G}^*}(s)^F$. We first explain how to get explicitly such a bijection when $C_{\mathbf{G}^*}(s)$ is a Levi subgroup of a parabolic subgroup of $\mathbf{G}^*$; this is in a sense the general case as "most" semi-simple elements have this property: we will show (see just after 12.1.12) that, modulo the centre, there is only a finite number of semi-simple conjugacy classes whose centraliser is not in a proper Levi subgroup of $\mathbf{G}$.

We first explain the correspondence between F-stable Levi subgroups of $\mathbf{G}$ and F^*-stable Levi subgroups of $\mathbf{G}^*$. We first classify $\mathbf{G}^F$-conjugacy classes of F-stable Levi subgroups.

Proposition 11.4.1   *Let* **G** *be a connected reductive group over* $\overline{\mathbb{F}}_p$ *with a Frobenius root* F. *Let* $\mathbf{T} \subset \mathbf{B}$ *be a pair of an* F-*stable maximal torus and an* F-*stable Borel subgroup and let* $(W(\mathbf{T}), S)$ *be the corresponding Coxeter system. Then the* $\mathbf{G}^F$-*classes of* F-*stable Levi subgroups of* **G** *are parametrised by* F-*conjugacy classes of cosets* $W_I w$ *where* $I \subset S$ *and* w *are such that* $^{wF}W_I = W_I$.

Proof   According to 4.2.14 the $\mathbf{G}^F$-classes of Levi subgroups conjugate under **G** to **L** are parametrised by $H^1(F, N_\mathbf{G}(\mathbf{L})/\mathbf{L})$. We may conjugate **L** to some $\mathbf{L}_I$ by conjugating a maximal torus of **L** to **T** but then, by a similar argument to the remark after 4.2.22, the action of F on **L** becomes that of $\dot{v}F$ on $\mathbf{L}_I$ where $\dot{v} \in N_\mathbf{G}(\mathbf{T})$ is a representative of some $v \in W(\mathbf{T})$; the condition that **L** is F-stable becomes $^{\dot{v}F}\mathbf{L}_I = \mathbf{L}_I$, which is equivalent to $^{vF}W_I = W_I$ as seen in the proof of 3.4.3. So the $\mathbf{G}^F$-classes of Levi subgroups conjugate under **G** to **L** are parametrised by $H^1(\dot{v}F, N_\mathbf{G}(\mathbf{L}_I)/\mathbf{L}_I) \simeq H^1(vF, N_W(W_I)/W_I)$, which is the same as the F-classes under $N_W(W_I)$ of cosets $W_I z v$ where $z \in N_W(W_I)$. This last condition is equivalent to $^{zvF}W_I = W_I$. We thus get the proposition by putting $w = zv$; the F-classes under $N_W(W_I)$ are replaced by F-classes under W since, if the F-class of $W_I w$ and that of $W_J w'$ are equal, then W_I is F-conjugate under W to W_J. $\square$

Note that, as the parametrisation given by 11.4.1 is the same in **G** and $\mathbf{G}^*$, the $\mathbf{G}^F$-classes of F-stable Levi subgroups of **G** correspond one-to-one to the similar classes in $\mathbf{G}^*$. If the class of **L** corresponds to the F-class under $W(\mathbf{T})$ of $W_I w$ (that is, to the $W(\mathbf{T})$-class of $W_I wF$), we make **L** correspond to a Levi subgroup $\mathbf{L}^*$ of $\mathbf{G}^*$ whose class is parametrised by $F^* w^* W_I^*$. As seen in the proof of 11.4.1, we may find in **L** a torus $\mathbf{T}_w$ of type w; that is, $(\mathbf{T}_w, F)$ is geometrically conjugate to $(\mathbf{T}, wF)$; similarly we may find in $\mathbf{L}^*$ a torus $\mathbf{T}_w^*$ such that $(\mathbf{T}_w^*, F^*)$ is geometrically conjugate to $(\mathbf{T}^*, F^* w^*)$. With these choices $(\mathbf{L}, \mathbf{T}_w, F)$ is dual to $(\mathbf{L}^*, \mathbf{T}_w^*, F^*)$, where the isomorphism $Y(\mathbf{T}_w) \xrightarrow{\sim} X(\mathbf{T}_w^*)$ is "transported" from the isomorphism $Y(\mathbf{T}) \xrightarrow{\sim} X(\mathbf{T}^*)$ by the chosen geometric conjugations. Given the duality between $(\mathbf{G}, \mathbf{T})$ and $(\mathbf{G}^*, \mathbf{T}^*)$, the duality thus defined between **L** and $\mathbf{L}^*$ is defined up to the automorphisms of **L** induced by **G**; that is, by $N_\mathbf{G}(\mathbf{L})$. It is easy to check that this duality is compatible with 11.1.15 and 11.1.16; that is, if $(\mathbf{T}, \theta)$ where $\mathbf{T} \subset \mathbf{L}$ corresponds to $(\mathbf{T}^*, s)$ where $\mathbf{T}^* \subset \mathbf{L}^*$ for this duality, then it also corresponds to $(\mathbf{T}^*, s)$ for the duality in **G**.

Definition 11.4.2   A semi-simple element s of a reductive group **G** is **quasi-isolated** if $C_\mathbf{G}(s)$ is not contained in any proper Levi subgroup of **G**.

We may then state the following result, which reduces the problem of parametrising $\mathcal{E}(\mathbf{G}^F, (s))$ to the case of quasi-isolated elements. In particular the case of elements such that $C_{\mathbf{G}^*}(s)$ is a Levi subgroup is reduced to the case of central elements.

Theorem 11.4.3 *Let s be a semi-simple element of $\mathbf{G}^{*F^*}$ and let $\mathbf{L}^*$ be an F^*-stable Levi subgroup of $\mathbf{G}^*$ which contains $C_{\mathbf{G}^*}(s)$. Let $\mathbf{L}$ be an F-stable Levi subgroup of $\mathbf{G}$ whose $\mathbf{G}^F$-class corresponds to the $\mathbf{G}^{*F^*}$-class of $\mathbf{L}^*$ as explained above, and let $\mathbf{U} \rtimes \mathbf{L}$ be the Levi decomposition of some parabolic subgroup of $\mathbf{G}$ containing $\mathbf{L}$; then:*

(i) *For any $\lambda \in \mathcal{E}(\mathbf{L}^F, (s))$ there exists an integer $i(\lambda)$ such that the space $H_c^i(\mathbf{X_U}) \otimes_{\overline{\mathbb{Q}}_\ell \mathbf{L}^F} \lambda$ is zero for $i \neq i(\lambda)$ and affords an irreducible representation of $\mathbf{G}^F$ for $i = i(\lambda)$; we thus have $(-1)^{i(\lambda)} R_{\mathbf{L}}^{\mathbf{G}}(\lambda) \in \mathrm{Irr}(\mathbf{G}^F)$. When $\mathbf{X_U}$ is known to be affine then $i(\lambda) = \dim \mathbf{X_U}$.*

(ii) *The functor $\varepsilon_{\mathbf{G}} \varepsilon_{\mathbf{L}} R_{\mathbf{L}}^{\mathbf{G}}$ induces a bijection from $\mathcal{E}(\mathbf{L}^F, (s))$ to $\mathcal{E}(\mathbf{G}^F, (s))$.*

Proof We begin with a lemma which translates for the group $\mathbf{G}$ the condition $C_{\mathbf{G}^*}(s) \subset \mathbf{L}^*$.

Lemma 11.4.4 *In the situation of 11.4.3*

(i) *Let $\mathbf{T}$ and $\mathbf{T}'$ be two F-stable maximal tori of $\mathbf{L}$ and let $\theta \in \mathrm{Irr}(\mathbf{T}^F)$ and $\theta' \in \mathrm{Irr}(\mathbf{T}'^F)$ be such that $(\mathbf{T}, \theta)$ and $(\mathbf{T}', \theta')$ are geometrically conjugate in $\mathbf{L}$ and their geometric class corresponds to $(s) \subset \mathbf{L}^{*F^*}$ by the duality between $\mathbf{L}$ and $\mathbf{L}^*$; then any $g \in \mathbf{G}$ which geometrically conjugates $(\mathbf{T}, \theta)$ to $(\mathbf{T}', \theta')$ is in $\mathbf{L}$.*

(ii) *Let $\mathbf{T}$, an F-stable maximal torus of $\mathbf{G}$ and $\theta \in \mathrm{Irr}(\mathbf{T}^F)$ be such that the geometric conjugacy class of $(\mathbf{T}, \theta)$ corresponds to the geometric class of s in $\mathbf{G}^{*F^*}$. Then $\mathbf{T}$ is $\mathbf{G}^F$-conjugate to some torus of $\mathbf{L}$.*

Proof To prove (i) it is clearly enough to consider the case where $\mathbf{T} = \mathbf{T}'$ and $\theta = \theta'$. By 11.1.8 g is then an element of $N_{\mathbf{G}}(\mathbf{T})$ which fixes θ seen as an element of $\mathrm{Hom}(Y, (\mathbb{Q}/\mathbb{Z})_{p'})$ as in 11.1.7(ii). The assumption does not change if we replace F by F^n. Thus we can assume that $\mathbf{T}$ is split and $g \in \mathbf{G}^F$. Then the Weyl group W of $\mathbf{G}$ is $N_{\mathbf{G}}(\mathbf{T})/\mathbf{T}$ and the image w of g in W fixes the element of $\mathrm{Hom}(Y, (\mathbb{Q}/\mathbb{Z})_{p'})$ defined by θ by 11.1.7(ii); that is, by the proof of 11.1.15, fixes the corresponding element of $Y(\mathbf{T}^*) \otimes (\mathbb{Q}/\mathbb{Z})_{p'}$ which is s by definition. Hence $w \in W(s) \subset W_{\mathbf{L}}$ so that g is in $\mathbf{L}$.

We now prove (ii). Let $\mathbf{T}_0 \subset \mathbf{L}$ and $\theta_0 \in \mathrm{Irr}(\mathbf{T}_0^F)$ be such that the geometric class in $\mathbf{L}$ of the pair $(\mathbf{T}_0, \theta_0)$ corresponds to the class in $\mathbf{L}^*$ of s. Then $(\mathbf{T}_0, \theta_0)$ also corresponds to (s) for the duality between $\mathbf{G}$ and $\mathbf{G}^*$, so $(\mathbf{T}, \theta)$ and $(\mathbf{T}_0, \theta_0)$ are geometrically conjugate in $\mathbf{G}$. Let $g \in \mathbf{G}$ be an element which geometrically conjugates $(\mathbf{T}, \theta)$ to $(\mathbf{T}_0, \theta_0)$. Then ${}^F g$ also geometrically conjugates $(\mathbf{T}, \theta)$ to $(\mathbf{T}_0, \theta_0)$, so $g.{}^F g^{-1}$ geometrically conjugates $(\mathbf{T}_0, \theta_0)$ to itself. Thus by (i) we have $g.{}^F g^{-1} \in \mathbf{L}$. Using the Lang–Steinberg theorem to write $g.{}^F g^{-1} = l^{-1}.{}^F l$ with $l \in \mathbf{L}$, we get an element lg which is F-stable and conjugates $\mathbf{T}$ to some torus of $\mathbf{L}$. $\qquad\square$

We now place ourselves in the setting of 9.2.1, where we take $\mathbf{L}' = \mathbf{L}$ and $\mathbf{P}' = \mathbf{P} = \mathbf{LU}$ and we work with the variety $\mathcal{L}^{-1}(\mathbf{U})$ instead of $\mathbf{X}_\mathbf{U}$ as is allowed by 9.1.5. The key step for 11.4.3 is Proposition 11.4.5.

Proposition 11.4.5 *Under the same assumptions as 11.4.3, let λ and λ' be two irreducible representations in $\mathcal{E}(\mathbf{L}^F,(s))$, then*

$$\lambda^{\mathrm{opp}} \otimes_{\overline{\mathbb{Q}}_\ell \mathbf{L}^F} H_c^i(\mathbf{Z},\overline{\mathbb{Q}}_\ell) \otimes_{\overline{\mathbb{Q}}_\ell \mathbf{L}^F} \lambda'$$

is 0 except for $i = 2d$, where $d = \dim \mathbf{U} + \dim(\mathbf{U} \cap {}^F\mathbf{U})$, where $\mathbf{Z}$ is as in 9.2.1 and λ^{opp} is as in the proof of 11.1.3. In addition, we have

$$\dim(\lambda^{\mathrm{opp}} \otimes_{\overline{\mathbb{Q}}_\ell \mathbf{L}^F} H_c^{2d}(\mathbf{Z},\overline{\mathbb{Q}}_\ell) \otimes_{\overline{\mathbb{Q}}_\ell \mathbf{L}^F} \lambda') = \langle \lambda,\overline{\lambda'}\rangle_{\mathbf{L}^F}.$$

Proof We may choose two F-stable maximal tori $\mathbf{T}$ and $\mathbf{T}'$ in $\mathbf{L}$, characters $\theta \in \mathrm{Irr}(\mathbf{T}^F)$ and $\theta' \in \mathrm{Irr}(\mathbf{T}'^F)$ and $\mathbf{TV}$ (resp. $\mathbf{T}'\mathbf{V}'$) (the Levi decomposition of) a Borel subgroup of $\mathbf{L}$ containing $\mathbf{T}$ (resp. $\mathbf{T}'$) such that there exists j (resp. k) for which λ (resp. λ') is a constituent of $H_c^j(\mathcal{L}^{-1}(\mathbf{V})) \otimes_{\overline{\mathbb{Q}}_\ell \mathbf{T}^F} \theta$ (resp. of $H_c^k(\mathcal{L}^{-1}(\mathbf{V}')) \otimes_{\overline{\mathbb{Q}}_\ell \mathbf{T}'^F} \theta'$); let $\mathbf{Z}_w$ be as defined after 9.2.1, then for all i, $\lambda^{\mathrm{opp}} \otimes_{\overline{\mathbb{Q}}_\ell \mathbf{L}^F} H_c^i(\mathbf{Z}_w,\overline{\mathbb{Q}}_\ell) \otimes_{\overline{\mathbb{Q}}_\ell \mathbf{L}^F} \lambda'$ is a subspace of

$$\theta^{\mathrm{opp}} \otimes_{\overline{\mathbb{Q}}_\ell \mathbf{T}^F} H_c^j(\mathcal{L}^{-1}(\mathbf{V}))^{\mathrm{opp}} \otimes_{\overline{\mathbb{Q}}_\ell \mathbf{L}^F} H_c^i(\mathbf{Z}_w,\overline{\mathbb{Q}}_\ell) \otimes_{\overline{\mathbb{Q}}_\ell \mathbf{L}^F} H_c^k(\mathcal{L}^{-1}(\mathbf{V}')) \otimes_{\overline{\mathbb{Q}}_\ell \mathbf{T}'^F} \theta'.$$

By similar arguments to those used in the proof of 9.1.8, the $\overline{\mathbb{Q}}_\ell \mathbf{T}^F$-module-$\overline{\mathbb{Q}}_\ell \mathbf{T}'^F$ given by

$$H_c^j(\mathcal{L}^{-1}(\mathbf{V}))^{\mathrm{opp}} \otimes_{\overline{\mathbb{Q}}_\ell \mathbf{L}^F} H_c^i(\mathbf{Z}_w,\overline{\mathbb{Q}}_\ell) \otimes_{\overline{\mathbb{Q}}_\ell \mathbf{L}^F} H_c^k(\mathcal{L}^{-1}(\mathbf{V}'))$$

is isomorphic to a submodule of $H_c^{i+j+k}(\bigcup_{w_1} \mathbf{Z}_{w_1}^1,\overline{\mathbb{Q}}_\ell)$, where $\mathbf{Z}_{w_1}^1$ is the variety analogous to $\mathbf{Z}_w$, relative to $\mathbf{T}$ and $\mathbf{T}'$, and where w_1 runs over elements of $\mathbf{T}\backslash\mathcal{S}(\mathbf{T},\mathbf{T}')/\mathbf{T}'$ having the same image in $\mathbf{L}\backslash\mathcal{S}(\mathbf{L},\mathbf{L})/\mathbf{L}$ as w. By an argument similar to the proofs of 11.1.4 and 11.1.5, the character $\theta \otimes \theta'^{\mathrm{opp}}$ does not occur in this module if none of the w_1 geometrically conjugate θ to θ'. By 11.4.4(i), this implies that w is in $\mathbf{L}$. Thus $\lambda^{\mathrm{opp}} \otimes_{\overline{\mathbb{Q}}_\ell \mathbf{L}^F} H_c^i(\mathbf{Z}_w,\overline{\mathbb{Q}}_\ell) \otimes_{\overline{\mathbb{Q}}_\ell \mathbf{L}^F} \lambda'$ is zero if w is not in $\mathbf{L}$. If w is in $\mathbf{L}$, that is $w = 1$ in $\mathbf{L}\backslash\mathcal{S}(\mathbf{L},\mathbf{L})/\mathbf{L}$, the variety $\mathbf{Z}_w = \mathbf{Z}_1$ may be simplified. We have

$$\mathbf{Z}_1 = \{(u,u',g) \in \mathbf{U} \times \mathbf{U} \times {}^{F^{-1}}\mathbf{P} \mid u^F g = gu'\},$$

so $(u,u',g) \in \mathbf{Z}_1$ implies $u^F g u'^{-1} = g \in \mathbf{P} \cap {}^{F^{-1}}\mathbf{P}$. Using the decomposition 3.4.8(ii) $\mathbf{P} \cap {}^{F^{-1}}\mathbf{P} = \mathbf{L}.(\mathbf{U} \cap {}^{F^{-1}}\mathbf{U})$, and taking projections to $\mathbf{L}$, we see that $g \in \mathbf{L}^F.(\mathbf{U} \cap {}^{F^{-1}}\mathbf{U})$; so the variety projects to $\mathbf{L}^F$ with all fibres isomorphic to $\mathbf{U} \times (\mathbf{U} \cap {}^F\mathbf{U})$. Thus, up to a shift of $2d$, the cohomology is that of the discrete variety $\mathbf{L}^F$, thus

$$\lambda^{\mathrm{opp}} \otimes_{\overline{\mathbb{Q}}_\ell \mathbf{L}^F} H_c^i(\mathbf{Z}_w, \overline{\mathbb{Q}}_\ell) \otimes_{\overline{\mathbb{Q}}_\ell \mathbf{L}^F} \lambda' =$$

$$\begin{cases} \lambda^{\mathrm{opp}} \otimes_{\overline{\mathbb{Q}}_\ell \mathbf{L}^F} \overline{\mathbb{Q}}_\ell \mathbf{L}^F \otimes_{\overline{\mathbb{Q}}_\ell \mathbf{L}^F} \lambda' & \text{if } w \in \mathbf{L} \text{ and } i = 2d, \\ 0 & \text{otherwise,} \end{cases}$$

and the dimension of $\lambda^{\mathrm{opp}} \otimes_{\overline{\mathbb{Q}}_\ell \mathbf{L}^F} \overline{\mathbb{Q}}_\ell \mathbf{L}^F \otimes_{\overline{\mathbb{Q}}_\ell \mathbf{L}^F} \lambda'$ is $\langle \lambda, \overline{\lambda'} \rangle_{\mathbf{L}^F}$. Using the arguments of 11.1.5 we get the required result for $\lambda^{\mathrm{opp}} \otimes_{\overline{\mathbb{Q}}_\ell \mathbf{L}^F} H_c^i(\mathbf{Z}, \overline{\mathbb{Q}}_\ell) \otimes_{\overline{\mathbb{Q}}_\ell \mathbf{L}^F} \lambda'$. □

As the scalar product $\langle R_\mathbf{L}^\mathbf{G} \lambda, R_\mathbf{L}^\mathbf{G} \lambda' \rangle_{\mathbf{G}^F}$ is equal (using 8.1.6) to the alternating sum of the dimensions of

$$\lambda^{\mathrm{opp}} \otimes_{\overline{\mathbb{Q}}_\ell \mathbf{L}^F} H_c^*(\mathcal{L}^{-1}(\mathbf{U}))^{\mathrm{opp}} \otimes_{\overline{\mathbb{Q}}_\ell \mathbf{G}^F} H_c^*(\mathcal{L}^{-1}(\mathbf{U})) \otimes_{\overline{\mathbb{Q}}_\ell \mathbf{L}^F} \overline{\lambda'},$$

the above proposition thus shows, using 9.2.1, that

$$\langle R_\mathbf{L}^\mathbf{G} \lambda, R_\mathbf{L}^\mathbf{G} \lambda' \rangle_{\mathbf{G}^F} = \langle \lambda, \lambda' \rangle_{\mathbf{L}^F}.$$

Remark 11.4.6 We note that under the assumption of 11.4.3, even if we do not assume $\mathbf{P} = \mathbf{P}'$, the preceding proof shows that $w \in \mathbf{L}$ and this together with Lemma 9.2.5 shows that the following statement (equivalent to the Mackey formula when λ or λ' is cuspidal) holds:

$$\langle R_{\mathbf{L} \subset \mathbf{P}}^\mathbf{G} \lambda, R_{\mathbf{L} \subset \mathbf{P}'}^\mathbf{G} \lambda' \rangle_\mathbf{G} = \langle \lambda, \lambda' \rangle_{\mathbf{L}^F}.$$

As seen in the proof of 5.3.1, this implies that $R_{\mathbf{L} \subset \mathbf{P}}^\mathbf{G} \lambda$ is independent of $\mathbf{P}$ in that case.

We now prove theorem 11.4.3. From 11.4.5 the dimension of

$$\oplus_{i+j=2d} \lambda^{\mathrm{opp}} \otimes_{\overline{\mathbb{Q}}_\ell \mathbf{L}^F} H_c^i(\mathcal{L}^{-1}(\mathbf{U}))^{\mathrm{opp}} \otimes_{\overline{\mathbb{Q}}_\ell \mathbf{G}^F} H_c^j(\mathcal{L}^{-1}(\mathbf{U})) \otimes_{\overline{\mathbb{Q}}_\ell \mathbf{L}^F} \overline{\lambda}$$

$$\simeq \lambda^{\mathrm{opp}} \otimes_{\overline{\mathbb{Q}}_\ell \mathbf{L}^F} H_c^{2d}(\mathbf{Z}) \otimes_{\overline{\mathbb{Q}}_\ell \mathbf{L}^F} \overline{\lambda}$$

is equal to 1, so all the summands have dimension 0 except one, say

$$\lambda^{\mathrm{opp}} \otimes_{\overline{\mathbb{Q}}_\ell \mathbf{L}^F} H_c^{i_0}(\mathcal{L}^{-1}(\mathbf{U}))^{\mathrm{opp}} \otimes_{\overline{\mathbb{Q}}_\ell \mathbf{G}^F} H_c^{2d-i_0}(\mathcal{L}^{-1}(\mathbf{U})) \otimes_{\overline{\mathbb{Q}}_\ell \mathbf{L}^F} \overline{\lambda}$$

which has dimension 1.

Suppose now that the $\overline{\mathbb{Q}}_\ell \mathbf{G}^F$-module given by

$$H_c^j(\mathcal{L}^{-1}(\mathbf{U})) \otimes_{\overline{\mathbb{Q}}_\ell \mathbf{L}^F} \lambda$$

is not 0. Let χ be one of its irreducible constituents. Then χ is in $\mathcal{E}(\mathbf{G}^F, (s))$, so $\overline{\chi}$ is in $\mathcal{E}(\mathbf{G}^F, (s^{-1}))$: indeed in the exact sequence of 11.1.7(i), if $\theta \in \mathrm{Irr}(\mathbf{T}^F)$ is the image of $x \in X$ then $\overline{\theta} = \theta^{-1}$ is the image of $-x$ and if $s \in \mathbf{T}^{*F^*}$ is the image of x in the exact sequence analog to 11.1.7(ii) for $\mathbf{T}^*$ then s^{-1} is the image of $-x$. By 11.4.4(ii), χ is a constituent of $R_\mathbf{T}^\mathbf{G}(\theta)$ where $\mathbf{T}$ is a maximal torus of $\mathbf{L}$

and θ is given by the geometric class of s. So $\overline{\chi}$ is a constituent of $R_{\mathbf{L}}^{\mathbf{G}}R_{\mathbf{T}}^{\mathbf{L}}(\overline{\theta})$ and in particular appears in some $R_{\mathbf{L}}^{\mathbf{G}}(\overline{\lambda'})$ with $\lambda' \in \mathcal{E}(\mathbf{L}^F,(s))$. Then $\overline{\chi}^{\mathrm{opp}}$ occurs in some $\lambda'^{\mathrm{opp}} \otimes_{\overline{\mathbb{Q}}_\ell \mathbf{L}^F} H_c^k(\mathcal{L}^{-1}(\mathbf{U}))^{\mathrm{opp}}$ and thus

$$E := \lambda^{\mathrm{opp}} \otimes_{\overline{\mathbb{Q}}_\ell \mathbf{L}^F} H_c^j(\mathcal{L}^{-1}(\mathbf{U}))^{\mathrm{opp}} \otimes_{\overline{\mathbb{Q}}_\ell \mathbf{G}^F} H_c^k(\mathcal{L}^{-1}(\mathbf{U})) \otimes_{\overline{\mathbb{Q}}_\ell \mathbf{L}^F} \lambda' \neq 0.$$

But E is a subspace of $\lambda^{\mathrm{opp}} \otimes_{\overline{\mathbb{Q}}_\ell \mathbf{L}^F} H_c^{j+k}(\mathbf{Z},\overline{\mathbb{Q}}_\ell) \otimes_{\overline{\mathbb{Q}}_\ell \mathbf{L}^F} \lambda'$, so this last space is not 0, which proves by 11.4.5 that $j + k = 2d$ and $\lambda' = \overline{\lambda}$ thus by definition of i_0 we have $j = i_0$. Since in that case E has dimension 1, we see that χ must be, in addition, the only irreducible constituent of $H_c^{i_0}(\mathcal{L}^{-1}(\mathbf{U})) \otimes_{\overline{\mathbb{Q}}_\ell \mathbf{L}^F} \lambda$. Whence (i) of the theorem with $i(\lambda) = i_0 - 2\dim(\mathbf{U} \cap {}^F\mathbf{U})$. When $\mathbf{X}_{\mathbf{U}}$ is affine, its cohomology is zero in degree less than $\dim \mathbf{X}_{\mathbf{U}}$, so that of $\mathcal{L}^{-1}(\mathbf{U})$ is zero in degree less than $\dim \mathbf{X}_{\mathbf{U}} + 2\dim(\mathbf{U} \cap {}^F\mathbf{U}) = d$, hence $i_0 \geq d$ and $2d - i_0 \geq d$, thus $i_0 = d$, whence $i(\lambda) = d - 2\dim(\mathbf{U} \cap {}^F\mathbf{U}) = \dim \mathbf{X}_{\mathbf{U}}$.

Let us show (ii). (i) implies that $R_{\mathbf{L}}^{\mathbf{G}}$ maps up to sign $\mathcal{E}(\mathbf{L}^F,(s))$ to $\mathcal{E}(\mathbf{G}^F,(s))$. That the sign is $\varepsilon_{\mathbf{G}}\varepsilon_{\mathbf{L}}$ follows immediately from 10.2.9. Let us show that $\varepsilon_{\mathbf{G}}\varepsilon_{\mathbf{L}}R_{\mathbf{L}}^{\mathbf{G}}$ is surjective. Let $\chi \in \mathcal{E}(\mathbf{G}^F,(s))$; by definition χ is a constituent of $R_{\mathbf{T}'}^{\mathbf{G}}(\theta')$ for some torus $\mathbf{T}'$ and some character θ' which is in the class defined by s. By 11.4.4 (ii), the torus $\mathbf{T}'$ has a $\mathbf{G}^F$-conjugate in $\mathbf{L}$. We may thus assume that $\mathbf{T}'$ is in $\mathbf{L}$. In that case $R_{\mathbf{T}'}^{\mathbf{G}}(\theta') = R_{\mathbf{L}}^{\mathbf{G}} \circ R_{\mathbf{T}'}^{\mathbf{L}}(\theta')$, so χ is indeed the $R_{\mathbf{L}}^{\mathbf{G}}$ of a character of $\mathcal{E}(\mathbf{L}^F,(s))$. $\qquad\square$

Exercise 11.4.7   (i) With the notation as in 11.4.3, let $\mathbf{B}_{\mathbf{L}}$ be an F-stable Borel subgroup of $\mathbf{L}$ and let $w \in W$ be such that the pair $(\mathbf{B}_{\mathbf{L}}\mathbf{U},F)$ is $\mathbf{G}$-conjugate to $(\mathbf{B}_0,wF)$ where $\mathbf{B}_0$ is an F-stable Borel subgroup of $\mathbf{G}$. Show that $\dim \mathbf{X}_{\mathbf{U}} = l(w)$.

(ii) Show that a quasi-split torus of $\mathbf{L}$ contained in $\mathbf{B}_{\mathbf{L}}$ is of type w in $\mathbf{G}$.

(iii) If $\mathbf{X}_{\mathbf{U}}$ is known to be affine, under the assumption of Theorem 11.4.3, show using (i) of the theorem without using 10.2.9 that $\varepsilon_{\mathbf{G}}\varepsilon_{\mathbf{L}}R_{\mathbf{L}}^{\mathbf{G}}$ maps a true character to a true character (hint: we have $\varepsilon_{\mathbf{G}}\varepsilon_{\mathbf{L}} = (-1)^{l(w)}$).

The special case of Theorem 11.4.3 where $\mathbf{L}^* = C_{\mathbf{G}^*}(s)$ shows that $R_{\mathbf{L}}^{\mathbf{G}}$ is then up to sign a bijection from $\mathcal{E}(C_{\mathbf{G}^*}(s)^{*F},(s))$ to $\mathcal{E}(\mathbf{G}^F,(s))$. Moreover,

Proposition 11.4.8   *Let s be an F^*-stable central element of $\mathbf{G}^*$. Then:*

(i) *There exists a linear character $\hat{s} \in \mathrm{Irr}(\mathbf{G}^F)$ such that, for any F-stable maximal torus $\mathbf{T}$ of $\mathbf{G}$, the pair $(\mathbf{T},\hat{s}|_{\mathbf{T}^F})$ is in the geometric conjugacy class defined by s.*

(ii) *Taking the tensor product with $\hat{s}$ defines a bijection*

$$\mathcal{E}(\mathbf{G}^F,1) \xrightarrow{\sim} \mathcal{E}(\mathbf{G}^F,(s)).$$

Proof We follow Deligne and Lusztig (1976, 5.11).

As $s \in Z(\mathbf{G}^{*F^*})$, its geometric class defines a character of $\mathbf{T}'^F$ for any F-stable maximal torus $\mathbf{T}' \subset \mathbf{G}$ since, if $(\mathbf{G}, \mathbf{T})$ is dual to $(\mathbf{G}^*, \mathbf{T}^*)$, the element s is in $\mathbf{T}^{*wF^*}$ for any w. Let $(\tilde{\mathbf{G}}, \tilde{\mathbf{T}})$ be a dual pair to $(\mathbf{G}^*/Z(\mathbf{G}^*), \mathbf{T}^*/Z(\mathbf{G}^*))$.

Proposition 11.4.9 *The pair* $(\mathbf{G}^*/Z(\mathbf{G}^*)^0, \mathbf{T}^*/Z(\mathbf{G}^*)^0)$ *is dual to the pair* $(\mathrm{Der}\,\mathbf{G}, \mathbf{T} \cap \mathrm{Der}\,\mathbf{G})$. *Let* π_0 *be the inclusion* $\mathrm{Der}\,\mathbf{G} \to \mathbf{G}$ *and* $f : \tilde{\mathbf{G}} \to \mathrm{Der}\,\mathbf{G}$ *be the dual isogeny to* $f^* : \mathbf{G}^*/Z(\mathbf{G}^*)^0 \to \mathbf{G}^*/Z(\mathbf{G}^*)$; *then the composed map* $\pi := \pi_0 \circ f$ *is an* F-*equivariant map with central kernel which induces an isomorphism on the* $\mathbf{U}_\alpha$ *and such that* $\pi(\tilde{\mathbf{T}}) \subset \mathbf{T}$.

Proof The groups $\mathbf{G}$ and $\mathrm{Der}\,\mathbf{G}$ have same root systems $(\Phi, \Phi^\vee)$ with respect to $\mathbf{T}$ and $\mathbf{T} \cap \mathrm{Der}\,\mathbf{G}$ respectively and the groups $\mathbf{G}^*$ and $\mathbf{G}^*/Z(\mathbf{G}^*)^0$ have same root systems $(\Phi^\vee, \Phi)$ with respect to $\mathbf{T}^*$ and $\mathbf{T}^*/Z(\mathbf{G}^*)^0$ respectively. We have $X(\mathbf{T}^*/Z(\mathbf{G}^*)^0) = ((\Phi^\vee)^\perp)^\perp = Y(\mathbf{T} \cap \mathrm{Der}\,\mathbf{G})$, the first equality by 2.3.10(i) and the second one by 2.3.10(ii). This shows that $(\mathbf{G}^*/Z(\mathbf{G}^*)^0, \mathbf{T}^*/Z(\mathbf{G}^*)^0)$ is dual to $(\mathrm{Der}\,\mathbf{G}, \mathbf{T} \cap \mathrm{Der}\,\mathbf{G})$.

The map f is F-equivariant since f^* is F^*-equivariant. The other statements are obvious since they are true for an isogeny and for the inclusion $\mathrm{Der}\,\mathbf{G} \subset \mathbf{G}$. □

We have $\pi(\tilde{\mathbf{G}}) = \mathrm{Der}\,\mathbf{G}$, and $\pi(\tilde{\mathbf{T}}) = \mathbf{T} \cap \mathrm{Der}\,\mathbf{G}$. We thus get $\pi(Z(\tilde{\mathbf{G}})) = Z(\mathrm{Der}\,\mathbf{G})$ since the centre consists of the elements in a maximal torus which act trivially on the $\mathbf{U}_\alpha$. Since the kernel of π is central, we deduce that π induces an isomorphism $\tilde{\mathbf{T}}/Z(\tilde{\mathbf{G}}) \simeq (\mathbf{T} \cap \mathrm{Der}\,\mathbf{G})/Z(\mathrm{Der}\,\mathbf{G})$.

Lemma 11.4.10 $\mathbf{T} \cap \pi(\tilde{\mathbf{G}}) = \pi(\tilde{\mathbf{T}})$ *and* $\mathbf{T}^F \cap \pi(\tilde{\mathbf{G}}^F) = \pi(\tilde{\mathbf{T}}^F)$.

Proof We have seen the first statement. The second comes from $\pi^{-1}(\mathbf{T}) = \tilde{\mathbf{T}}$, since $\mathrm{Ker}\,\pi \subset \tilde{\mathbf{T}}$, thus an element of $\mathbf{T}^F \cap \pi(\tilde{\mathbf{G}}^F)$ is the image of an element in $\tilde{\mathbf{G}}^F \cap \pi^{-1}(\mathbf{T}) = \tilde{\mathbf{T}}^F$. □

Lemma 11.4.11 *The inclusion* $\mathbf{T} \subset \mathbf{G}$ *induces an isomorphism* $\mathbf{T}^F/\pi(\tilde{\mathbf{T}}^F) \simeq \mathbf{G}^F/\pi(\tilde{\mathbf{G}}^F)$.

Proof We have $\mathbf{G} = \mathbf{T}\,\mathrm{Der}\,\mathbf{G} = \mathbf{T}\pi(\tilde{\mathbf{G}})$, the first equality by 2.3.8. For $g \in \mathbf{G}^F$, let us write $g = t\pi(\tilde{g})$ with $t \in \mathbf{T}$ and $\tilde{g} \in \tilde{\mathbf{G}}$. Since g is F-stable we have $t^{-1}{}^F t = \pi(\tilde{g}^F \tilde{g}^{-1}) \in \mathbf{T} \cap \pi(\tilde{\mathbf{G}}) = \pi(\tilde{\mathbf{T}})$, the last equality by the previous lemma. Since the kernel of π is contained in $\tilde{\mathbf{T}}$ this implies $\tilde{g}^F \tilde{g}^{-1} \in \tilde{\mathbf{T}}$. By the Lang–Steinberg theorem we can write $\tilde{g}^F \tilde{g}^{-1} = \tilde{t}_1^{-1}{}^F \tilde{t}_1$ with $\tilde{t}_1 \in \tilde{\mathbf{T}}$. Then $\tilde{t}_1 \tilde{g}$ is in $\tilde{\mathbf{G}}^F$ and $t\pi(\tilde{t}_1^{-1}) = g\pi(\tilde{t}_1 \tilde{g})^{-1}$ is in $\mathbf{T}^F$. Thus we have $g = t\pi(\tilde{t}_1^{-1})\pi(\tilde{t}_1 \tilde{g}) \in \mathbf{T}^F\pi(\tilde{\mathbf{G}}^F)$, hence $\mathbf{G}^F = \mathbf{T}^F\pi(\tilde{\mathbf{G}}^F)$. Using the second assertion of the previous lemma we get $\mathbf{G}^F/\pi(\tilde{\mathbf{G}}^F) \simeq \mathbf{T}^F/\pi(\tilde{\mathbf{T}}^F)$. □

Proposition 11.4.12 *There is a natural bijection between $Z(\mathbf{G}^*)^{F^*}$ and the characters of the abelian group $\mathbf{G}^F/\pi(\tilde{\mathbf{G}}^F)$.*

Proof We have $Y(\tilde{\mathbf{T}}) = X(\mathbf{T}^*/Z(\mathbf{G}^*)) = ((\Phi^\vee)^\perp_{\mathbf{T}^*})^\perp_{X(\mathbf{T}^*)}$ where $(\Phi, \Phi^\vee)$ is the root system of $\mathbf{G}$. A character $\theta \in \mathrm{Irr}(\mathbf{T}^F)$ vanishes on $\pi(\tilde{\mathbf{T}}^F)$ that is on the images of the coroots if and only if the corresponding element $s \in \mathbf{G}^{*F^*}$ is in $(\Phi^\vee)^\perp_{\mathbf{T}^*}$ that is $s \in Z(\mathbf{G}^*)^{F^*}$. Conversely for any F-stable $\mathbf{T}$, an element $s \in Z(\mathbf{G}^*)^{F^*}$ defines a character of $\mathbf{T}^F$ which vanishes on $\pi(\tilde{\mathbf{T}}^F)$ hence a character of $\mathbf{G}^F/\pi(\tilde{\mathbf{G}}^F)$. This character is independent of the torus $\mathbf{T}$ since the characters obtained in various tori are geometrically conjugate, and

Lemma 11.4.13 *Geometrically conjugate characters of F-stable maximal tori induce the same character of $\mathbf{G}^F/\pi(\tilde{\mathbf{G}}^F)$.*

Proof Let $\mathbf{T}$ and $\mathbf{T}'$ be two F-stable maximal tori of $\mathbf{G}$ and let n be such that there exists $x \in \mathbf{G}^{F^n}$ such that $\mathbf{T}' = {}^x\mathbf{T}$. It is sufficient to show that for $t \in \mathbf{T}^{F^n}$ the elements $N_{F^n/F}(t)$ and $N_{F^n/F}({}^xt)$ have same image in $\mathbf{G}^F/\pi(\tilde{\mathbf{G}}^F)$, where $N_{F^n/F} : \mathbf{G} \to \mathbf{G}$ is the map $y \mapsto y^F y \ldots {}^{F^{n-1}}y$. Since $\mathbf{G} = Z(\mathbf{G})^0\pi(\tilde{\mathbf{G}})$ we may assume $x = \pi(\tilde{x}) \in \pi(\tilde{\mathbf{G}})$ and take n large enough so that $\tilde{x} \in \tilde{\mathbf{G}}^{F^n}$.

Let $\phi_{\mathbf{G}} : \mathbf{T} \to \mathbf{G}$ be the map $t \mapsto N_{F^n/F}(t)^{-1}N_{F^n/F}({}^xt)$, and let $\phi_{\tilde{\mathbf{G}}}$ be the analogous map $\tilde{\mathbf{T}} \to \tilde{\mathbf{G}}$ defined using $\tilde{x}$.

Take now $t \in \mathbf{T}^{F^n}$. Lift it to $\tilde{t} \in \tilde{\mathbf{G}}$ such that $\pi(\tilde{t}) = t$; then $\tilde{t}^{-1 F^n}\tilde{t} \in \mathrm{Ker}\,\pi \subset Z(\tilde{\mathbf{G}})$. Using that for any y in an F-stable torus we have ${}^FN_{F^n/F}(y) = y^{-1 F^n}yN_{F^n/F}(y)$, we get ${}^F\phi_{\tilde{\mathbf{G}}}(\tilde{t}) = (\tilde{t}^{-1 F^n}\tilde{t})^{-1}\phi_{\tilde{\mathbf{G}}}(\tilde{t})^{\tilde{x}}(\tilde{t}^{-1 F^n}\tilde{t}) = \phi_{\tilde{\mathbf{G}}}(\tilde{t})$, the last equality since $\tilde{t}^{-1 F^n}\tilde{t}$ is central. Thus $\phi_{\mathbf{G}}(t) = \pi(\phi_{\tilde{\mathbf{G}}}(\tilde{t}))$ is in $\pi(\tilde{\mathbf{G}}^F)$.   □

□

We have seen in the proof of 11.4.12 that the character of $\mathbf{T}^F$ corresponding to $s \in Z(\mathbf{G}^*)^{F^*}$ is in the geometric class defined by s, whence 11.4.8(i).

Let us now prove 11.4.8(ii). Taking the tensor product with $\hat{s}$ permutes the irreducible characters, and by 10.1.2(i) we have $R_{\mathbf{T}}^{\mathbf{G}}(\theta)\hat{s} = R_{\mathbf{T}}^{\mathbf{G}}(\theta\hat{s})$ for any $\mathbf{T}$. As the isomorphism $\mathrm{Irr}(\mathbf{T}^{wF}) \simeq \mathbf{T}^{*F^*w^*}$ is compatible with multiplication, if the geometric class of $(\mathbf{T}, \theta)$ corresponds to that of $s_1 \in \mathbf{G}^{*F^*}$, then the class of $(\mathbf{T}, \theta\hat{s})$ corresponds to that of s_1s, whence 11.4.8(ii).   □

Remark 11.4.14 Actually $\mathbf{G}^F/\pi(\tilde{\mathbf{G}}^F)$ is the semi-simple quotient of the abelian quotient of $\mathbf{G}^F$ (for quasi-simple groups, the abelian quotient of $\mathbf{G}^F$ has no unipotent elements except for the cases mentioned in 4.3.6). Indeed, the theorem of Steinberg recalled above 4.3.6 says that $\tilde{\mathbf{G}}^F$ is generated by its unipotent elements, thus the same holds for $\pi(\tilde{\mathbf{G}}^F)$; thus, by 11.4.11, $\pi(\tilde{\mathbf{G}}^F)$ is generated by the unipotent elements of $\mathbf{G}^F$. The result then follows from the theorem of Tits recalled above 4.3.6.

Proposition 11.4.8 together with 11.4.3(ii) gives the special case of Theorem 11.5.1 below where $C_{\mathbf{G}^*}(s)$ is a Levi subgroup, if we show that the unipotent characters of a group and of its dual are in one-to-one correspondence; this results from 11.3.8 when the root systems of $\mathbf{G}$ and $\mathbf{G}^*$ have same type; that is, unless $\mathbf{G}$ has a quasi-simple component of type B_n or C_n; it was proved in this last case in Lusztig (1977).

11.5 Lusztig's Jordan Decomposition of Characters: The General Case

We now give a statement of Lusztig's "Jordan decomposition of characters". The statement is more complicated when $Z(\mathbf{G})$ is not connected; to give a general statement, we first generalise the definition of Deligne–Lusztig characters to a disconnected group $\mathbf{G}$ by setting $R_{\mathbf{T}^0}^{\mathbf{G}}(\theta) = \mathrm{Ind}_{\mathbf{G}^{0F}}^{\mathbf{G}^{F}}(R_{\mathbf{T}^0}^{\mathbf{G}^0}(\theta))$, where $\mathbf{T}^0$ is an F-stable maximal torus of $\mathbf{G}^0$ – we will not use this, but $R_{\mathbf{T}^0}^{\mathbf{G}}(\theta)$ is a sum of some characters $R_{\mathbf{T}}^{\mathbf{G}}(\theta')$ which may be defined using a "quasi-torus" $\mathbf{T} = N_{\mathbf{G}}(\mathbf{T}^0, \mathbf{B})$ where $\mathbf{B}$ is a Borel subgroup which contains $\mathbf{T}^0$, see Digne and Michel (1994).

This definition allows us to extend to the case of disconnected $\mathbf{G}$ the definition of unipotent characters, keeping the same definition; that is, $\mathcal{E}(\mathbf{G}^F, 1)$ is defined as the set of irreducible constituents of the $R_{\mathbf{T}^0}^{\mathbf{G}}(1)$. With these definitions, and using the definition 12.4.3 of $\mathbf{G}^{*F^*}$-Lusztig series $\mathcal{E}(\mathbf{G}^F, (s)_{\mathbf{G}^{*F^*}})$, we may state the following theorem.

Theorem 11.5.1 *Let $\mathbf{G}$ be a connected reductive group with a Frobenius root F; for any semi-simple element $s \in \mathbf{G}^{*F^*}$, there is a bijection from $\mathcal{E}(\mathbf{G}^F, (s)_{\mathbf{G}^{*F^*}})$ to $\mathcal{E}(C_{\mathbf{G}^*}(s)^{F^*}, 1)$. This bijection may be chosen such that, extended by linearity to virtual characters, it sends $\varepsilon_{\mathbf{G}} R_{\mathbf{T}^*}^{\mathbf{G}}(s)$ (see remark after 11.1.16) to $\varepsilon_{C_{\mathbf{G}^*}(s)^0} R_{\mathbf{T}^*}^{C_{\mathbf{G}^*}(s)}(1)$ for any F^*-stable maximal torus $\mathbf{T}^*$ of $C_{\mathbf{G}^*}(s)$.*

Comments and Proof. When the centre of $\mathbf{G}$ is connected, the group $C_{\mathbf{G}^*}(s)$ is connected for any semi-simple element $s \in \mathbf{G}^*$ – see 11.2.2(ii) – and the remarks above about disconnected groups are not needed.

When F is a Frobenius endomorphism and $\mathbf{G}$ has a connected centre Theorem 11.5.1 is the main topic of Lusztig (1984a). The description of the bijection in Lusztig (1984a) leaves some ambiguity, but this ambiguity may be lifted – see Digne and Michel (1990, Theorem 7.1). The bijection was extended in Lusztig (1988) to the disconnected centre case, but it is not explicit. We will first in 11.5.2 and 11.5.3 state the theorem of Lusztig (1988) and explain how to deduce our statement 11.5.1 when F is a Frobenius morphism.

An easy computation shows that conjugation by $g \in \mathbf{G}$ preserves $\mathbf{G}^F$ if and only if $\mathcal{L}_F(g) \in C_{\mathbf{G}}(\mathbf{G}^F)$; the induced automorphisms of $\mathbf{G}^F$ obtained this way are called **diagonal automorphisms** of $\mathbf{G}^F$. Now $C_{\mathbf{G}}(\mathbf{G}^F) = Z(\mathbf{G})$ – see 12.2.17 – and the condition $\mathcal{L}_F(g) \in Z(\mathbf{G})$ is equivalent to the image $\bar{g} \in \mathbf{G}/Z(\mathbf{G})$ being F-stable. So we have an action of $(\mathbf{G}/Z(\mathbf{G}))^F$ on $\mathbf{G}^F$ by diagonal automorphisms. Further, we have the Galois cohomology exact sequence 4.3.5

$$1 \to Z(\mathbf{G})^F \to \mathbf{G}^F \to (\mathbf{G}/Z(\mathbf{G}))^F \to H^1(F, Z(\mathbf{G})) \to 1$$

thus the image of the diagonal automorphisms in the outer automorphisms of $\mathbf{G}^F$ factors through $H^1(F, Z(\mathbf{G}))$. We thus get an action of the group $H^1(F, Z(\mathbf{G}))$ (see 12.2.13) on $\mathrm{Irr}(\mathbf{G}^F)$. To simplify notation, let $H = C_{\mathbf{G}^*}(s)^{F^*}$ and $H^0 = C_{\mathbf{G}^*}(s)^{0F^*}$. We denote by $C_H(\rho)$ the inertia group in H of $\rho \in \mathrm{Irr}(H^0)$. Lusztig (1988, 5.1) can then be stated as follows.

Proposition 11.5.2 *Assume that F is a Frobenius endomorphism. There is a bijection between $H^1(F, Z(\mathbf{G}))$-orbits on $\mathcal{E}(\mathbf{G}^F, (s)_{\mathbf{G}^{*F^*}})$ for the action through diagonal automorphisms and H-orbits on $\mathcal{E}(H^0, 1)$. If the orbits $O \subset \mathcal{E}(\mathbf{G}^F, (s)_{\mathbf{G}^{*F^*}})$ and $O' \subset \mathcal{E}(H^0, 1)$ correspond, then $|O| = |C_H(\rho')/H^0|$ for any $\rho' \in O'$. Further, for any $\rho \in O$ and any F^*-stable torus $\mathbf{T}^*$ of $C_{\mathbf{G}^*}(s)^0$ we have*

$$\varepsilon_{\mathbf{G}}\langle \rho, R_{\mathbf{T}^*}^{\mathbf{G}}(s) \rangle_{\mathbf{G}^F} = \varepsilon_{C_{\mathbf{G}^*}(s)} \langle \sum_{\rho' \in O'} \rho', R_{\mathbf{T}^*}^{C_{\mathbf{G}^*}(s)^0}(1) \rangle_{H^0}.$$

By 11.4.3(ii) it is enough to prove 11.5.1 when s is quasi-isolated. In this case the statement, when F is a Frobenius endomorphism, follows from Lusztig's statement and from the following.

Proposition 11.5.3 *Assume that F is a Frobenius endomorphism. Let $s \in \mathbf{G}^{*F^*}$ be quasi-isolated, and let $H = C_{\mathbf{G}^*}(s)^{F^*}$ and $H^0 = C_{\mathbf{G}^*}(s)^{0F^*}$. The restriction of any $\tilde{\rho} \in \mathcal{E}(H, 1)$ to H^0 is of the form $\sum_{\rho' \in O'} \rho'$ where O' is an H-orbit on $\mathcal{E}(H^0, 1)$. There are $|C_H(\rho')/H^0|$ elements of $\mathcal{E}(H, 1)$ with a given restriction to H^0.*

Proof By Clifford's theory – see for example Curtis and Reiner (1981, §11) – all statements result from the first one, and the first one is equivalent to the claim that any $\rho' \in \mathcal{E}(H^0, 1)$ can be extended to its inertia group $C_H(\rho')$.

The following lemma reduces the proof of 11.5.3 to the case where $Z(\mathbf{G}^*) = \{1\}$.

Lemma 11.5.4 *Let $(\mathbf{H}, F)$ and $(\mathbf{H}_1, F_1)$ be two reductive groups with Frobenius roots and let $f : \mathbf{H} \to \mathbf{H}_1$ be a morphism such that the morphism $f_0 : \mathbf{H}^0 \to \mathbf{H}_1^0$ induced on the identity components satisfies the assumptions of 11.3.8; that is, $\mathrm{Ker} f_0 \subset Z(\mathbf{H}^0)$, and $f_0 \circ F = F_1 \circ f_0$, and $f_0(\mathbf{H}^0)$ contains $\mathrm{Der}\,\mathbf{H}_1^0$. Assume* moreover that all unipotent characters of $\mathbf{H}_1^{0F_1}$ can be extended to their inertia group in $\mathbf{H}_1^{F_1}$. Then all unipotent characters of $\mathbf{H}^{0F}$ can be extended to their inertia group in $\mathbf{H}^F$.

Proof By 11.3.8 the unipotent characters of $\mathbf{H}^{0F}$ are of the form $\chi \circ f_0$ where χ is a unipotent character of $\mathbf{H}_1^{0F_1}$. We have $f(C_{\mathbf{H}^F}(\chi \circ f_0)) \subset C_{\mathbf{H}_1^{F_1}}(\chi)$. Hence, if $\tilde{\chi}$ is an extension of χ to $C_{\mathbf{H}_1^{F_1}}(\chi)$, then $\tilde{\chi} \circ f$ is an extension of $\chi \circ f_0$ to $C_{\mathbf{H}^F}(\chi \circ f_0)$. □

Let $\bar{s}$ be the image of s in $\overline{\mathbf{G}}^* := \mathbf{G}^*/Z(\mathbf{G}^*)$. Note that $\bar{s}$ is still quasi-isolated. If we prove that all unipotent characters of $C_{\overline{\mathbf{G}}^*}(\bar{s})^{0F^*}$ can be extended to their inertia group in $C_{\overline{\mathbf{G}}^*}(\bar{s})^{F^*}$, we will get the extension property, thus 11.5.3, in $\mathbf{G}^*$ by applying 11.5.4 with $\mathbf{H} = C_{\mathbf{G}^*}(s)$ and $\mathbf{H}_1 = C_{\overline{\mathbf{G}}^*}(\bar{s})$.

We thus assume now $Z(\mathbf{G}^*) = \{1\}$. We next reduce to the case where $\mathbf{G}^*$ (thus $\mathbf{G}$ also) is simple. Since $Z(\mathbf{G}^*) = 1$ we have $\mathbf{G}^* = \prod_{i=1}^k \mathbf{G}_i^*$, a direct product of simple groups; we can write $s = s_1 s_2 \ldots s_k$ with $s_i \in \mathbf{G}_i^*$, and the centraliser of s is the direct product of the centralisers of s_i in $\mathbf{G}_i^*$. Splitting the product into orbits under F^*, we get $C_{\mathbf{G}^*}(s)^{F^*} \simeq \prod_{i \in I} C_{\mathbf{G}_i^*}(s_i)^{F^{*k_i}}$, with $I \subset \{1 \ldots k\}$ such that $\{\mathbf{G}_i^* \mid i \in I\}$ is a set of representatives of the orbits of F^* and where k_i is the size of the corresponding orbit. Thus it is sufficient to prove the result for one of these simple factors.

We assume now $\mathbf{G}^*$ simple. The first statement of 11.5.3 holds when H/H^0 is cyclic, since it is always possible to extend characters to their inertia group in this case; see Curtis and Reiner (1981, 11.47). By 11.2.1(iii) the group H/H^0 is isomorphic to a subgroup of $\mathrm{Irr}(Z(\mathbf{G}))$ (we have $Z(\mathbf{G})^0 = \{1\}$); by 11.2.6 and table 11.1, this group is cyclic unless $\mathbf{G}^*$ is adjoint of type D_{2n}, in which case it is isomorphic to $(\mathbb{Z}/2\mathbb{Z})^2$. Thus the result is proved unless s is a quasi-isolated element of $\mathbf{G} = \mathbf{PSO}_{4n}$ such that $C_{\mathbf{G}}(s)/C_{\mathbf{G}}(s)^0 \simeq (\mathbb{Z}/2\mathbb{Z})^2$. These elements are completely described by Bonnafé (2005b, Theorem 5.1): let $\tilde{\Pi}$ be an extended basis of the root system of type D_{2n}. By Bonnafé's theorem quasi-isolated elements such that $C_{\mathbf{G}}(s)/C_{\mathbf{G}}(s)^0 \simeq (\mathbb{Z}/2\mathbb{Z})^2$ correspond to orbits of Ω on $\tilde{\Pi}$, and the group $C_{\mathbf{G}}(s)^0$ is generated by $\mathbf{T}$ and $\mathbf{U}_\alpha$ for $\alpha \in \tilde{\Pi} - \Omega$. We thus find two cases; in the first case $C_{\mathbf{G}}(s)^0$ is of type A_{2n-3}; $C_{\mathbf{G}}(s)$ is generated, in addition, by two elements of W which act respectively by an automorphism of type ${}^2A_{2n-3}$ and an automorphism of the radical. We can reduce to the adjoint quotient of $C_{\mathbf{G}}(s)^0$ by applying Lemma 11.5.4 with $\mathbf{H} = C_{\mathbf{G}}(s)$ and $\mathbf{H}_1 = C_{\mathbf{G}}(s)/Z(C_{\mathbf{G}}(s)^0)$. Then $\mathbf{H}_1$ is the direct product of $\mathbb{Z}/2\mathbb{Z}$ by a reductive group $\mathbf{H}_2$ such that $\mathbf{H}_2/\mathbf{H}_1^0$ is cyclic, and the extendibility of characters follows.

In the second case $C_{\mathbf{G}}(s)^0$ is of type $D_{n-i} \times D_{n-i} \times A_{2i-1}$ and $C_{\mathbf{G}}(s)$ is generated, in addition, by the representatives of two elements of W:

- An element σ which acts by an automorphism of type $^2D_{n-i}$ on both components of type D.
- An element τ which interchanges these two components and acts by an automorphism of type $^2A_{2i-1}$ on the other component.

By Adams and He (2017, Theorem A), we can lift the Weyl group to $N_{\mathbf{G}}(\mathbf{T})$, in particular we can assume that σ and τ are commuting involutions. We can as in the first case reduce to the adjoint quotient and we get $\mathbf{H}_1^0 = \mathbf{PSO}_{2n-2i} \times \mathbf{PSO}_{2n-2i} \times \mathbf{PGL}_{2i}$. A character τ and σ-invariant is of the form $\rho \otimes \rho \otimes \rho'$ where ρ is σ-invariant and ρ' is τ-invariant. We can extend it first to σ as $\tilde\rho \otimes \tilde\rho \otimes \rho'$, taking care to take the same extension of both components ρ. Such a character can then be extended to τ by extending ρ' and making τ act on $\tilde\rho \otimes \tilde\rho$ by the interchange of the two factors of the tensor product. $\qquad\square$

Note that Proposition 11.5.3 just determines the existence of an (arbitrary) bijection between O and O' in 11.5.2; the existence of a "canonical" bijection is an open problem.

We finally indicate how to proceed to prove 11.5.1 when F is a Frobenius root which is not a Frobenius morphism. Using our arguments, we can reduce to the case where $\mathbf{G}$ is simple and using 11.4.3 and 11.4.8 we can reduce to the case where s is a non-central quasi-isolated element. Such elements do not exist in $^2B_2$ and $^2G_2$; in $^2F_4$ there is a single conjugacy class of such elements where s is of order 3 and $C_{\mathbf{G}^*}(s)$ is connected of type $A_2 \times A_2$ where F^* interchanges the two components. The corresponding characters were computed by Malle, see Malle (1990) and Geck et al. (1996a). $\qquad\square$

We still call **uniform functions** in a disconnected group the class functions which are linear combination of Deligne–Lusztig characters. Note that, since the Deligne–Lusztig characters of a disconnected group $\mathbf{G}$ are defined as induced from the Deligne–Lusztig characters of $\mathbf{G}^{0F}$, the induced function of a uniform function on $\mathbf{G}^{0F}$ is uniform; in particular – by 10.2.6 – $\mathrm{reg}_{\mathbf{G}^F}$ is uniform. We generalise 9.3.1 to disconnected groups in Lemma 11.5.5.

Lemma 11.5.5

(i) *Let $\mathbf{T}$ and $\mathbf{T}'$ be two F-stable maximal tori of the (possibly disconnected) reductive group $\mathbf{G}$; for $\theta \in \mathrm{Irr}(\mathbf{T}^F)$ and $\theta' \in \mathrm{Irr}(\mathbf{T}'^F)$ we have*

$$\langle R_{\mathbf{T}}^{\mathbf{G}}(\theta), R_{\mathbf{T}'}^{\mathbf{G}}(\theta') \rangle_{\mathbf{G}^F} = |\mathbf{T}^F|^{-1} \#\{n \in \mathbf{G}^F \mid {}^n\mathbf{T} = \mathbf{T}' \text{ and } {}^n\theta = \theta'\}.$$

(ii) *The $R_{\mathbf{T}}^{\mathbf{G}}(\theta)$ where the pairs $(\mathbf{T}, \theta)$ are taken up to $\mathbf{G}^F$-conjugacy form an orthogonal basis of the space of uniform functions.*

Proof We have $\langle R_{\mathbf{T}}^{\mathbf{G}}(\theta), R_{\mathbf{T}'}^{\mathbf{G}}(\theta')\rangle_{\mathbf{G}^F} = \langle \operatorname{Res}_{\mathbf{G}^{\circ F}}^{\mathbf{G}^F} R_{\mathbf{T}}^{\mathbf{G}^\circ}(\theta), R_{\mathbf{T}'}^{\mathbf{G}^\circ}(\theta')\rangle_{\mathbf{G}^{\circ F}}$. By the ordinary Mackey formula this is equal to

$$\sum_{x \in \mathbf{G}^F/\mathbf{G}^{0F}} \langle R_{x\mathbf{T}}^{\mathbf{G}^0}({}^x\theta), R_{\mathbf{T}'}^{\mathbf{G}^0}(\theta')\rangle_{\mathbf{G}^{\circ F}}.$$

By 9.3.1(i) the summand indexed by x is equal to $|\mathbf{T}^F|^{-1}|\{g \in \mathbf{G}^{0F} \mid {}^{gx}(\mathbf{T}, \theta) = (\mathbf{T}', \theta')\}|$, whence (i).

(ii) is an immediate consequence of (i). □

Proposition 11.5.6 *If ψ_s is any bijection as in 11.5.1, then for any $\chi \in \mathcal{E}(\mathbf{G}^F, (s))$ we have*

$$\chi(1) = \frac{|\mathbf{G}^F|_{p'}}{|C_{\mathbf{G}^*}(s)^{F^*}|_{p'}} \psi_s(\chi)(1).$$

Proof Since the characteristic function of the identity is uniform, see 10.2.6, χ has the same dimension as its projection $p(\chi)$ to the space of uniform functions. By 11.5.5(ii) this projection is equal to

$$p(\chi) = \sum_{(\mathbf{T}, \theta)} \frac{\langle \chi, R_{\mathbf{T}}^{\mathbf{G}}(\theta)\rangle_{\mathbf{G}^F}}{\langle R_{\mathbf{T}}^{\mathbf{G}}(\theta), R_{\mathbf{T}}^{\mathbf{G}}(\theta)\rangle_{\mathbf{G}^F}} R_{\mathbf{T}}^{\mathbf{G}}(\theta), \qquad (11.5.1)$$

where $(\mathbf{T}, \theta)$ is taken up to $\mathbf{G}^F$-conjugacy. Since $\chi \in \mathcal{E}(\mathbf{G}^F, (s))$ the only non-zero terms in this sum are those such that the pair $(\mathbf{T}, \theta)$ corresponds to $(\mathbf{T}^*, s)$ under the bijection of 11.1.16. Hence

$$p(\chi) = \sum_{(\mathbf{T}^*)} \frac{\langle \chi, R_{\mathbf{T}^*}^{\mathbf{G}}(s)\rangle_{\mathbf{G}^F}}{\langle R_{\mathbf{T}^*}^{\mathbf{G}}(s), R_{\mathbf{T}^*}^{\mathbf{G}}(s)\rangle_{\mathbf{G}^F}} R_{\mathbf{T}^*}^{\mathbf{G}}(s)$$

where $\mathbf{T}^*$ runs over $C_{\mathbf{G}^*}(s)^{F^*}$-classes of F^*-stable maximal tori of $C_{\mathbf{G}^*}(s)$. By 9.3.1 applied in $\mathbf{G}$ and 11.5.5 applied in $C_{\mathbf{G}^*}(s)$, we have $\langle R_{\mathbf{T}^*}^{\mathbf{G}}(s), R_{\mathbf{T}^*}^{\mathbf{G}}(s)\rangle_{\mathbf{G}^F} = \langle R_{\mathbf{T}^*}^{C_{\mathbf{G}^*}(s)}(1), R_{\mathbf{T}^*}^{C_{\mathbf{G}^*}(s)}(1)\rangle_{C_{\mathbf{G}^*}(s)^{F^*}}$, whence

$$p(\chi) = \sum_{(\mathbf{T}^*)} \frac{\langle \chi, R_{\mathbf{T}^*}^{\mathbf{G}}(s)\rangle_{\mathbf{G}^F}}{\langle R_{\mathbf{T}^*}^{C_{\mathbf{G}^*}(s)}(1), R_{\mathbf{T}^*}^{C_{\mathbf{G}^*}(s)}(1)\rangle_{C_{\mathbf{G}^*}(s)^{F^*}}} R_{\mathbf{T}^*}^{\mathbf{G}}(s).$$

We have

$$R_{\mathbf{T}^*}^{\mathbf{G}}(s)(1) = \varepsilon_{\mathbf{G}} \varepsilon_{C_{\mathbf{G}^*}(s)^0} \frac{|\mathbf{G}^F|_{p'}}{|C_{\mathbf{G}^*}(s)^{F^*}|_{p'}} R_{\mathbf{T}^*}^{C_{\mathbf{G}^*}(s)}(1)(1)$$

by 10.2.2 in both sides, using that by 11.2.1(iii) $C_{\mathbf{G}^*}(s)^{F^*}/C_{\mathbf{G}^*}(s)^{0F^*}$ is a p'-group. By 11.5.1 we have

$$\langle \chi, R_{\mathbf{T}^*}^{\mathbf{G}}(s)\rangle_{\mathbf{G}^F} = \varepsilon_{\mathbf{G}} \varepsilon_{C_{\mathbf{G}^*}(s)^0} \langle \psi_s(\chi), R_{\mathbf{T}^*}^{C_{\mathbf{G}^*}(s)}(1)\rangle_{C_{\mathbf{G}^*}(s)^{F^*}}.$$

We deduce

$$\chi(1) = \frac{|\mathbf{G}^F|_{p'}}{|C_{\mathbf{G}^*}(s)^{F^*}|_{p'}}$$

$$\sum_{(\mathbf{T}^*)} \frac{\langle \psi_s(\chi), R_{\mathbf{T}^*}^{C_{\mathbf{G}^*}(s)}(\mathbf{1}) \rangle_{C_{\mathbf{G}^*}(s)^{F^*}}}{\langle R_{\mathbf{T}^*}^{C_{\mathbf{G}^*}(s)}(\mathbf{1}), R_{\mathbf{T}^*}^{C_{\mathbf{G}^*}(s)}(\mathbf{1}) \rangle_{C_{\mathbf{G}^*}(s)^{F^*}}} R_{\mathbf{T}^*}^{C_{\mathbf{G}^*}(s)}(\mathbf{1})(1).$$

By (11.5.1) applied in $C_{\mathbf{G}^*}(s)$, which is valid in a disconnected group by 11.5.5(ii), and using that by definition of unipotent characters of $C_{\mathbf{G}^*}(s)^{F^*}$, the character $\psi_s(\chi)$ is orthogonal to $R_{\mathbf{T}^*}^{C_{\mathbf{G}^*}(s)}(\theta)$ for $\theta \neq \mathbf{1}$, the right-hand side is equal to the dimension of the uniform projection of $\psi_s(\chi)$ that is to $\psi_s(\chi)(1)$ since the characteristic function of the identity is uniform (see the remark above 11.5.5). We thus get $\chi(1) = \frac{|\mathbf{G}^F|_{p'}}{|C_{\mathbf{G}^*}(s)^{F^*}|_{p'}} \psi_s(\chi)(1)$.   □

11.6 More about Unipotent Characters

Theorem 11.5.1 reduces the problem of decomposing $R_{\mathbf{T}}^{\mathbf{G}}(\theta)$ to the case of unipotent characters. We give here a construction which is a first step for decomposing $R_{\mathbf{T}}^{\mathbf{G}}(\mathbf{1})$. This construction solves the problem for general linear and unitary groups.

Let W be the Weyl group of a quasi-split torus. We consider the coset WF of W in $\tilde{W} := W \rtimes \langle F \rangle$, where $\langle F \rangle$ is the group generated by the automorphism of finite order induced by F on W.

Definition 11.6.1   Let $C(WF)$ be the space of functions $WF \to \mathbb{C}$ invariant by W-conjugacy. Then for $f \in C(WF)$ we define

$$R_f := |W|^{-1} \sum_{w \in W} f(wF) R_{\mathbf{T}_w}^{\mathbf{G}}(\mathbf{1}) = \sum_{w \in H^1(F,W)} |C_W(wF)|^{-1} R_{\mathbf{T}_w}^{\mathbf{G}}(\mathbf{1}).$$

In particular, if π_{wF} is the normalised characteristic function of the class of wF, equal to $|C_W(wF)|$ on that class and 0 outside it, we have $R_{\pi_{wF}} = R_{\mathbf{T}_w}^{\mathbf{G}}(\mathbf{1})$.

Lemma 11.6.2   *If we define a scalar product on $C(WF)$ by*

$$\langle f, f' \rangle_{WF} = |W|^{-1} \sum_{w \in W} f(wF) \overline{f'(wF)},$$

then the map $f \mapsto R_f$ is an isometry between $C(WF)$ and unipotent uniform functions on $\mathbf{G}^F$.

Proof   This is an immediate consequence of 9.3.2.   □

By abuse of notation, we will still write R_f when f is a class function on $\tilde{W}$, where we interpret f as its restriction to the coset WF.

Proposition 11.6.3

(i) $\chi \in \mathrm{Irr}(\tilde{W})$ *vanishes on* WF *unless* $\mathrm{Res}_W^{\tilde{W}} \chi \in \mathrm{Irr}(W)$.

(ii) *Let* $\chi, \psi \in \mathrm{Irr}(\tilde{W})$ *such that* $\mathrm{Res}_W^{\tilde{W}} \chi$ *and* $\mathrm{Res}_W^{\tilde{W}} \psi$ *are in* $\mathrm{Irr}(W)$. *Then*

$$\langle \chi, \psi \rangle_{WF} = \begin{cases} 0 & unless\ \mathrm{Res}_W^{\tilde{W}} \chi = \mathrm{Res}_W^{\tilde{W}} \psi, \\ 1 & if\ \chi = \psi. \end{cases}$$

Proof Let us prove (i). If $\chi \in \mathrm{Irr}(\tilde{W})$ does not have an irreducible restriction to W then by 11.3.9 there exists a non-trivial character ζ of $\tilde{W}/W = \langle F \rangle$ such that $\zeta \chi = \chi$. Thus $\lambda = \zeta(F)$ is a non-trivial root of unity; let o be its order. Then for $w \in W$ we have $o\chi(wF) = \sum_{i=0}^{o-1} \zeta^i \chi(wF) = (1 + \lambda + \cdots + \lambda^{o-1})\chi(wF) = 0$ which proves (i).

Now let $\delta = |\langle F \rangle|$, let ζ be a character of $\langle F \rangle$ such that $\lambda = \zeta(F)$ is a primitive root of unity of order δ and for $\chi \in \mathrm{Irr}(\tilde{W})$ set $p(\chi) = \chi + \lambda^{-1}\zeta\chi + \cdots + \lambda^{1-\delta}\zeta^{\delta-1}\chi$. Then an easy computation shows that for $w \in W$ we have $p(\chi)(wF^i) = \begin{cases} \delta\chi(wF) & \text{if } i = 1 \\ 0 & \text{otherwise} \end{cases}$. Thus if $\chi, \psi \in \mathrm{Irr}(\tilde{W})$ we have $\langle p(\chi), p(\psi) \rangle_{\tilde{W}} = \delta \langle \chi, \psi \rangle_{WF}$. Now, if χ and ψ have irreducible restrictions to W, using again 11.3.9 we get

$$\langle p(\chi), p(\psi) \rangle_{\tilde{W}} = \begin{cases} \delta & \text{if } \mathrm{Res}_W^{\tilde{W}} \chi = \mathrm{Res}_W^{\tilde{W}} \psi, \\ 0 & \text{otherwise,} \end{cases}$$

whence the result. $\qquad\square$

It follows from 11.6.2 and 11.6.3 that:

Corollary 11.6.4 *Let $\mathcal{E}$ be a set consisting of one extension to $\tilde{W}$ of each character in* $\mathrm{Irr}(W)^F$. *Then* $\{R_{\tilde{\chi}}\}_{\tilde{\chi} \in \mathcal{E}}$ *is an orthonormal basis of the space of unipotent uniform functions on* $\mathbf{G}^F$.

Some $R_{\tilde{\chi}}$ are irreducible characters, for instance 10.2.5(i) and 10.2.5(ii) together with 7.1.6 can be interpreted respectively as $R_1 = \mathbf{1}_{\mathbf{G}^F}$ and $R_{\mathrm{sgn}} = \mathrm{St}_{\mathbf{G}^F}$, where sgn is the sign character $w \mapsto (-1)^{l(w)}$ of W, extended trivially to WF. We will see that all $R_{\tilde{\chi}}$ are irreducible characters for the general linear groups. In general they are only $\mathbb{Q}$-linear combinations of irreducible characters; but to decompose the $R_{\mathbf{T}_w}^{\mathbf{G}}(1)$ it is enough to decompose the $R_{\tilde{\chi}}$ since:

Lemma 11.6.5 $R_{\mathbf{T}_w}^{\mathbf{G}}(1) = \sum_{\tilde{\chi} \in \mathcal{E}} \overline{\tilde{\chi}(wF)} R_{\tilde{\chi}}$.

Proof   The formula is a consequence of the "second orthogonality relation for $C(WF)$"

$$\sum_{\chi \in \mathcal{E}} \overline{\chi(wF)}\chi(w'F) = \begin{cases} |C_W(wF)| & \text{if } wF \text{ and } w'F \text{ are } W\text{-conjugate} \\ 0 & \text{otherwise,} \end{cases}$$

which is itself a consequence of the second orthogonality relation for $\tilde{W}$:

$$\sum_{\chi \in \mathrm{Irr}(\tilde{W})} \overline{\chi(wF)}\chi(w'F) = \begin{cases} |C_{\tilde{W}}(wF)| & \text{if } wF \text{ and } w'F \text{ are conjugate} \\ 0 & \text{otherwise,} \end{cases}$$

since:

- wF and $w'F$ are conjugate in $\tilde{W}$ if and only if they are W-conjugate.
- In the formula only the characters with an irreducible restriction to W contribute and the contribution of two characters with the same restriction is the same.
- There are $|\langle F \rangle|$ characters with the same restriction and $|C_{\tilde{W}}(wF)| = |\langle F \rangle||C_W(wF)|$.

$\square$

Let W' be a standard parabolic subgroup of W which is $w'F$-stable for some $w' \in W$. By restricting $\mathrm{Ind}_{W' \rtimes \langle w'F \rangle}^{W \rtimes \langle F \rangle}$ to cosets we define an induction between cosets given, for $f \in C(W'w'F)$, by the formula

$$(\mathrm{Ind}_{W'w'F}^{WF} f)(wF) = |W'|^{-1} \sum_{\{h \in W \mid hwFh^{-1} \in W'w'F\}} f(hwFh^{-1}).$$

With this definition we can make the following proposition.

Proposition 11.6.6   *Let* $\mathbf{L}$ *be an F-stable Levi subgroup of a parabolic subgroup of* $\mathbf{G}$, *which as in 11.4.1 is parametrised by a coset $W'w$ where W' is a $w'F$-stable standard parabolic subgroup of W. Then $R_{\mathbf{L}}^{\mathbf{G}}(R_f) = R_{\mathrm{Ind}_{W'w'F}^{WF}(f)}$.*

Proof   It is enough to prove the proposition for the functions $\pi_{ww'F}^{W'}$, where the exponent shows the considered group. It is straightforward using the above definition that $\mathrm{Ind}_{W'w'F}^{WF}(\pi_{ww'F}^{W'}) = \pi_{ww'F}^{W}$; the corresponding functions on the reductive group side are $R_{\mathbf{T}_{ww'}}^{\mathbf{L}}(1)$ and $R_{\mathbf{T}_{ww'}}^{\mathbf{G}}(1)$, thus the proposition is a consequence of transitivity of the Lusztig induction: $R_{\mathbf{L}}^{\mathbf{G}}(R_{\mathbf{T}_{ww'}}^{\mathbf{L}}(1)) = R_{\mathbf{T}_{ww'}}^{\mathbf{G}}(1)$. $\square$

11.7 The Irreducible Characters of $\mathbf{GL}_n^F$ and $\mathbf{U}_n^F$

We will express all irreducible characters of the general linear and unitary groups as explicit combinations of Deligne–Lusztig characters, following the method of Lusztig and Srinivasan (1977). Note that this will prove in particular that all class functions are uniform in these groups.

We will denote by $\mathbf{G}$ the group $\mathbf{GL}_n$ and by F a Frobenius endomorphism which may be either the split one or that of the unitary group (called F' in 7.1.8). The torus of diagonal matrices is F-stable in both cases; it is split for general linear groups, but in the unitary case the Borel subgroup $\mathbf{B}$ of upper triangular matrices is sent by F to the Borel subgroup $^{w_0}\mathbf{B}$ (see example 7.1.8), thus the torus of diagonal matrices is of type w_0, which implies that the Frobenius acts by conjugation by w_0 on its Weyl group $W = \mathfrak{S}_n$. Thus in both cases the set $\mathcal{E}$ of 11.6.4 consists in one extension of each irreducible character of W, and for $\chi \in \mathrm{Irr}(W)$, we set $\tilde\chi(wF) = \chi(w)$ in the case of $\mathbf{GL}_n$ and $\tilde\chi(wF) = \chi(ww_0)$ in the case of $\mathbf{U}_n$. In both cases we write R_χ for $R_{\tilde\chi}$. Lemma 11.6.5 becomes:

Lemma 11.7.1

$$R_{\mathbf{T}_w}^{\mathbf{G}}(1) = \begin{cases} \sum_{\chi \in \mathrm{Irr}(W)} \chi(w) R_\chi & \textit{for } \mathbf{GL}_n, \\ \sum_{\chi \in \mathrm{Irr}(W)} \chi(ww_0) R_\chi & \textit{for } \mathbf{U}_n. \end{cases}$$

We may now prove:

Theorem 11.7.2 *In the general linear and unitary groups, the set* $\mathcal{E}(\mathbf{G}^F, 1)$ *is equal (up to signs in the unitary case) to* $\{R_\chi\}_{\chi \in \mathrm{Irr}(W)}$.

Proof If we prove that the R_χ are virtual characters, they will be up to sign distinct irreducible characters (since they form an orthonormal set) and by 11.7.1 they will span the unipotent characters.

We know that $R_1 = \mathbf{1}$ is irreducible. It is also well known that in the symmetric group any irreducible character is a linear combination with integral coefficients of the character induced from the trivial character of various parabolic subgroups; that is, any $\chi \in \mathrm{Irr}(W)$ is a linear combination with integral coefficients of $\mathrm{Ind}_{W_{\mathbf{L}}}^W(1)$ where $\mathbf{L}$ runs over diagonal Levi subgroups of $\mathbf{G}$. In the unitary case this correspond to cosets $W_I w_0$ where W_I runs over the standard parabolic subgroups of W. Applying the isometry 11.6.2 and proposition 11.6.6, we get an expression of R_χ as an integral linear combination of $R_{\mathbf{L}}^{\mathbf{G}}(1)$, which proves indeed that it is a virtual character.

To prove that the R_χ are actual characters for general linear groups, we note that, in that case, as the split torus $\mathbf{T}_1$ of diagonal matrices is contained in the

F-stable Borel subgroup **B** of upper triangular matrices, $R_{\mathbf{T}_1}^{\mathbf{G}}(1) = \mathrm{Ind}_{\mathbf{B}^F}^{\mathbf{G}^F}(1)$ is an actual character; since $R_{\mathbf{T}_1}^{\mathbf{G}}(1) = \sum_\chi \chi(1)R_\chi$, and $\chi(1) > 0$, this proves the result.   □

In the unitary case, it is not difficult to obtain a formula for the dimension of R_χ and deduce from it, for each χ, which of R_χ or $-R_\chi$ is an irreducible character. The same argument as in $\mathbf{GL}_n$, using that $R_{\mathbf{T}_1}^{\mathbf{G}} = \sum_\chi \chi(w_0)R_\chi$ is an actual character, shows that $\chi(w_0)R_\chi$ is an actual character. So, when $\chi(w_0) \neq 0$, the sign we have to give to R_χ to make it an actual character is that of $\chi(w_0)$. For the sign in general, we will see in 13.1.14(ii) that $R_\chi(1)$ is a polynomial in q, and the sign is $(-1)^{B_\chi}$ where B_χ is the degree of that polynomial.

Once we know the unipotent characters of $\mathbf{G}^F$, we can easily get all characters.

Theorem 11.7.3 *The irreducible characters of the general linear and unitary groups are (up to sign) the*

$$R_\chi(s) := |W_I|^{-1} \sum_{w \in W_I} \tilde{\chi}(ww_1)R_{\mathbf{T}_{ww_1}}^{\mathbf{G}}(s),$$

where (s) *runs over F-stable semi-simple conjugacy classes of $\mathbf{G}$; then $C_{\mathbf{G}}(s)$ is a Levi subgroup parametrised by a coset $W_I w_1$ as in 11.4.1. The character χ runs over w_1-stable irreducible characters of W_I and $\tilde{\chi}$ stands for a real extension to $W_I.\langle w_1 \rangle$ of χ.*

Proof We can take $(\mathbf{G}^*, F^*)$ to be $(\mathbf{G}, F)$; see examples 11.1.13. Moreover, by the description of parabolic subgroups in $\mathbf{GL}_n$ – see 3.2.5 – and that of centralisers of semi-simple elements – see 3.5.4(ii) which applies also to $\mathbf{GL}_n$ – the centraliser of a semi-simple element is a Levi subgroup conjugate to a group of block-diagonal matrices. If $s \in \mathbf{G}^F$ is semi-simple, by 11.4.1 the action of F on $C_{\mathbf{G}}(s)$ permutes blocks of equal size. An orbit of blocks is of the form $\mathbf{G}_1^k$ where F cyclically permutes the copies of $\mathbf{G}_1$ and F^k still acts on $\mathbf{G}_1$ as a split or unitary type Frobenius endomorphism (on a bigger field). If W_1 is the Weyl group of $\mathbf{G}_1$, an F-invariant character of W_1^k is of the form $\chi \otimes \cdots \otimes \chi$ and it has a natural real extension which on $(w_1, \ldots, w_k)F$ takes the value $\chi(w_1 \cdots w_k F^k)$. With this extension Theorem 11.7.2 can be extended easily to such groups. Then, using 11.4.8 and the fact that by 11.4.3 the functor $R_{C_{\mathbf{G}}(s)}^{\mathbf{G}}$ is an isometry from the series $\mathcal{E}(C_{\mathbf{G}}(s)^F, (s))$ to $\mathcal{E}(\mathbf{G}^F, (s))$, we get the theorem.   □

Example 11.7.4   The character table of $\mathbf{GL}_2(\mathbb{F}_q)$. (See Table 11.2).

The elements of $\mathbf{GL}_2(\mathbb{F}_q)$ are the invertible 2×2 matrices over $\mathbb{F}_q$; the Frobenius F raises the matrix entries to their qth power. The diagonal matrices **T** form a split F-stable maximal torus. The Weyl group $W = W_{\mathbf{G}}(\mathbf{T})$ can be identified as the two permutation matrices, which we denote by 1 and s; there are thus two $\mathbf{G}^F$-conjugacy classes of F-stable maximal tori, that of $\mathbf{T}$ and of $\mathbf{T}_s$ such that $(\mathbf{T}_s, F)$ is $\mathbf{G}$-conjugate to $(\mathbf{T}, sF)$.

We have $\mathbf{T}^F \simeq \mathbb{F}_q^\times \times \mathbb{F}_q^\times$, thus an element of $\mathrm{Irr}(\mathbf{T}^F)$ is given by a pair (α, β) of characters of $\mathbb{F}_q^\times$, and s interchanges α and β. When $\alpha = \beta$ the character may be written $\alpha \circ \det$ thus extends to a linear character of $\mathbf{G}^F$. By 9.3.1, the Deligne–Lusztig characters $R_{\mathbf{T}}^{\mathbf{G}}(\alpha, \beta)$, which are true characters since $R_{\mathbf{T}}^{\mathbf{G}}$ is a Harish-Chandra induction in this case (see remark before 9.1.5), are distinct irreducible characters when $\{\alpha, \beta\}$ runs over (non-ordered) pairs of distinct characters of $\mathbb{F}_q^\times$; the $R_{\mathbf{T}}^{\mathbf{G}}(\alpha \circ \det)$, where α runs over characters of $\mathbb{F}_q^\times$, are pairwise orthogonal and are each one sum of two distinct irreducible constituents.

We have $\mathbf{T}_s^F \simeq \mathbf{T}^{sF} = \{\mathrm{diag}(t_1, t_1^q) \mid t_1 \in \mathbb{F}_{q^2}^\times\} \simeq \mathbb{F}_{q^2}^\times$. The element s acts on $\mathrm{Irr}(\mathbb{F}_{q^2}^\times)$ by $\omega \mapsto \omega^q$. Thus by 9.3.1 the Deligne–Lusztig characters $R_{\mathbf{T}_s}^{\mathbf{G}}(\omega)$ are pairwise orthogonal when ω runs over representatives of $\mathrm{Irr}(\mathbb{F}_{q^2}^\times)$ mod $\omega \equiv \omega^q$; these characters are the negative (by 10.2.9) of irreducible characters unless $\omega = \omega^q$; when they are not irreducible they have two irreducible constituents.

We now decompose the non-irreducible Deligne–Lusztig characters. Note that $\omega \in \mathrm{Irr}(\mathbb{F}_{q^2}^\times)$ such that $\omega = \omega^q$ is equal to $\alpha \circ N_{\mathbb{F}_{q^2}/\mathbb{F}_q}$ for some $\alpha \in \mathrm{Irr}(\mathbb{F}_q^\times)$, and ω is the restriction to $\mathbf{T}_s^F$ of $\alpha \circ \det$. Thus by 10.1.6 we have $R_{\mathbf{T}}^{\mathbf{G}}(\alpha \circ \det) = R_{\mathbf{T}}^{\mathbf{G}}(1) \cdot \alpha \circ \det$ and $R_{\mathbf{T}_s}^{\mathbf{G}}(\omega) = R_{\mathbf{T}_s}^{\mathbf{G}}(1) \cdot \alpha \circ \det$ and 11.7.1 gives $R_{\mathbf{T}}^{\mathbf{G}}(1) = R_1 + R_\varepsilon$ and $R_{\mathbf{T}_s}^{\mathbf{G}}(1) = R_1 - R_\varepsilon$, which using the remarks after 11.6.4 becomes $R_{\mathbf{T}}^{\mathbf{G}}(1) = 1_{\mathbf{G}^F} + \mathrm{St}_{\mathbf{G}^F}$ and $R_{\mathbf{T}_s}^{\mathbf{G}}(1) = 1_{\mathbf{G}^F} - \mathrm{St}_{\mathbf{G}^F}$.

The list of irreducible characters of $\mathbf{GL}_2(\mathbb{F}_q)$ is thus:

- $R_{\mathbf{T}}^{\mathbf{G}}(\alpha, \beta)$ where $\{\alpha, \beta\} \subset \mathrm{Irr}(\mathbb{F}_q^\times)$ is of size 2.
- $-R_{\mathbf{T}_s}^{\mathbf{G}}(\omega)$ attached to $\{\omega, \omega^q\} \subset \mathrm{Irr}(\mathbb{F}_{q^2}^\times)$ of size 2.
- $1.(\alpha \circ \det)$ and $\mathrm{St}_{\mathbf{G}^F}.(\alpha \circ \det)$ where $\alpha \in \mathrm{Irr}(\mathbb{F}_q^\times)$.

We now give the list of conjugacy classes of $\mathbf{GL}_2(\mathbb{F}_q)$. In general, to list the conjugacy classes of a group $\mathbf{G}^F$, we may first describe the geometric semi-simple classes and their centralisers, and get the $\mathbf{G}^F$-classes by 4.2.15(ii). We are then reduced for each centraliser $C_{\mathbf{G}}(s)$ to describe the unipotent classes of $C_{\mathbf{G}}(s)$, which are actually in $C_{\mathbf{G}}(s)^0$ by 3.5.3. By 3.5.4(i), geometric classes coincide with the $\mathbf{G}^F$-classes for $\mathbf{GL}_n^F$. In $\mathbf{GL}_2$, the only non-semi-simple classes are of the form zu with $z \in Z(\mathbf{GL}_2)$ and u a non-trivial unipotent element; the only non-trivial Jordan form in $\mathbf{GL}_2$ is $u := \begin{pmatrix} 1 & 1 \\ 0 & 1 \end{pmatrix}$; thus we get $q - 1$ classes for $z \in Z(\mathbf{G}^F)$ and $q - 1$ classes for zu.

Table 11.2. *Character Table of* $\mathbf{GL}_2(\mathbb{F}_q)$ *(note that* $|\mathbf{GL}_2(\mathbb{F}_q)| = q(q-1)^2(q+1)$*)*

Classes	$\begin{pmatrix} a & 0 \\ 0 & a \end{pmatrix}$ $a \in \mathbb{F}_q^\times$	$\begin{pmatrix} a & 0 \\ 0 & b \end{pmatrix}$ $a,b \in \mathbb{F}_q^\times$ $a \neq b$	$\begin{pmatrix} x & 0 \\ 0 & {}^F x \end{pmatrix}$ $x \in \mathbb{F}_{q^2}^\times$ $x \neq {}^F x$	$\begin{pmatrix} a & 1 \\ 0 & a \end{pmatrix}$ $a \in \mathbb{F}_q^\times$
Number of classes of this type	$q-1$	$\dfrac{(q-1)(q-2)}{2}$	$\dfrac{q(q-1)}{2}$	$q-1$
Cardinal of the class	1	$q(q+1)$	$q(q-1)$	q^2-1
$R_{\mathbf{T}}^{\mathbf{G}}(\alpha,\beta)$ $\alpha,\beta \in \mathrm{Irr}(\mathbb{F}_q^\times)$ $\alpha \neq \beta$	$(q+1)\alpha(a)\beta(a)$	$\alpha(a)\beta(b)+$ $\alpha(b)\beta(a)$	0	$\alpha(a)\beta(a)$
$-R_{\mathbf{T}_s}^{\mathbf{G}}(\omega)$ $\omega \in \mathrm{Irr}(\mathbb{F}_{q^2}^\times)$ $\omega \neq \omega^q$	$(q-1)\omega(a)$	0	$-\omega(x) - \omega({}^F x)$	$-\omega(a)$
$\alpha \circ \det$ $\alpha \in \mathrm{Irr}(\mathbb{F}_q^\times)$	$\alpha(a^2)$	$\alpha(ab)$	$\alpha(x.{}^F x)$	$\alpha(a^2)$
$\mathrm{St}_{\mathbf{G}^F}.(\alpha \circ \det)$ $\alpha \in \mathrm{Irr}(\mathbb{F}_q^\times)$	$q\alpha(a^2)$	$\alpha(ab)$	$-\alpha(x.{}^F x)$	0

In $\mathbf{GL}_n^F$ the semi-simple elements of $\mathbf{T}_w^F \simeq \mathbf{T}^{wF}$ are conjugate in $\mathbf{GL}_n(\overline{\mathbb{F}}_q)$ to $\mathrm{diag}(t_1,\ldots,t_n)$ where $t_{w(i)} = t_i^q$. Two such elements are conjugate in $\mathbf{GL}_n$ if and only if they are conjugate under W, see 1.3.6(iii) and (iv). The semi-simple classes in $\mathbf{GL}_2$ with distinct eigenvalues are:

- In $\mathbf{T}^F$, the classes $\mathrm{diag}(a,b)$, determined by $\{a,b\} \in \mathbb{F}_q^\times$ of size 2.
- In $\mathbf{T}_s^F$, the classes which over $\overline{\mathbb{F}}_q$ diagonalise to $\mathrm{diag}(x,{}^F x)$ where $\{x,{}^F x\} \subset \mathbb{F}_{q^2}^\times$ is of size 2. Conjugates in $\mathbf{G}^F$ are $\begin{pmatrix} 0 & 1 \\ -x.{}^F x & x + {}^F x \end{pmatrix}$.

All the entries in the character table of $\mathbf{GL}_2(\mathbb{F}_q)$ (Table 11.2) may be obtained by applying the character formula 10.1.2(i), once we know the value of $R_{\mathbf{T}}^{\mathbf{G}}(1)$ and $R_{\mathbf{T}_s}^{\mathbf{G}}(1)$ on non-trivial unipotent elements. This is 1, since $R_{\mathbf{T}}^{\mathbf{G}}(1) = 1_{\mathbf{G}^F} + \mathrm{St}_{\mathbf{G}^F}$ and $R_{\mathbf{T}_s}^{\mathbf{G}}(1) = 1_{\mathbf{G}^F} - \mathrm{St}_{\mathbf{G}^F}$, and $\mathrm{St}_{\mathbf{G}^F}$ vanishes on non-trivial unipotents.

Notes

The statements in this chapter about geometric conjugacy of $(\mathbf{T}, \theta)$ and its interpretation in the dual group are in Deligne and Lusztig (1976). Langlands had introduced before a dual group which has the same root system as $\mathbf{G}^*$ but is over the complex field. Most of 11.4.3 is already in Lusztig (1976b). Subsequent works Lusztig (1976a), Lusztig (1977), Lusztig (1984a) first gave the definition of series, and then parametrised, initially unipotent characters, then all characters, and gave the decomposition of Deligne–Lusztig characters into irreducible constituents.

A good survey on the representation theory of reductive groups with disconnected centre is Bonnafé (2006).

12

Regular Elements; Gelfand–Graev Representations; Regular and Semi-Simple Characters

The aim of this chapter is to expound the main results about Gelfand–Graev representations. As the properties of these representations involve the notion of regular elements, we begin with their definition and some of their properties, in particular of regular unipotent elements.

12.1 Regular Elements

Definition 12.1.1   An element x of an algebraic group $\mathbf{G}$ is said to be **regular** if the dimension of its centraliser is minimal.

If $\mathbf{G}$ is reductive, this minimal dimension is $\mathrm{rank}(\mathbf{G})$.

Proposition 12.1.2   *Let $\mathbf{G}$ be a connected linear algebraic group.*

(i) *For any element $x \in \mathbf{G}$ we have $\dim C_{\mathbf{G}}(x) \geq \mathrm{rank}(\mathbf{G})$.*

(ii) *Assume $\mathbf{G}$ reductive; then in any torus $\mathbf{T}$ of $\mathbf{G}$ there exists an element t such that $C_{\mathbf{G}}(t) = \mathbf{T}$.*

Proof Let $\mathbf{B}$ be a Borel subgroup of $\mathbf{G}$ containing x and let $\mathbf{U}$ be its unipotent radical. The conjugacy class of x under $\mathbf{B}$ is contained in $x\mathbf{U}$, so its dimension is at most $\dim(\mathbf{U}) = \dim(\mathbf{B}) - \mathrm{rank}(\mathbf{G})$. As this dimension is equal to $\dim(\mathbf{B}) - \dim C_{\mathbf{B}}(x)$, we get $\dim C_{\mathbf{B}}(x) \geq \mathrm{rank}(\mathbf{G})$, whence $\dim C_{\mathbf{G}}(x) \geq \mathrm{rank}(\mathbf{G})$. We now prove (ii). If we apply 1.2.7 to the action of $\mathbf{T}$ on $\mathbf{G}$ by conjugation, we get an element $t \in \mathbf{T}$ such that $C_{\mathbf{G}}(t) = C_{\mathbf{G}}(\mathbf{T})$. But by 2.3.2(iii) this means that $C_{\mathbf{G}}(t) = \mathbf{T}$, whence the result. □

Examples 12.1.3   Take $\mathbf{G} = \mathbf{GL}_n$; its rank is n (see 1.5.1). Any diagonal matrix with diagonal entries all distinct is regular, as its centraliser is the torus which consists of all diagonal matrices. The unipotent matrix

$$\begin{pmatrix} 1 & 1 & & & \\ & 1 & \ddots & & \\ & & \ddots & 1 & \\ & & & 1 \end{pmatrix}$$

is regular, as its centraliser is the group of all upper triangular matrices with constant diagonals.

From now on $\mathbf{G}$ will denote a connected reductive group over an algebraically closed field. The following result gives a characterisation of regular elements.

Theorem 12.1.4   *Let $x = su$ be the Jordan decomposition of an element of $\mathbf{G}$. The element x is regular if and only if u is regular in $C_\mathbf{G}(s)^0$.*

Proof   We have $C_\mathbf{G}(su) = C_\mathbf{G}(s) \cap C_\mathbf{G}(u)$, so $\dim C_\mathbf{G}(su) = \dim C_{C_\mathbf{G}(s)^0}(u)$. But su is regular if and only if this dimension is equal to $\text{rank}(\mathbf{G}) = \text{rank}(C_\mathbf{G}(s)^0)$. By 12.1.2 applied in $C_\mathbf{G}(s)^0$ this is equivalent to u being regular in $C_\mathbf{G}(s)^0$. $\qquad\qquad\square$

By 12.1.2 regular semi-simple elements exist. The existence of regular unipotent elements is much more difficult, and will be proved later. Once this existence is known, Theorem 12.1.4 shows that there are regular elements with arbitrary semi-simple parts.

Exercise 12.1.5   Show that in $\mathbf{GL}_n$ or $\mathbf{SL}_n$ a matrix is regular if and only if its characteristic polynomial is equal to its minimal polynomial.

We first study semi-simple regular elements.

Proposition 12.1.6

(i)   *Let $\mathbf{T}$ be a maximal torus of $\mathbf{G}$; an element s of $\mathbf{T}$ is regular if and only if no root of $\mathbf{G}$ relative to $\mathbf{T}$ is trivial on s.*

(ii) *A semi-simple element is regular if and only if it is contained in only one maximal torus.*

(iii) *A semi-simple element s is regular if and only if $C_\mathbf{G}(s)^0$ is a torus.*

Proof   Assertions (i) and (iii) are straightforward from Proposition 3.5.1. Assertion (ii) is a direct consequence of assertion (iii). $\qquad\qquad\square$

Corollary 12.1.7   *Let* **B** *be a Borel subgroup of* **G**. *A semi-simple element of* **B** *is regular in* **B** *if and only if it is regular in* **G**. *Regular semi-simple elements are dense in* **B**.

Proof   Let $\mathbf{B} = \mathbf{U} \rtimes \mathbf{T}$ be a Levi decomposition of **B**. The first assertion results from 12.1.6(i) and (iii) which are valid in **B** as well as in **G**: if Φ is the root system of **G** with respect to **T**, and Φ^+ is the positive subsystem defined by **B**, the condition for $t \in \mathbf{T}$ that $\alpha(t) \neq 1$ for any $\alpha \in \Phi$ is equivalent to the same condition for $\alpha \in \Phi^+$; that is, for $\mathbf{U}_\alpha \subset \mathbf{B}$.

Now consider the map $f : (u,t) \mapsto {}^u t : \mathbf{U} \times \mathbf{T}_{\mathrm{reg}} \to \mathbf{B}$ where $\mathbf{T}_{\mathrm{reg}}$ is the set of regular elements of **T**. The image of f is the set of regular semi-simple elements of **B** since all maximal tori of **B** are conjugate under **U** to **T**. By 12.1.6(i) $\mathbf{T}_{\mathrm{reg}}$ is an open (hence dense) subset of **T**; further the fibres of f are reduced to a point since an equality ${}^u t = {}^{u'} t'$ implies $u'^{-1} u^t (u'^{-1} u)^{-1} \in \mathbf{T}$, thus $u'^{-1} u = {}^t(u'^{-1} u)$ which implies $u' = u$ since t is regular. Thus the image of f has same dimension as **B**, hence f is dominant since **B** is irreducible, so the image of f is dense.   □

Remark 12.1.8   The method of the above proof shows that if $t \in \mathbf{T}_{\mathrm{reg}}$, the image of the map $u \mapsto u^{-1} {}^t u : \mathbf{U} \to \mathbf{U}$ is dense in **U**; actually it can be proved, using the method of the proof of 4.2.9, that this map is surjective.

Corollary 12.1.9   *Regular semi-simple elements are dense in* **G**.

Proof   This is clear from 12.1.7 as **G** is the union of all Borel subgroups by 1.3.3(ii).   □

We now consider regular unipotent elements. The proof of their existence (due to Steinberg) is not easy. We shall deduce it from the fact that there is only a finite number of unipotent classes in **G**. This result was proved by Kostant and Dynkin in characteristic 0 and by Richardson in good characteristic – see definition below 12.2.5 – by classifying the unipotent classes. It can be shown that to prove it in characteristic p it is enough to prove it over $\overline{\mathbb{F}}_p$. The general finiteness property was first proved in Lusztig (1976b) whose proof we follow; it is a good illustration of the use of the theory of representations of finite groups of Lie type. As the quotient morphism $\mathbf{G} \to \mathbf{G}/Z(\mathbf{G})$ induces a bijection on unipotent elements and also on unipotent classes it is enough to show the finiteness when **G** has a connected centre. The idea of the proof is to use the representation theory of $\mathbf{G}^F$, where F is a Frobenius endomorphism associated to the integer q, to separate the unipotent classes of $\mathbf{G}^F$ by a finite set of class functions whose cardinality is bounded independently of q, which gives the result as $\mathbf{G} = \bigcup_{n \in \mathbb{N}^*} \mathbf{G}^{F^n}$.

Now **G** will be a connected reductive group over $\overline{\mathbb{F}}_q$, with a Frobenius root F, and $(\mathbf{G}^*, \mathbf{T}^*, F^*)$ are dual to $(\mathbf{G}, \mathbf{T}, F)$. Note that an F-stable semi-simple element is quasi-isolated if and only if $C_{\mathbf{G}}(s)$ is not contained in any proper F-stable Levi subgroup. Indeed, if s is not quasi-isolated – thus $C_{\mathbf{G}}(s) \subset \mathbf{L}$ for some proper Levi subgroup $\mathbf{L}$ – then $\cap_i {}^{F^i}\mathbf{L}$ is an F-stable Levi subgroup containing $C_{\mathbf{G}}(s)$ by 3.4.8(i) since all ${}^{F^i}\mathbf{L}$ contain a given F-stable maximal torus of $C_{\mathbf{G}}(s)$.

We define $\mathcal{E}(\mathbf{G}^F)$ to be the linear span of the Lusztig series $\mathcal{E}(\mathbf{G}^F, (s))$ when s runs over the set $\mathcal{S}(\mathbf{G}^*)$ of semi-simple elements of $\mathbf{G}^{*F^*}$ such that there exists an F^*-stable Levi subgroup $\mathbf{L}^*$ of $\mathbf{G}^*$ such that $s \in \mathrm{Der}\,\mathbf{L}^*$ and such that s is quasi-isolated in $\mathbf{L}^*$.

We denote by $\mathcal{E}(\mathbf{G}^F)_u$ the space of restrictions of $\mathcal{E}(\mathbf{G}^F)$ to unipotent elements. The main lemma is the following.

Lemma 12.1.10 *Let $\mathbf{G}$ be a connected reductive group over $\overline{\mathbb{F}}_q$ with a connected centre and a Frobenius root F; the restriction of any class function f on $\mathbf{G}^F$ to unipotent elements is in $\mathcal{E}(\mathbf{G}^F)_u$.*

Proof The proof is by induction on the dimension of $\mathbf{G}$. If $Z(\mathbf{G}) \neq \{1\}$, the map $\mathbf{G} \xrightarrow{\pi} \mathbf{G}/Z(\mathbf{G})$ induces a bijection on F-stable unipotent elements and their conjugacy classes, so the restriction of f to unipotent elements is equal to $\overline{f} \circ \pi$ for a class function $\overline{f}$ on unipotent elements of $(\mathbf{G}/Z(\mathbf{G}))^F$; this last group is equal to $\mathbf{G}^F/Z(\mathbf{G})^F$ since $Z(\mathbf{G})$ is connected. By the induction hypothesis we have $\overline{f} \in \mathcal{E}(\mathbf{G}^F/Z(\mathbf{G})^F)_u$. By 11.3.10 we have $R_{\mathbf{T}/Z(\mathbf{G})}^{\mathbf{G}/Z(\mathbf{G})}(\theta) \circ \pi = R_{\mathbf{T}}^{\mathbf{G}}(\theta \circ \pi)$ and the dual group of $\mathbf{G}/Z(\mathbf{G})$ is isomorphic to $\mathrm{Der}(\mathbf{G}^*)$ – see 11.4.9 – thus the semi-simple element defining the series of $(\mathbf{T}, \theta \circ \pi)$ is in $\mathrm{Der}(\mathbf{G}^*)$ whence $f \in \mathcal{E}(\mathbf{G}^F)_u$ in this case.

We now assume that $Z(\mathbf{G}) = 1$, so $\mathbf{G}$ and $\mathbf{G}^*$ are semi-simple. Let ρ be an irreducible character of $\mathbf{G}^F$; it is in some Lusztig series $\mathcal{E}(\mathbf{G}^F, (s))$. Assume first that the centraliser of s is contained in some proper F^*-stable Levi subgroup $\mathbf{L}^*$ of $\mathbf{G}^*$. Then, by 11.4.3(ii), we have $\rho = R_{\mathbf{L}}^{\mathbf{G}}(\lambda)$, where λ is, up to sign, an element of $\mathcal{E}(\mathbf{L}^F, (s))$. By the induction hypothesis applied to $\mathbf{L}$, the restriction of λ to the unipotent elements of $\mathbf{L}$ is in $\mathcal{E}(\mathbf{L}^F)_u$. As $R_{\mathbf{L}}^{\mathbf{G}}$ commutes with the restriction to the unipotent elements (see 10.1.6 and 7.3.2(ii)) and maps $\mathcal{E}(\mathbf{L}^F)$ into $\mathcal{E}(\mathbf{G}^F)$, the restriction of $R_{\mathbf{L}}^{\mathbf{G}}(\lambda)$ to the unipotent elements is in $\mathcal{E}(\mathbf{G}^F)_u$, whence the result in this case. Assume now that the centraliser of s is not contained in a proper F-stable Levi subgroup. We have $s \in \mathrm{Der}(\mathbf{G}^*) = \mathbf{G}^*$, so $\mathcal{E}(\mathbf{G}^F, (s)) \subset \mathcal{E}(\mathbf{G}^F)$ and we get the result in this last case. $\square$

Theorem 12.1.11 *In a connected reductive algebraic group $\mathbf{G}$ over $\overline{\mathbb{F}}_q$, the number of unipotent classes is finite.*

Proof We will show that the dimension of $\mathcal{E}(\mathbf{G}^F)$ is bounded independently of q, which proves the theorem, using 12.1.10. For that, we first show that

$S(\mathbf{G}^*)$ is a finite union of conjugacy classes, in number bounded independently of q. By 11.2.1(iii), in the dual of a group with connected centre the semi-simple elements have connected centralisers. So the condition that the s is quasi-isolated in $\mathbf{L}^*$ can be seen on the root system of $C_{\mathbf{G}^*}(s)$ (with respect to some torus containing s). Explicitly, an element $s \in \mathbf{G}^{*F^*}$ is in $S(\mathbf{G}^*)$ if and only if there exists a Levi subgroup $\mathbf{L}^*$ such that $s \in \mathrm{Der}\,\mathbf{L}^*$ and that the root system of $C_{\mathbf{L}^*}(s)$ is not contained in a proper parabolic subsystem of the root system of $\mathbf{L}^*$. But in a given root system there is only a finite number of root subsystems, whence only a finite number of $\mathbf{G}^*$-conjugacy classes of possible groups $C_{\mathbf{G}^*}(s)$, depending only on the type of the root system of $\mathbf{G}$. Any one of these groups is the centraliser of a finite number of elements – bounded independently of q – of the corresponding $\mathrm{Der}\,\mathbf{L}^*$ by the following lemma applied to $\mathbf{H} = C_{\mathbf{G}^*}(s) \cap \mathrm{Der}\,\mathbf{L}^*$ and $\mathbf{G} = \mathrm{Der}\,\mathbf{L}^*$ and by 11.2.6.

Lemma 12.1.12   *Let* $\mathbf{G}$ *be a connected reductive group and let* $\mathbf{H}$ *be a connected subgroup of maximal rank of* $\mathbf{G}$. *Assume that the root system of* $\mathbf{H}$ *(relative to some maximal torus* $\mathbf{T}$*) is not contained in any parabolic subsystem of the root system of* $\mathbf{G}$ *relative to* $\mathbf{T}$*; then* $Z(\mathbf{H})^0 = Z(\mathbf{G})^0$. *In particular, if* $\mathbf{G}$ *is semi-simple, then so is* $\mathbf{H}$.

Proof   The centraliser in $\mathbf{G}$ of $Z(\mathbf{H})^0$ is a Levi subgroup of $\mathbf{G}$ containing $\mathbf{H}$ (see 3.4.7), so is equal to $\mathbf{G}$ and we have $Z(\mathbf{H})^0 \subset Z(\mathbf{G})^0$. As $\mathbf{H}$ has maximal rank it contains $Z(\mathbf{G})^0$, whence the result.   □

We have thus proved that $\mathcal{E}(\mathbf{G}^F)$ is the union of a finite number of series $\mathcal{E}(\mathbf{G}^F, (s))$ in number bounded independently of q. We shall be done if we bound independently of q the cardinality of $\mathcal{E}(\mathbf{G}^F, (s))$; we claim $|W|^2$ is such a bound: each Deligne–Lusztig character has at most $|W|$ irreducible constituents by 9.3.1, and the number of Deligne–Lusztig characters whose irreducible constituents lie in a Lusztig series $\mathcal{E}(\mathbf{G}^F, (s))$ is the number of classes of F-stable maximal tori in the centraliser of s, thus is bounded by $|H^1(F, W(s))| \le |W|$.   □

We now recall a well-known result from algebraic geometry which will be used in the proofs of 12.2.2 and of 12.2.9.

Lemma 12.1.13   *The dimension of any fibre of a surjective morphism from an irreducible variety* $\mathbf{X}$ *to another* $\mathbf{Y}$ *is at least* $\dim(\mathbf{X}) - \dim(\mathbf{Y})$*; moreover this inequality is an equality for an open dense subset of* $\mathbf{Y}$.

Proof   For the proof of the inequality see Hartshorne (1977, II, Exercise 3.22). The second statement is proved in Steinberg (1974, Appendix to 2.11, Prop. 2).
                    □

12.2 Regular Unipotent Elements

We can now state the properties of regular unipotent elements. We first introduce a definition.

Definition 12.2.1 For $x \in \mathbf{G}$, we define the variety $\mathcal{B}_x$ (called the **Springer fibre** of x) as the variety of Borel subgroups containing x; equivalently if we fix a Borel subgroup $\mathbf{B}$ of $\mathbf{G}$ it identifies with the fixed points of x for left translation in $\mathbf{G}/\mathbf{B}$.

The Springer fibres are projective varieties as closed subvarieties of a projective variety.

Proposition 12.2.2 *Let u be a unipotent element of $\mathbf{G}$; the following properties are equivalent:*

(i) *u is regular.*

(ii) *$\mathcal{B}_u$ is finite.*

(iii) *$\mathcal{B}_u$ consists of only one Borel subgroup.*

(iv) *There exists a Borel subgroup $\mathbf{B}$ containing u and a maximal torus $\mathbf{T}$ of $\mathbf{B}$ such that in the decomposition $u = \prod_{\alpha \in \Phi^+} \mathbf{u}_\alpha(x_\alpha)$, where Φ^+ is the set of positive roots of $\mathbf{G}$ relative to $\mathbf{T} \subset \mathbf{B}$ (see 2.3.1(vi)), none of the x_α for α simple is equal to 0.*

(v) *Property (iv) is true for any Borel subgroup $\mathbf{B}$ containing u and any maximal torus of $\mathbf{B}$.*

Proof It is obvious that (iii) implies (ii) and that (v) implies (iv).

We prove that (ii) implies (v). Let $\mathbf{B} = \mathbf{U} \rtimes \mathbf{T}$ be a Levi decomposition of a Borel subgroup containing u and let $\prod_{\alpha \in \Phi^+} \mathbf{u}_\alpha(x_\alpha)$ be the decomposition of u. If $x_\alpha = 0$ for a simple α then for any $v \in \mathbf{U}_\alpha$ the element $^v u$ has also a trivial component in $\mathbf{U}_\alpha$ by the commutation formulae 2.3.1(vii). Let $n \in N_{\mathbf{G}}(\mathbf{T})$ be a representative of the simple reflection s_α of $W(\mathbf{T})$ corresponding to α; then $^{nv} u$ is in $\mathbf{B}$ since the only root subgroup of $\mathbf{B}$ which is mapped onto a negative root subgroup under the action of n is $\mathbf{U}_\alpha$, thus u is in $^{vs_\alpha}\mathbf{B}$ for any $v \in \mathbf{U}_\alpha$. The Borel subgroups $^{vs_\alpha}\mathbf{B}$ are all distinct since an equality $^{vs_\alpha}\mathbf{B} = {}^{v's_\alpha}\mathbf{B}$ is equivalent to $vs_\alpha\mathbf{B} = v's_\alpha\mathbf{B}$ and by the unique decomposition 3.2.7 this is possible only if $v = v'$. Thus we have got an infinite number (at least an affine line) of Borel subgroups containing u.

We show now that (iv) implies (iii). Let $\mathbf{B} = \mathbf{U} \rtimes \mathbf{T}$, be as in (iv), let $\mathbf{B}_1$ be another Borel subgroup containing u and let $\mathbf{T}_1$ be a maximal torus contained in $\mathbf{B} \cap \mathbf{B}_1$ (see 3.1.4); then $\mathbf{T} = {}^v\mathbf{T}_1$ with $v \in \mathbf{U}$, so that $^v\mathbf{B}_1$ is a Borel subgroup containing $\mathbf{T}$ and so equals $^w\mathbf{B}$ for some $w \in W(\mathbf{T})$. Thus we see that $^w\mathbf{B}$ contains $^v u$. By the commutation formulae 2.3.1(vii), the element $^v u$ still satisfies (iv), but if α is a simple root such that $^{w^{-1}}\alpha$ is negative, then the root subgroup $\mathbf{U}_\alpha$ has a trivial intersection with ${}^w\mathbf{B}$, so the component of ${}^v u$ in $\mathbf{U}_\alpha$ is necessarily trivial. This is a contradiction if w is not 1; that is, if $\mathbf{B}_1 \neq \mathbf{B}$.

So far we have proved the equivalence of (ii), (iii), (iv) and (v).

We shall show now that (i) is equivalent to these four properties. First we prove that (i) implies (iv). Let u be a regular unipotent element and $\mathbf{B} = \mathbf{U} \rtimes \mathbf{T}$ be a Borel subgroup containing u, and let $u = \prod_{\alpha \in \Phi^+} \mathbf{u}_\alpha(x_\alpha)$ be the corresponding decomposition. Assume that for some simple α we have $x_\alpha = 0$. The element u is in the parabolic subgroup $\mathbf{P} = \mathbf{B} \cup \mathbf{B}s_\alpha\mathbf{B}$ where s_α is the reflection with respect to α. In that parabolic subgroup the conjugacy class of u contains only elements of $\mathbf{U}$ whose component in $\mathbf{U}_\alpha$ is 1, so has dimension at most $\dim(\mathbf{U}) - 1$. Hence the dimension of the centraliser of u in $\mathbf{P}$ is at least $\dim(\mathbf{P}) - (\dim(\mathbf{U}) - 1)$. As $\dim(\mathbf{P}) = \dim(\mathbf{U}) + \dim(\mathbf{T}) + 1$, we get $C_\mathbf{P}(u) \geq \mathrm{rank}(\mathbf{G}) + 2$, which is a contradiction by 12.1.2.

We will now prove that some element satisfying (iv) is regular. By 12.1.11, the variety $\mathbf{G}_u$ of unipotent elements is a finite union of conjugacy classes, so has the same dimension as one of them. We know $\mathrm{codim}(\mathbf{G}_u) \geq \mathrm{rank}(\mathbf{G})$, as the codimension of any conjugacy class is at least $\mathrm{rank}(\mathbf{G})$ by 12.1.2 (i). Consider the morphism $\mathbf{G} \times \mathbf{U} \to \mathbf{G}$ given by $(g,u) \mapsto {}^g u$. Its image is $\mathbf{G}_u$. Let u be as in (iii). Then (g,v) is in the fibre of u if and only if $g \in N_\mathbf{G}(\mathbf{B}) = \mathbf{B}$ and $v = {}^{g^{-1}}u$; thus the fibre of u is isomorphic to $\mathbf{B}$. By 12.1.13 we have $\dim(\mathbf{B}) \geq \dim(\mathbf{G}) + \dim(\mathbf{U}) - \dim(\mathbf{G}_u)$. As $\dim(\mathbf{B}) - \dim(\mathbf{U}) = \dim(\mathbf{T}) = \mathrm{rank}(\mathbf{G})$, we get $\mathrm{codim}(\mathbf{G}_u) \leq \mathrm{rank}(\mathbf{G})$. Thus $\mathrm{codim}(\mathbf{G}_u) = \mathrm{rank}(\mathbf{G})$, so a class having same dimension as $\mathbf{G}_u$ is regular (and this class satisfies (iv)).

Let us prove now that the elements satisfying (iv) form a single conjugacy class under $\mathbf{B}$ which will prove that every element satisfying (iv) is regular. Let u satisfy (iv) and be regular, and let v satisfy (iv); replacing v by a $\mathbf{T}$-conjugate, we may assume that the components of v and u in $\mathbf{U}_\alpha$ for α simple are equal (see 2.3.1(i) and 1.2.11), so that vu^{-1} is in $\mathbf{U}^* = \prod_{\alpha \in \Phi^+ - \Pi} \mathbf{U}_\alpha$. As the orbits of a unipotent group acting on a variety are closed, see Steinberg (1974, 2.5, Proposition), the set $[x,\mathbf{U}] := \{[x,u] \mid x \in \mathbf{U}\}$ is a closed subvariety of $\mathbf{U}^*$ of dimension $\dim(\mathbf{U}) - \dim(C_\mathbf{U}(u))$. If we prove $\dim([x,\mathbf{U}]) = \dim(\mathbf{U}^*)$, we shall get $\mathbf{U}^* = [x,\mathbf{U}]$, thus the class $(u)_\mathbf{U}$ of u under $\mathbf{U}$ is equal to $u\mathbf{U}^*$, thus $v \in u\mathbf{U}^*$ is $\mathbf{U}$-conjugate to u. Thus we have to prove $\dim(C_\mathbf{U}(u)) = \dim(\mathbf{U}) - \dim(\mathbf{U}^*) = \mathrm{rank}(\mathbf{G}) - \dim(Z(\mathbf{G}))$. Since u is regular, this follows from the following lemma.

Lemma 12.2.3   *If u satisfies (iv) then $C_\mathbf{G}(u) = Z(\mathbf{G}).C_\mathbf{U}(u)$.*

Proof   If g centralises u then u is in ${}^{g^{-1}}\mathbf{B}$, thus since u satisfies (iii) we get $g \in N_\mathbf{G}(\mathbf{B}) = \mathbf{B}$, thus $g \in C_\mathbf{B}(u)$. Let $g = tv$ be the Jordan decomposition of g

with t semi-simple and v unipotent, both in $C_{\mathbf{B}}(u)$; the fact that u is of the form (iv) implies that $t \in Z(\mathbf{G})$, whence the lemma. □

□

We have seen in the proof of 12.2.2

Corollary 12.2.4 *Regular unipotent elements exist and form a single conjugacy class of* **G**.

We have also proved that any element of $\mathbf{U}^*$ is a commutator of a regular element and another element of $\mathbf{U}$, so in particular the following proposition is valid.

Proposition 12.2.5 *We have* $\operatorname{Der} \mathbf{U} = \prod_{\alpha \in \Phi^+ - \Pi} \mathbf{U}_\alpha$.

To state the next result we need to define good and bad characteristics.

Definition 12.2.6 The characteristic is **good** for **G** if it does not divide the coefficients of the highest root of the positive root system associated to a Borel subgroup of **G**.

Bad (that is, not good) characteristics are $p = 2$ for a root system of type B_n, C_n or D_n, $p = 2, 3$ for types G_2, F_4, E_6, E_7 and $p = 2, 3, 5$ for type E_8, see Bourbaki (1968, VI §4).

Proposition 12.2.7 *Let u be a regular unipotent element of* **G***; if the characteristic is good for* **G** *then $C_{\mathbf{U}}(u)$ is connected, otherwise $C_{\mathbf{U}}(u)/C_{\mathbf{U}}^0(u)$ is a non-trivial cyclic group generated by the image of u.*

References See Springer (1966b, 4.11, 4.12) and Lou (1968). Most of the proof is by a case-by-case check. □

From the above results on unipotent and semi-simple regular elements, we can deduce results for all regular elements.

Corollary 12.2.8 *An element x is regular if and only if $\mathcal{B}_x$ is finite, and then $|\mathcal{B}_x| = |W/W^0(s)|$ where s is the semi-simple part of x.*

Proof By 12.1.4 if su is the Jordan decomposition of x, then x is regular if and only if u is regular in $C_{\mathbf{G}}(s)^0$, which is equivalent to u being contained in a finite number of Borel subgroups (in fact one) of $C_{\mathbf{G}}(s)^0$ by 12.2.2(iii). But a Borel subgroup of $C_{\mathbf{G}}(s)^0$ is of the form $\mathbf{B} \cap C_{\mathbf{G}}(s)^0$ where $\mathbf{B}$ is a Borel subgroup of $\mathbf{G}$ containing a maximal torus of $C_{\mathbf{G}}(s)^0$ (see 3.4.10(i)). As the number of Borel subgroups containing a given maximal torus in a connected reductive group is finite, we get the first assertion. This number is equal to the cardinality of the Weyl group, so is $|W|$ in **G** and $|W^0(s)|$ in $C_{\mathbf{G}}(s)^0$; moreover any two Borel subgroups of $C_{\mathbf{G}}(s)^0$ are contained in the same number of Borel subgroups of $\mathbf{G}$ since they are conjugate, whence the second assertion.   □

We now give a general result about centralisers. This result is due to Springer (1966a). We follow here the proof given in Steinberg (1974, 3.5, remark following Proposition 1).

Proposition 12.2.9   *The centraliser of any element of* $\mathbf{G}$ *contains an abelian algebraic subgroup of dimension at least* rank($\mathbf{G}$).

Proof   The result is clear for semi-simple elements. The idea of the proof is to use the fact that semi-simple elements are dense in $\mathbf{G}$ (see 12.1.9). Consider the variety

$$\mathbf{S}_n = \{(x_1, x_2, \ldots, x_n) \in \mathbf{G}^n \mid \text{all } x_i \text{ belong to the same torus}\};$$

then the projection $(x_1, x_2, \ldots, x_n) \mapsto (x_1, x_2, \ldots, x_{n-1})$ is surjective from $\mathbf{S}_n$ to $\mathbf{S}_{n-1}$. Moreover, if $\mathbf{T}$ is a maximal torus of $\mathbf{G}$, as $\mathbf{S}_n$ is the image of $\mathbf{G} \times \mathbf{T}^n$ by the map $(g, t_1, \ldots, t_n) \mapsto ({}^g t_1, \ldots, {}^g t_n)$, it is irreducible, so its closure $\overline{\mathbf{S}}_n$ is also irreducible. It follows that the projection is also surjective from $\overline{\mathbf{S}}_n$ to $\overline{\mathbf{S}}_{n-1}$. Note also that the components of an element of $\mathbf{S}_n$ commute with each other, so the same property is true for $\overline{\mathbf{S}}_n$. The density of semi-simple elements means that $\overline{\mathbf{S}}_1 = \mathbf{G}$. Now let x be an arbitrary element of $\mathbf{G}$ and choose n and $(x_1, x_2, \ldots, x_{n-1}) \in \mathbf{G}^{n-1}$ such that $(x, x_1, \ldots, x_{n-1}) \in \overline{\mathbf{S}}_n$ and that $C_{\mathbf{G}}(x, x_1, \ldots, x_{n-1})$ is minimal among all such sequences when n and x_i vary, which is possible since $\mathbf{G}$ is a Noetherian topological space. Then any $z \in \mathbf{G}$ such that $(x, x_1, \ldots, x_{n-1}, z) \in \overline{\mathbf{S}}_{n+1}$, satisfies

$$C_{\mathbf{G}}(x, x_1, \ldots, x_{n-1}, z) = C_{\mathbf{G}}(x, x_1, \ldots, x_{n-1}),$$

so is in the centre of $C_{\mathbf{G}}(x, x_1, \ldots, x_{n-1})$. The set of such elements z is thus an abelian subgroup of $C_{\mathbf{G}}(x)$, so we shall be done if we prove that the fibres of the projection $\overline{\mathbf{S}}_{n+1} \to \overline{\mathbf{S}}_n$ have dimension at least rank($\mathbf{G}$). By 12.1.13, which we can use since the morphism is surjective and $\overline{\mathbf{S}}_{n+1}$ is irreducible, the dimension of any fibre is at least $\dim(\overline{\mathbf{S}}_{n+1}) - \dim(\overline{\mathbf{S}}_n)$. But the fibre of an element of $\mathbf{S}_n$ contains at least a maximal torus, so has dimension at least rank($\mathbf{G}$) whence $\dim(\overline{\mathbf{S}}_{n+1}) - \dim(\overline{\mathbf{S}}_n) \geq \text{rank}(\mathbf{G})$, which gives the result.   □

Corollary 12.2.10   *If* x *is regular then* $C_{\mathbf{G}}(x)^0$ *is abelian.*

Proof   We know that the dimension of $C_{\mathbf{G}}(x)^0$ is equal to rank($\mathbf{G}$), but by 12.2.9, $C_{\mathbf{G}}(x)^0$ contains an abelian subgroup of that dimension, whence the result.   □

We now study $\mathbf{G}^F$-classes of regular unipotent elements, where F is a Frobenius root on $\mathbf{G}$.

Remark 12.2.11   Note that all the $\mathbf{G}^F$-classes of regular unipotent elements have the same cardinality: indeed by 12.2.3, 12.2.7 and 12.2.10 the group $C_{\mathbf{G}}(u)$ is abelian, and we have the following lemma.

Lemma 12.2.12   *If $g \in \mathbf{G}^F$ is such that $C_{\mathbf{G}}(g)$ is abelian, then all $\mathbf{G}^F$-classes in the geometric conjugacy class of g have same cardinality.*

Proof   By 4.2.15(ii) an element of $\mathbf{G}^F$ geometrically conjugate to g is of the form ${}^x g$ where $x \in \mathbf{G}$ is such that $x^{-1}F(x) \in C_{\mathbf{G}}(g)$. But, since $C_{\mathbf{G}}(g)$ is abelian, for any $h \in C_{\mathbf{G}^F}(g)$ we have ${}^x h \in \mathbf{G}^F$, thus $\mathrm{ad}\, x$ is a bijection from $C_{\mathbf{G}^F}(g)$ to $C_{\mathbf{G}^F}({}^x g)$. $\qquad\square$

Remark 12.2.13   Note that if H is a finite abelian group with an automorphism F we have $H^1(F,H) \simeq H/\mathcal{L}(H)$; as in this case $\mathcal{L}$ is a group morphism with kernel H^F, we get $|H^1(F,H)| = |H^F|$. Thus for an algebraic group $\mathbf{H}$ with a Frobenius root F and such that $\mathbf{H}/\mathbf{H}^0$ is abelian we get $|H^1(F,\mathbf{H})| = |H^1(F,\mathbf{H}/\mathbf{H}^0)| = |(\mathbf{H}/\mathbf{H}^0)^F| = |\mathbf{H}^F|/|\mathbf{H}^{0F}|$.

We use this remark in the proof of the following proposition.

Proposition 12.2.14   *The number of regular unipotent elements in $\mathbf{G}^F$ is $|\mathbf{G}^F|/(|Z(\mathbf{G})^{0F}|q^l)$ where l is the semi-simple rank of $\mathbf{G}$.*

Proof   A regular unipotent element lies in only one Borel subgroup, so the number of regular unipotent elements in $\mathbf{G}^F$ is equal to the product of the number of F-stable Borel subgroups – which is $|\mathbf{G}^F/\mathbf{B}^F|$ for any F-stable Borel subgroup $\mathbf{B}$ – with the number of such elements in a given Borel subgroup. By 12.2.2(v) and 12.2.5, the number of regular unipotent elements in $\mathbf{B}^F$, where $\mathbf{B} = \mathbf{T}\mathbf{U}$, is $|(\mathrm{Der}\,\mathbf{U})^F|$ times the number of elements of $\mathbf{U}^F/(\mathrm{Der}\,\mathbf{U})^F$ whose every component is non-trivial in the decomposition $\mathbf{U}/\mathrm{Der}\,\mathbf{U} \simeq \prod \mathbf{U}_\alpha$; we shall say that these elements are regular in $\mathbf{U}/\mathrm{Der}\,\mathbf{U}$. By 2.3.1(i) and 1.2.11 any two regular elements of $\mathbf{U}/\mathrm{Der}\,\mathbf{U}$ are conjugate by an element of $\mathbf{T}$, but the centraliser of a regular element in $\mathbf{T}$ is $Z(\mathbf{G})$, so by 4.2.14 the number of $\mathbf{T}^F$-orbits of regular elements in $\mathbf{U}^F/(\mathrm{Der}\,\mathbf{U})^F$ is $|H^1(F,Z(\mathbf{G}))|$. In each such orbit there are $|\mathbf{T}^F/Z(\mathbf{G})^F|$ elements, so the number of regular elements in $\mathbf{G}^F$ is $|\mathbf{G}^F|/(|\mathbf{T}^F||\mathbf{U}^F|)|(\mathrm{Der}\,\mathbf{U})^F||H^1(F,Z(\mathbf{G}))||\mathbf{T}^F|/|Z(\mathbf{G})^F|$. Now $\mathbf{U}^F/(\mathrm{Der}\,\mathbf{U})^F$ has cardinality q^l (by 4.4.2 applied to $\mathbf{U}^F$ and $(\mathrm{Der}\,\mathbf{U})^F$) and $|H^1(F,Z(\mathbf{G}))| = |Z(\mathbf{G})^F/Z(\mathbf{G})^{0F}|$ since $Z(\mathbf{G})/Z(\mathbf{G})^0$ is abelian. The result follows. $\qquad\square$

Note that by 4.2.14(i) and the above proof $\mathcal{L}_{\mathbf{T}}^{-1}(Z(\mathbf{G}))$ permutes transitively the regular elements of $\mathbf{U}^F/(\mathrm{Der}\,\mathbf{U})^F$.

Proposition 12.2.15   *If the characteristic is good for $\mathbf{G}$, the $\mathbf{G}^F$-conjugacy classes of regular unipotent elements are parametrised by the F-conjugacy classes $H^1(F,Z(\mathbf{G})/Z(\mathbf{G})^0)$.*

Proof If u is an F-stable regular unipotent element, by 4.2.15(ii) the $\mathbf{G}^F$-classes inside its $\mathbf{G}$-class are parametrised by $H^1(F, C_\mathbf{G}(u)/C_\mathbf{G}(u)^0)$. If the characteristic is good, by 12.2.7 we have $C_\mathbf{G}(u)/C_\mathbf{G}(u)^0 = Z(\mathbf{G})/Z(\mathbf{G})^0$, whence the result. □

The following corollary is clear.

Corollary 12.2.16 *If the characteristic is good for $\mathbf{G}$ and $Z(\mathbf{G})$ is connected, there is a single class of regular unipotent elements in $\mathbf{G}^F$.*

We can also use regular unipotent elements to prove the following proposition.

Proposition 12.2.17 $C_\mathbf{G}(\mathbf{G}^F) = Z(\mathbf{G})$.

Proof Let u be a regular unipotent element of $\mathbf{G}^F$. Let $\mathbf{B}$ be the unique Borel subgroup of $\mathbf{G}$ containing u, which is necessarily F-stable, and let $\mathbf{B} = \mathbf{U} \rtimes \mathbf{T}$ be an F-stable Levi decomposition of $\mathbf{B}$. We have $C_\mathbf{G}(\mathbf{G}^F) \subset C_\mathbf{G}(u) = Z(\mathbf{G}) \cdot C_\mathbf{U}(u)$, where the right equality is by 12.2.3. Let now $w_0 \in W_\mathbf{G}(\mathbf{T})$ be the longest element which, being unique, is F-stable and thus has an F-stable representative $\dot{w}_0 \in N_\mathbf{G}(\mathbf{T})$. Since $\mathbf{U} \cap {}^{w_0}\mathbf{U} = \prod_{\{\alpha \in \Phi^+ \mid w_0(\alpha) \in \Phi^+\}} \mathbf{U}_\alpha = \{1\}$, the Borel subgroup ${}^{w_0}\mathbf{B}$ satisfies ${}^{w_0}\mathbf{B} \cap \mathbf{B} = \mathbf{T}$ – such a Borel subgroup is said to be **opposed** to $\mathbf{B}$. Since ${}^{\dot{w}_0}u$ is also regular, it follows that $C_\mathbf{G}(\mathbf{G}^F) \subset C_\mathbf{G}({}^{\dot{w}_0}u) \cap C_\mathbf{G}(u) = Z(\mathbf{G})$, whence the result. □

Note that the parametrisation of 12.2.15 is well-defined only once we have chosen a particular regular unipotent element $u_1 \in \mathbf{G}^F$ which corresponds to the trivial element of $H^1(F, Z(\mathbf{G})/Z(\mathbf{G})^0)$. From now on we fix a regular unipotent $u_1 \in \mathbf{G}^F$. In arbitrary characteristic, by 12.2.3 the $\mathbf{G}^F$-classes of regular unipotent elements are parametrised by

$$H^1(F, Z(\mathbf{G})/Z(\mathbf{G})^0 \times C_\mathbf{U}(u_1)/C_\mathbf{U}^0(u_1));$$

the first projection maps these F-classes onto the F-classes of $Z(\mathbf{G})/Z(\mathbf{G})^0$. We shall denote by $\mathcal{U}_z$ the set of regular unipotent elements of $\mathbf{G}^F$ in a class parametrised by the fibre of $z \in H^1(F, Z(\mathbf{G})/Z(\mathbf{G})^0)$; by construction we have $u_1 \in \mathcal{U}_1$. If $\mathbf{B} = \mathbf{U} \rtimes \mathbf{T}$ is an F-stable Levi decomposition of some Borel subgroup, the sets $\mathcal{U}_z$ are characterised by the following property.

Proposition 12.2.18 *With the above notation, two regular elements u and u' of $\mathbf{U}^F$ are in the same set $\mathcal{U}_z$ if and only if there exists $t \in \mathbf{T}^F$ such that $u^{-1 \, t}u'$ is in Der($\mathbf{U}$).*

Proof There is an F-stable conjugate of u_1 in $\mathbf{U}$, which we denote again by u_1. Any element which conjugates u_1 to u is in $\mathbf{B}$ by 12.2.2(iii). So we have

$u = {}^{sv}u_1$ with $s \in \mathbf{T}$ and $v \in \mathbf{U}$. The class of u is parametrised by the F-class of $v^{-1}s^{-1}{}^F s^F v$ in $H^1(F, Z(\mathbf{G})/Z(\mathbf{G})^0 \times C_\mathbf{G}(u_1)/C_\mathbf{G}(u_1)^0)$ – we know by 12.2.3 that $s^{-1}{}^F s$ is in $Z(\mathbf{G})$. The element u' is in the same $\mathcal{U}_z$ as u if and only if it can be written ${}^{s'v'}u_1$ with $s^{-1}{}^F s$ and $s'^{-1}{}^F s'$ F-conjugate by some $z_1 \in Z(\mathbf{G})$. This implies that we have ${}^{ts'v'}u_1 = {}^{sv}u_1$ with $t = s z_1 s'^{-1} \in \mathbf{T}^F$. So u and ${}^t u'$ are conjugate under $\mathbf{U}$, which implies by 2.3.1(vii) that they have the same image in $\mathbf{U}/\operatorname{Der}\mathbf{U}$. Conversely, if ${}^{sv}u_1$ and ${}^{ts'v'}u_1$ have the same image in $\mathbf{U}/\operatorname{Der}\mathbf{U}$, the element $s^{-1}ts'$ must be central by 12.2.2(v) and 2.3.1(i), so $s^{-1}{}^F s$ and $s'^{-1}{}^F s'$ are in the same F-class of $Z(\mathbf{G})$, whence the result. □

12.3 Gelfand–Graev Representations

In the following we fix an F-stable Levi decomposition $\mathbf{B} = \mathbf{U} \rtimes \mathbf{T}$ of an F-stable Borel subgroup $\mathbf{B}$. We denote by Π the corresponding basis of the root system of $\mathbf{G}$ relative to $\mathbf{T}$. Then F defines a permutation σ of Π (see 4.2.5 and 2.4.8). For any orbit O of σ in Π, we denote by $\mathbf{U}_O$ the image of $\prod_{\alpha \in O} \mathbf{U}_\alpha$ in $\mathbf{U}/\operatorname{Der}\mathbf{U}$. It is a commutative group isomorphic to $\mathbb{G}_a^{|O|}$. The action of F maps $\mathbf{U}_\alpha$ on $\mathbf{U}_{\sigma\alpha}$ for each $\alpha \in \Pi$ – and is an isomorphism of abstract groups. The action of $F^{|O|}$ stabilises each $\mathbf{U}_\alpha$ for $\alpha \in O$, so the group $\mathbf{U}_O^F$ is isomorphic to $\mathbf{U}_\alpha^{F^{|O|}}$ for any $\alpha \in O$. We choose such an α; by the Lang–Steinberg theorem we can choose the isomorphism $\mathbb{G}_a \xrightarrow{\mathbf{u}_\alpha} \mathbf{U}_\alpha$ such that $\mathbf{u}_\alpha(1) \in \mathbf{U}_\alpha^{F^{|O|}}$; using 4.1.9 we then get $\mathbf{U}_O^F \simeq \mathbb{F}_{q^{|O|}}^+$ – the additive group. In the following we shall assume that we have made such choices for each orbit O.

Definition 12.3.1   A linear character $\phi \in \operatorname{Irr}(\mathbf{U}^F)$ is **regular** if its restriction to $\operatorname{Der}(\mathbf{U})^F$ is trivial and its restriction to $\mathbf{U}_O^F$ is not trivial for any orbit O of σ in Π.

Note that $(\operatorname{Der}\mathbf{U})^F$ contains $\operatorname{Der}(\mathbf{U}^F)$ – actually it is almost always equal to that derived group, the only exceptions for quasi-simple groups being groups of type B_l, C_l, G_2 or F_4 over $\mathbb{F}_2$ and groups of type G_2 over $\mathbb{F}_3$; see Howlett (1974, Lemma 7) where the proof is given in the case of split groups, for non-split groups the proof is similar. Note also that the quotient map is an isomorphism of varieties $\prod_{\alpha \in \Pi} \mathbf{U}_\alpha \xrightarrow{\sim} \mathbf{U}/\operatorname{Der}\mathbf{U}$ compatible with the actions of $\mathbf{T}$ and that this isomorphism induces a bijection on F-stable elements. We thus have an action of $\mathbf{T}^F$ on regular characters.

Proposition 12.3.2   *The $\mathbf{T}^F$-orbits of regular characters of $\mathbf{U}^F$ are in one-to-one correspondence with $H^1(F, Z(\mathbf{G}))$.*

Proof   By the remark after 12.2.14 $\mathcal{L}_\mathbf{T}^{-1}(Z(\mathbf{G}))$ acts transitively on regular elements of $\mathbf{U}^F/(\operatorname{Der}\mathbf{U})^F$. By the remarks before 12.3.1, a regular character of

$\mathbf{U}^F$ is identified with an element of $\prod_{O \in \Pi/\sigma} \{\mathrm{Irr}(\mathbb{F}^+_{q^{|O|}}) - \{\mathbf{1}\}\}$ and a regular element of $\mathbf{U}^F/(\mathrm{Der}\,\mathbf{U})^F$ is identified with an element of $\prod_{O \in \Pi/\sigma} \{\mathbb{F}_{q^{|O|}} - \{0\}\}$. So we see that $\mathcal{L}_{\mathbf{T}}^{-1}(Z(\mathbf{G}))$ acts transitively on the set of regular characters of $\mathbf{U}^F$. The stabiliser in $\mathcal{L}_{\mathbf{T}}^{-1}(Z(\mathbf{G}))$ of a regular character of $\mathbf{U}^F$ is $Z(\mathbf{G})$ and $\mathcal{L}_{\mathbf{T}}^{-1}(Z(\mathbf{G}))^F = \mathbf{T}^F$. So, by 4.2.14(ii) applied with $\mathbf{G}$ replaced by $\mathcal{L}_{\mathbf{T}}^{-1}(Z(\mathbf{G}))$ we get the result (in 4.2.14(ii) $\mathbf{G}$ is assumed to be connected but the only property actually needed is that the stabiliser of an element is contained in the image of the Lang map, so we can apply 4.2.14(ii) to $\mathcal{L}_{\mathbf{T}}^{-1}(Z(\mathbf{G}))$ in our case. □

Note that, by the above proof, the regular characters of $\mathbf{U}^F$ are parametrised by $\mathcal{L}_{\mathbf{T}}^{-1}(Z(\mathbf{G}))/Z(\mathbf{G})$. As in the case of regular unipotent elements, this parametrisation is well-defined once we have chosen a regular character. From now on we shall fix such a character which will be denoted by ψ_1. We make the choice in the following way: we fix a non-trivial additive character χ_0 of $\mathbb{F}_p^+$; by composing with the trace, for any N we get a character χ_N of $\mathbb{F}^+_{q^N}$. We take for ψ_1, parametrised by $1 \in H^1(F, Z(\mathbf{G}))$, the product of the characters $\chi_{|O|}$ of the additive groups $\mathbb{F}^+_{q^{|O|}}$. The orbit of regular characters parametrised by $z \in H^1(F, Z(\mathbf{G}))$ will be denoted by Ψ_z. If $\dot{z}$ is the image in $\mathcal{L}_{\mathbf{T}}^{-1}(Z(\mathbf{G}))/Z(\mathbf{G})$ of an element $t \in \mathcal{L}_{\mathbf{T}}^{-1}(Z(\mathbf{G}))$ such that the F-conjugacy class of $\mathcal{L}(t)$ is equal to z, then the character $\psi_{\dot{z}} := {}^t\psi_1$ is in Ψ_z.

Definition 12.3.3   For $z \in H^1(F, Z(\mathbf{G}))$ we define the **Gelfand–Graev representation** Γ_z (or $\Gamma_z^{\mathbf{G}}$) by $\Gamma_z = \mathrm{Ind}_{\mathbf{U}^F}^{\mathbf{G}^F}(\psi_{\dot{z}})$.

Note that $\Gamma_z = {}^t\Gamma_1$ if $\dot{z}$ is the image of t as above. The following result is due to Steinberg:

Theorem 12.3.4   *The Gelfand–Graev representations are multiplicity free.*

Proof   We shall not give the proof here. It may be found, for example, in Carter (1985, 8.1.3); the proof given there is written under the assumption that $Z(\mathbf{G})$ is connected and F is a Frobenius, but it applies without any change to our case. □

We now study the effect of Harish-Chandra restriction and of duality on the Gelfand–Graev representations. For this we need some notation and preliminary results. If $\mathbf{L}$ is an F-stable Levi subgroup containing $\mathbf{T}$ of an F-stable parabolic subgroup of $\mathbf{G}$, for parametrising the regular characters of $\mathbf{U}^F \cap \mathbf{L}^F$ by $H^1(F, Z(\mathbf{L}))$ we have to fix a particular regular character: we choose $\mathrm{Res}_{\mathbf{U}^F \cap \mathbf{L}^F}^{\mathbf{U}^F} \psi_1$, which corresponds for $\mathbf{L}$ to the same definition we gave of ψ_1 for $\mathbf{G}$. The next lemma will allow us to compare the parametrisations in $\mathbf{G}$ and in $\mathbf{L}$.

Lemma 12.3.5 *The inclusion* $Z(\mathbf{G}) \subset Z(\mathbf{L})$ *induces a surjective map*

$$h_{\mathbf{L}} : H^1(F, Z(\mathbf{G})) \to H^1(F, Z(\mathbf{L})).$$

Proof The inclusion $Z(\mathbf{G}) \subset Z(\mathbf{L})$ gives rise to the Galois cohomology exact sequence, see 4.2.10 and the remark following for normal subgroups

$$1 \to Z(\mathbf{G})^F \to Z(\mathbf{L})^F \to (Z(\mathbf{L})/Z(\mathbf{G}))^F \to$$

$$\to H^1(F, Z(\mathbf{G})) \xrightarrow{h_{\mathbf{L}}} H^1(F, Z(\mathbf{L})) \to H^1(F, Z(\mathbf{L})/Z(\mathbf{G})).$$

Thus it is enough to show that the quotient $Z(\mathbf{L})/Z(\mathbf{G})$ is connected, which results from 11.2.1(ii) as – by 2.3.4(ii) – $Z(\mathbf{L})/Z(\mathbf{G})$ is the centre of the Levi subgroup $\mathbf{L}/Z(\mathbf{G})$ of the group $\mathbf{G}/Z(\mathbf{G})$. □

Proposition 12.3.6 *With the above notation we have*

$$^*R_{\mathbf{L}}^{\mathbf{G}}(\Gamma_z^{\mathbf{G}}) = \Gamma_{h_{\mathbf{L}}(z)}^{\mathbf{L}}.$$

Proof We have to show that, for any class function f on $\mathbf{L}^F$, we have

$$\langle R_{\mathbf{L}}^{\mathbf{G}}(f), \Gamma_z^{\mathbf{G}} \rangle_{\mathbf{G}^F} = \langle f, \Gamma_{h_{\mathbf{L}}(z)}^{\mathbf{L}} \rangle_{\mathbf{L}^F}.$$

Up to $\mathbf{G}^F$-conjugacy, we may assume that $\mathbf{L}$ is the standard Levi subgroup $\mathbf{L}_I$, and we may compute the Harish-Chandra induction using the parabolic subgroup $\mathbf{P}$ having $\mathbf{L}$ as a Levi subgroup and containing $^{n_0}\mathbf{B}$, where n_0 is a representative of the longest element of W. We have $R_{\mathbf{L}}^{\mathbf{G}}(f) = \mathrm{Ind}_{\mathbf{P}^F}^{\mathbf{G}^F} \tilde{f}$ where the value of $\tilde{f} = \mathrm{Inf}_{\mathbf{L}^F}^{\mathbf{P}^F} f$ is given by $\tilde{f}(lu) = f(l)$ for $u \in \mathbf{V}$, where $\mathbf{V} = R_u(\mathbf{P})$. So the left-hand side is equal to

$$\langle \mathrm{Ind}_{\mathbf{P}^F}^{\mathbf{G}^F} \tilde{f}, \mathrm{Ind}_{\mathbf{U}^F}^{\mathbf{G}^F} \psi_z \rangle_{\mathbf{G}^F} = \sum_{g \in \mathbf{P}^F \backslash \mathbf{G}^F / \mathbf{U}^F} \langle \tilde{f}, {}^g \psi_z \rangle_{\mathbf{P}^F \cap {}^g \mathbf{U}^F},$$

this equality is by the ordinary Mackey formula. Now – by 5.2.2(ii) – $(\mathbf{L} \backslash \mathcal{S}(\mathbf{L}, \mathbf{T}) / \mathbf{T})^F \xrightarrow{\sim} \mathbf{P}^F \backslash \mathbf{G}^F / \mathbf{B}^F$. Since $\mathbf{L} \backslash \mathcal{S}(\mathbf{L}, \mathbf{T}) / \mathbf{T} = W_I \backslash W$ is represented by the I-reduced element, we get – using the unicity of an I-reduced element in its coset – that the F-stable I-reduced elements in W^F form a set of representatives of $\mathbf{P}^F \backslash \mathbf{G}^F / \mathbf{B}^F$. So we can choose as a cross section of $\mathbf{P}^F \backslash \mathbf{G}^F / \mathbf{U}^F$ a set of representatives in $N_{\mathbf{G}}(\mathbf{T})^F$ of I-reduced elements of W^F. If n is such an element – a representative of $w \in W^F$ – we have

$$\mathbf{P}^F \cap {}^n\mathbf{U}^F = (\mathbf{L}^F \cap {}^n\mathbf{U}^F).(\mathbf{V}^F \cap {}^n\mathbf{U}^F),$$

and the restriction of $\tilde{f}$ to $\mathbf{P}^F \cap {}^n\mathbf{U}^F$ is equal to $\mathrm{Inf}_{\mathbf{L}^F \cap {}^n\mathbf{U}^F}^{\mathbf{P}^F \cap {}^n\mathbf{U}^F} \circ \mathrm{Res}_{\mathbf{L}^F \cap {}^n\mathbf{U}^F}^{\mathbf{L}^F}(f)$ in that decomposition. So we have

$$\langle \tilde{f}, {}^n\psi_z \rangle_{\mathbf{P}^F \cap {}^n\mathbf{U}^F} = \langle \mathrm{Res}_{\mathbf{L}^F \cap {}^n\mathbf{U}^F}^{\mathbf{L}^F} f, {}^n\psi_z \rangle_{\mathbf{L}^F \cap {}^n\mathbf{U}^F} \langle 1, \psi_z \rangle_{n^{-1}\mathbf{V}^F \cap \mathbf{U}^F}.$$

But, by the definition of $\psi_{\dot{z}}$, the scalar product $\langle 1, \psi_{\dot{z}} \rangle_{n^{-1}\mathbf{V}^F \cap \mathbf{U}^F}$ is not zero (and is equal to 1) if and only if $^{n^{-1}}\mathbf{V} \cap \mathbf{U}$ contains no $\mathbf{U}_\alpha$ for $\alpha \in \Pi$. However, as w is I-reduced, using 2.3.1(v) we get

$$^{n^{-1}}\mathbf{V} \cap \mathbf{U} = {}^{n^{-1}n_0}\mathbf{U} \cap \mathbf{U}.$$

This group contains no $\mathbf{U}_\alpha$ with $\alpha \in \Pi$ if and only if $w = 1$; that is, $n \in \mathbf{T}^F$. We then have $\mathbf{L} \cap {}^n\mathbf{U} = \mathbf{L} \cap \mathbf{U}$, so

$$\langle R_{\mathbf{L}}^{\mathbf{G}}(f), \Gamma_z \rangle_{\mathbf{G}^F} = \langle \operatorname{Res}_{\mathbf{L}^F \cap \mathbf{U}^F}^{\mathbf{L}^F} f, {}^n\psi_{\dot{z}} \rangle_{\mathbf{L}^F \cap \mathbf{U}^F}.$$

So we shall have proved the proposition if we show that the $\mathbf{T}^F$-orbit of ${}^n\psi_{\dot{z}}|_{\mathbf{L}^F \cap \mathbf{U}^F}$, that is of $\psi_{\dot{z}}|_{\mathbf{L}^F \cap \mathbf{U}^F}$, is the $\mathbf{T}^F$-orbit $\Psi_{h_{\mathbf{L}}(z)}$ of regular characters of $\mathbf{L}^F \cap \mathbf{U}^F$. But we have $\psi_{\dot{z}} = {}^t\psi_1$ for any $t \in \mathcal{L}_{\mathbf{T}}^{-1}(Z(\mathbf{G}))$ such that $\dot{z} = tZ(\mathbf{G})$ and – by the definition of $h_{\mathbf{L}}$– the character $\operatorname{Res}_{\mathbf{U}^F \cap \mathbf{L}^F}^{\mathbf{U}^F} \psi_{\dot{z}} = {}^t(\operatorname{Res}_{\mathbf{U}^F \cap \mathbf{L}^F}^{\mathbf{U}^F} \psi_1)$ is in $\Psi_{h_{\mathbf{L}}(z)}$. □

Remark 12.3.7   Proposition 12.3.6 can be extended to Lusztig induction when the characteristic p is good for $\mathbf{G}$ and q is large enough, in the form ${}^*R_{\mathbf{L}}^{\mathbf{G}}(\Gamma_z^{\mathbf{G}}) = \varepsilon_{\mathbf{G}}\varepsilon_{\mathbf{L}}\Gamma_{h_{\mathbf{L}}(z)}^{\mathbf{L}}$, see Bonnafé (2005a, 15.2).

We now give two results on the dual of a Gelfand–Graev representation.

Proposition 12.3.8

(i) *The dual of any Gelfand–Graev representation is zero outside regular unipotent elements.*

(ii) *We have* $\langle \Gamma_z, (-1)^{S/F} D_{\mathbf{G}} \Gamma_{z'} \rangle_{\mathbf{G}^F} = |Z(\mathbf{G})^F| \delta_{z,z'}.$

Proof   By 12.3.6 and 7.1.12 we have

$$D_{\mathbf{G}} \Gamma_z = \sum_{J \subset S/F} (-1)^{|J|} R_{\mathbf{L}_J}^{\mathbf{G}}(\Gamma_{h_{\mathbf{L}_J}(z)}).$$

By the definition of $R_{\mathbf{L}_J}^{\mathbf{G}}$, and using $\operatorname{Res}_{\mathbf{U}^F \cap \mathbf{L}_J^F}^{\mathbf{U}^F}(\psi_{\dot{z}}) \in \Psi_{h_{\mathbf{L}_J}(z)}$, which we saw at the end of the proof of 12.3.6, we have

$$R_{\mathbf{L}_J}^{\mathbf{G}}(\Gamma_{h_{\mathbf{L}_J}(z)}) = \operatorname{Ind}_{\mathbf{P}_J^F}^{\mathbf{G}^F} \circ \operatorname{Inf}_{\mathbf{L}_J^F}^{\mathbf{P}_J^F} \circ \operatorname{Ind}_{\mathbf{U}^F \cap \mathbf{L}_J^F}^{\mathbf{L}_J^F} \circ \operatorname{Res}_{\mathbf{U}^F \cap \mathbf{L}_J^F}^{\mathbf{U}^F}(\psi_{\dot{z}}).$$

If $v \in R_u(\mathbf{P}_J)^F$ and $l \in \mathbf{L}_J^F$, for any $f \in C(\mathbf{U}^F \cap \mathbf{L}_J^F)$ we have

$$(\operatorname{Ind}_{\mathbf{U}^F}^{\mathbf{P}_J^F} \circ \operatorname{Inf}_{\mathbf{U}^F \cap \mathbf{L}_J^F}^{\mathbf{U}^F} f)(vl) = |\mathbf{U}^F|^{-1} \sum_{\{p \in \mathbf{P}_J^F \mid {}^p(vl) \in \mathbf{U}^F\}} (\operatorname{Inf}_{\mathbf{U}^F \cap \mathbf{L}_J^F}^{\mathbf{U}^F} f)({}^p(vl)),$$

where the inflation in the left-hand side is defined using $\mathbf{U}^F = R_u(\mathbf{P}_J)^F \rtimes (\mathbf{U}^F \cap \mathbf{L}_J^F)$. The condition ${}^p(vl) \in \mathbf{U}^F$ is equivalent to ${}^{l'}l \in \mathbf{U}^F$ where l' is the component in $\mathbf{L}_J^F$ of p in the decomposition $\mathbf{P}_J^F = R_u(\mathbf{P}_J)^F \rtimes \mathbf{L}_J^F$ and

$(\mathrm{Inf}^{\mathbf{U}^F}_{\mathbf{U}^F \cap \mathbf{L}^F_J} f)(^p(vl)) = f(^{l'}l)$ when the condition is satisfied. Thus we get $\mathrm{Inf}^{\mathbf{P}^F_J}_{\mathbf{L}^F_J} \circ \mathrm{Ind}^{\mathbf{L}^F_J}_{\mathbf{U}^F \cap \mathbf{L}^F_J} = \mathrm{Ind}^{\mathbf{P}^F_J}_{\mathbf{U}^F} \circ \mathrm{Inf}^{\mathbf{U}^F}_{\mathbf{U}^F \cap \mathbf{L}^F_J}$. By transitivity of induction we get

$$R^{\mathbf{G}}_{\mathbf{L}_J}(\Gamma_{h_{L_J}(z)}) = \mathrm{Ind}^{\mathbf{G}^F}_{\mathbf{U}^F} \circ \mathrm{Inf}^{\mathbf{U}^F}_{\mathbf{U}^F \cap \mathbf{L}^F_J} \circ \mathrm{Res}^{\mathbf{U}^F}_{\mathbf{U}^F \cap \mathbf{L}^F_J}(\psi_{\hat{z}}).$$

Thus

$$D_{\mathbf{G}}(\Gamma_z) = \mathrm{Ind}^{\mathbf{G}^F}_{\mathbf{U}^F}\left(\sum_{J \subset S/F} (-1)^{|J|} \mathrm{Inf}^{\mathbf{U}^F}_{\mathbf{U}^F \cap \mathbf{L}^F_J} \circ \mathrm{Res}^{\mathbf{U}^F}_{\mathbf{U}^F \cap \mathbf{L}^F_J} \psi_{\hat{z}}\right).$$

To prove assertion (i) of the proposition it is sufficient to show that

$$\sum_{J \subset S/F} (-1)^{|J|} \mathrm{Inf}^{\mathbf{U}^F}_{\mathbf{U}^F \cap \mathbf{L}^F_J} \circ \mathrm{Res}^{\mathbf{U}^F}_{\mathbf{U}^F \cap \mathbf{L}^F_J} \psi_{\hat{z}}$$

is zero outside regular unipotent elements of $\mathbf{U}^F$. We use the notation of 12.3.1. Let u be in $\mathbf{U}^F$ and let $\prod_{O \in \Pi/\sigma} u_O$ denote the projection of u on $\prod_O \mathbf{U}^F_O$; if we put $\psi_O = \mathrm{Res}^{\mathbf{U}^F}_{\mathbf{U}^F_O} \psi_{\hat{z}}$, we have $\psi_{\hat{z}}(u) = \prod_{O \in \Pi/\sigma} \psi_O(u_O)$ whence

$$\sum_{J \subset S/F} (-1)^{|J|} (\mathrm{Inf}^{\mathbf{U}^F}_{\mathbf{U}^F \cap \mathbf{L}^F_J} \circ \mathrm{Res}^{\mathbf{U}^F}_{\mathbf{U}^F \cap \mathbf{L}^F_J} \psi_{\hat{z}})(u) = \prod_{O \in \Pi/\sigma} (1 - \psi_O(u_O))$$

which is zero outside regular unipotent elements, as u is regular if and only if $u_O \neq 1$ for all O.

We now prove (ii). By the above computations we have $D_{\mathbf{G}}\Gamma_z = \mathrm{Ind}^{\mathbf{G}^F}_{\mathbf{U}^F} \varphi$, where

$$\varphi = \sum_{J \subset S/F} (-1)^{|J|} \mathrm{Inf}^{\mathbf{U}^F}_{\mathbf{U}^F \cap \mathbf{L}^F_J} \circ \mathrm{Res}^{\mathbf{U}^F}_{\mathbf{U}^F \cap \mathbf{L}^F_J} \psi_{\hat{z}},$$

which is zero outside regular unipotent elements. By the ordinary Mackey formula, using the fact that $N_{\mathbf{G}}(\mathbf{T})^F$ is a cross section for $\mathbf{U}^F \backslash \mathbf{G}^F / \mathbf{U}^F$, we have

$$\langle D_{\mathbf{G}}\Gamma_z, \Gamma_{z'} \rangle_{\mathbf{G}^F} = \sum_{n \in N_{\mathbf{G}}(\mathbf{T})^F} \langle \varphi, {}^n\psi_{\hat{z}'} \rangle_{\mathbf{U}^F \cap {}^n\mathbf{U}^F}.$$

But $\mathbf{U}^F \cap {}^n\mathbf{U}^F$ does not contain any regular unipotent element if n is not in $\mathbf{T}$ by 12.2.2(iii), and $\langle \mathrm{Inf}^{\mathbf{U}^F}_{\mathbf{U}^F \cap \mathbf{L}^F_J} \circ \mathrm{Res}^{\mathbf{U}^F}_{\mathbf{U}^F \cap \mathbf{L}^F_J} \psi_{\hat{z}}, \psi_{\hat{z}'} \rangle_{\mathbf{U}^F}$ is equal to zero if $J \neq S/F$. So we have

$$\langle D_{\mathbf{G}}\Gamma_z, \Gamma_{z'} \rangle_{\mathbf{G}^F} = (-1)^{|S/F|} \sum_{t \in \mathbf{T}^F} \langle \psi_{\hat{z}}, {}^t\psi_{\hat{z}'} \rangle_{\mathbf{U}^F}.$$

The characters $\psi_{\hat{z}}$ and ${}^t\psi_{\hat{z}'}$ can be equal only if $z = z'$. There are then $|Z(\mathbf{G})^F|$ values of t such that $\psi_{\hat{z}} = {}^t\psi_{\hat{z}'}$, whence the result. □

Using the Gelfand–Graev representations, we can give the value of any character on a regular unipotent element at least when the characteristic is good.

Definition 12.3.9   We define $\sigma_z := \sum_{\psi \in \Psi_{z^{-1}}} \psi(u_1)$, where u_1 is the element we fixed after 12.2.17.

Theorem 12.3.10

(i) *For any $\chi \in \mathrm{Irr}(\mathbf{G}^F)$ we have*

$$|\mathcal{U}_z|^{-1} \sum_{u \in \mathcal{U}_z} \chi(u) = \sum_{z' \in H^1(F, Z(\mathbf{G}))} \sigma_{zz'^{-1}} \langle (-1)^{|S/F|} D_{\mathbf{G}}(\chi), \Gamma_{z'} \rangle_{\mathbf{G}^F}.$$

(ii) *We have*

$$\langle \sum_{z \in H^1(F, Z(\mathbf{G}))} \Gamma_z, \sum_{z \in H^1(F, Z(\mathbf{G}))} \Gamma_z \rangle_{\mathbf{G}^F} = |H^1(F, Z(\mathbf{G}))| |Z(\mathbf{G})^F| q^l,$$

where l is the semi-simple rank of $\mathbf{G}$.

Proof   In the following we shall denote by γ_z the class function on $\mathbf{G}^F$ whose value is zero outside $\mathcal{U}_z$ and $|\mathbf{G}^F|/|\mathcal{U}_z|$ on $\mathcal{U}_z$. We have seen in the proof of 12.3.8 that $D_{\mathbf{G}}\Gamma_z = \mathrm{Ind}_{\mathbf{U}^F}^{\mathbf{G}^F} \varphi$, where φ is a function on $\mathbf{U}^F/(\mathrm{Der}\,\mathbf{U})^F$ which is zero outside the set of regular elements. So by 12.2.18 $D_{\mathbf{G}}\Gamma_z$ is constant on $\mathcal{U}_{z'}$, and there exist coefficients $c_{z,z'}$ such that $D_{\mathbf{G}}\Gamma_z = \sum_{z' \in H^1(F, Z(\mathbf{G}))} c_{z,z'} \gamma_{z'}$. By 12.3.8 (ii) the matrix $(c_{z,z'})_{z,z'}$ is invertible and its inverse is

$$((-1)^{|S/F|} |Z(\mathbf{G})^F|^{-1} \langle \Gamma_{z'}, \gamma_z \rangle_{\mathbf{G}^F})_{z,z'}.$$

Let $t \in \mathcal{L}_{\mathbf{T}}^{-1}(Z(\mathbf{G}))$ be such that the F-class of $\mathcal{L}(t)$ is equal to z; we have

$$\langle \Gamma_{z'}, \gamma_z \rangle_{\mathbf{G}^F} = |\mathcal{U}_z|^{-1} \sum_{u \in \mathcal{U}_z} \Gamma_{z'}(u) = \Gamma_{z'}(^t u_1),$$

the last equality because $\Gamma_{z'}$ is constant on $\mathcal{U}_z$, as it is induced from $\psi_{z'}$ which factorises through $(\mathbf{U}/\mathrm{Der}\,\mathbf{U})^F$. So

$$\langle \Gamma_{z'}, \gamma_z \rangle_{\mathbf{G}^F} = \Gamma_{z'}(^t u_1) = \Gamma_{z'z^{-1}}(u_1).$$

We now apply the following lemma.

Lemma 12.3.11   *We have $\sigma_z = |Z(\mathbf{G})^F|^{-1} \Gamma_{z^{-1}}(u_1)$.*

Proof   By definition

$$\Gamma_z(u_1) = |\mathbf{U}^F|^{-1} \sum_{\{g \in \mathbf{G}^F | {}^g u_1 \in \mathbf{U}^F\}} \psi_{\hat{z}}(^g u_1).$$

As u_1 is regular, we must have $g \in \mathbf{B}^F$ in the above summation and, if $g = tv$ with $t \in \mathbf{T}^F$ and $v \in \mathbf{U}^F$, then $\psi_{\hat{z}}(^g u_1) = \psi_{\hat{z}}(^t u_1)$, whence, as $Z(\mathbf{G})^F$ is the kernel of the $\mathbf{T}^F$-action on Ψ_z, we get

$$|Z(\mathbf{G})^F|^{-1} \Gamma_z(u_1) = \sum_{\psi \in \Psi_z} \psi(u_1) = \sigma_{z^{-1}}.$$

$\square$

So we get

$$|Z(\mathbf{G})^F|^{-1}\langle \Gamma_{z'}, \gamma_z \rangle_{\mathbf{G}^F} = \sigma_{zz'^{-1}}.$$

So $(-1)^{|S/F|}(\sigma_{zz'^{-1}})_{z,z'}$ is the inverse matrix of $(c_{z,z'})_{z,z'}$; that is,

$$\gamma_z = \sum_{z' \in H^1(F, Z(\mathbf{G}))} \sigma_{zz'^{-1}}(-1)^{|S/F|} D_{\mathbf{G}}\Gamma_{z'},$$

whence 12.3.10(i) by taking the scalar product of both sides with χ.

Let us prove (ii). We have $\langle \sum_z \Gamma_z, \sum_z \Gamma_z \rangle_{\mathbf{G}^F} = \langle \sum_z D_{\mathbf{G}}\Gamma_z, \sum_z D_{\mathbf{G}}\Gamma_z \rangle_{\mathbf{G}^F}$ and, with the above notation, it follows that

$$\langle \sum_z D_{\mathbf{G}}\Gamma_z, \sum_z D_{\mathbf{G}}\Gamma_z \rangle_{\mathbf{G}^F} = \langle \sum_{z,z'} c_{z,z'}\gamma_{z'}, \sum_{z,z'} c_{z,z'}\gamma_{z'} \rangle_{\mathbf{G}^F}$$

$$= \sum_{z'} \langle \sum_z c_{z,z'}\gamma_{z'}, \sum_z c_{z,z'}\gamma_{z'} \rangle_{\mathbf{G}^F}.$$

But

$$\sum_{z'} c_{z,z'} = \langle D_{\mathbf{G}}\Gamma_z, 1 \rangle_{\mathbf{G}^F} = \langle \Gamma_z, \mathrm{St}_{\mathbf{G}^F} \rangle_{\mathbf{G}^F} = |\mathbf{G}^F|^{-1}\Gamma_z(1)\mathrm{St}_{\mathbf{G}^F}(1)$$

$$= |\mathbf{G}^F|^{-1}(|\mathbf{G}^F|/|\mathbf{U}^F|)|\mathbf{G}^F|_p = 1 \text{ (see 4.4.1 (iv))},$$

whence $\sum_z c_{z,z'} = 1$ as $c_{z,z'} = c_{1,z^{-1}z'}$. Thus

$$\langle \sum_z D_{\mathbf{G}}\Gamma_z, \sum_z D_{\mathbf{G}}\Gamma_z \rangle_{\mathbf{G}^F} = \sum_{z'} \langle \gamma_{z'}, \gamma_{z'} \rangle_{\mathbf{G}^F} = \sum_{z'} |\mathbf{G}^F|/|\mathcal{U}_{z'}|.$$

All the sets $\mathcal{U}_z$ have the same cardinality as they are geometrically conjugate, so their cardinality is $|H^1(F, Z(\mathbf{G}))|^{-1}$ times the number of regular unipotent elements. Hence the above sum is equal to $|Z(\mathbf{G})^{0F}|q^l|H^1(F, Z(\mathbf{G}))|^2$ by 12.2.14. As $Z(\mathbf{G})/Z(\mathbf{G})^0$ is abelian we have

$$|H^1(F, Z(\mathbf{G}))| = |H^1(F, Z(\mathbf{G})/Z(\mathbf{G})^0)| = |Z(\mathbf{G})^F/Z(\mathbf{G})^{0F}|,$$

whence the result. $\qquad\square$

Corollary 12.3.12 *If the centre of $\mathbf{G}$ is connected and Γ denotes the unique Gelfand–Graev representation, then $D_{\mathbf{G}}\Gamma$ is equal to $|Z(\mathbf{G})^F|q^l$ on regular unipotent elements and to zero elsewhere. If moreover the characteristic is good we have $D_{\mathbf{G}}\Gamma = \pi_u^{\mathbf{G}^F}$, where $u \in \mathbf{G}^F$ is regular unipotent.*

Proof In that case there is only one $c_{z,z'}$ whose value has to be 1 by the proof of 12.3.10(ii). If moreover the characteristic is good the set of regular unipotent elements is a single conjugacy class of $\mathbf{G}^F$ by 12.2.16. $\qquad\square$

Corollary 12.3.13   *If the characteristic is good for* $\mathbf{G}$, *for any* $\chi \in \mathrm{Irr}(\mathbf{G}^F)$ *and any* $u \in \mathcal{U}_z$ *we have*

$$\chi(u) = \sum_{z' \in H^1(F, Z(\mathbf{G}))} \sigma_{zz'^{-1}} \langle (-1)^{|S/F|} D_{\mathbf{G}}(\chi), \Gamma_{z'} \rangle_{\mathbf{G}^F}.$$

Proof   This is 12.3.10(i) since when the characteristic is good the set $\mathcal{U}_z$ is a single conjugacy class by 12.2.7.   □

Note that, since for bad characteristic the distinct conjugacy classes into which $\mathcal{U}_z$ splits have all the same cardinality (see 12.2.11), the left-hand side of 12.3.10(i) is the mean of the values of χ at these classes.

Finally, note that there are convenient formulas to compute σ_z in terms of Gauss sums, see Digne et al. (1997, 2.4) and the computation at the end of Section 12.5.

12.4 Regular and Semi-Simple Characters

The final part of this chapter is devoted to the study of irreducible constituents of Gelfand–Graev representations and of their dual. We shall prove in particular that when the centre of $\mathbf{G}$ is connected these representations are uniform.

Definition 12.4.1   An irreducible character of $\mathbf{G}^F$ is **regular** if it is a constituent of some Gelfand–Graev character. An irreducible character whose dual is (up to sign) a regular irreducible character will be called **semi-simple**.

For example the character $\mathrm{St}_{\mathbf{G}^F}$ is regular as $\mathrm{Res}^{\mathbf{G}^F}_{\mathbf{U}^F} \mathrm{St}_{\mathbf{G}^F} = \mathrm{reg}_{\mathbf{U}^F}$ (see 7.4.4 and 4.4.1(iv)), so by Frobenius reciprocity $\mathrm{St}_{\mathbf{G}^F}$ is a constituent of any Gelfand–Graev representation. The character $\mathbf{1}_{\mathbf{G}^F}$ is thus semi-simple. The following definition will allow us to describe the regular and semi-simple characters (see 12.4.10 and 12.4.12 below).

Definition 12.4.2   If s is a semi-simple element of $\mathbf{G}^{*F^*}$ and if $\mathbf{T}_1^*$ is quasi-split maximal torus of $C_{\mathbf{G}^*}(s)^0$, we define a class function $\chi_{(s)}$ on $\mathbf{G}^F$ by

$$\chi_{(s)} = |W^0(s)|^{-1} \sum_{w \in W^0(s)} \varepsilon_{\mathbf{G}} \varepsilon_{\mathbf{T}_w^*} R^{\mathbf{G}}_{\mathbf{T}_w^*}(s),$$

where $W^0(s)$ is as in 3.5.2 and $\mathbf{T}_w^*$ is a maximal torus of $C_{\mathbf{G}^*}(s)^0$ of type w (see 4.2.22).

Note that $\varepsilon_{\mathbf{G}} \varepsilon_{\mathbf{T}_w^*} = \varepsilon_{\mathbf{T}_1^*}(-1)^{l(w)}$, see remark after 10.3.6.

Definition 12.4.3   A $\mathbf{G}^{*F^*}$-**Lusztig series** of characters, denoted by $\mathcal{E}(\mathbf{G}^F, (s)_{\mathbf{G}^{*F^*}})$, is the set of irreducible constituents of the $R^{\mathbf{G}}_{\mathbf{T}^*}(s)$ where the semi-simple class $(s)_{\mathbf{G}^{*F^*}}$ of s in $\mathbf{G}^{*F^*}$ is fixed.

Thus $\mathcal{E}(\mathbf{G}^F, (s)_{\mathbf{G}^{*F^*}})$ is a subset of $\mathcal{E}(\mathbf{G}^F, (s))$. These series, in the case where F^* is a Frobenius endomorphism and thus the $\mathbf{G}^{*F^*}$-classes are the "rational classes", are also called in the literature "rational Lusztig series". We see that $\chi_{(s)}$ is in $\mathcal{E}(\mathbf{G}^F, (s)_{\mathbf{G}^{*F^*}})$. Note that by 11.2.2(ii) and 4.2.14(ii), if the centre of $\mathbf{G}$ is connected, the $\mathbf{G}^{*F^*}$-Lusztig series and the geometric series are the same. This is not true in general, as can be seen from the following proposition.

Proposition 12.4.4 *The $\mathbf{G}^{*F^*}$-Lusztig series of characters form a partition of* $\mathrm{Irr}(\mathbf{G}^F)$.

This result will be proved after 12.4.12. We shall also use the next result.

Proposition 12.4.5 *The number of geometric semi-simple classes of $\mathbf{G}^{*F^*}$ is* $|Z(\mathbf{G})^{0F}|q^l$, *where l is the semi-simple rank of $\mathbf{G}$.*

Proof Let $\mathbf{T}^*$ be a quasi-split maximal torus of $\mathbf{G}^*$. Any semi-simple class of $\mathbf{G}^*$ has a representative in $\mathbf{T}^*$ and two elements of $\mathbf{T}^*$ are conjugate in $\mathbf{G}^*$ if and only if they are conjugate by the Weyl group W of $\mathbf{G}^*$ with respect to $\mathbf{T}^*$, see 1.3.6(iv). So semi-simple conjugacy classes of $\mathbf{G}^*$ are in one-to-one correspondence with $\mathbf{T}^*/W$ and the F^*-stable ones – which are in bijection with the geometric semi-simple classes of $\mathbf{G}^{*F^*}$ – are parametrised by $(\mathbf{T}^*/W)^{F^*}$. Now $t \in \mathbf{T}^*$ has an F-stable image in $\mathbf{T}^*/W$ if and only if $t \in \mathbf{T}_1 := \{t \in \mathbf{T}^* \mid \exists w \in W, {}^{wF}t = t\}$. Thus if $O(t)$ denotes the W-orbit of $t \in \mathbf{T}_1$ we get

$$|(\mathbf{T}^*/W)^F| = |\mathbf{T}_1/W| = \sum_{t \in \mathbf{T}_1} \frac{1}{|O(t)|}$$

$$= \frac{1}{|W|} \sum_{t \in \mathbf{T}_1} |\{w \in W \mid {}^wt = t\}| = \frac{1}{|W|} \sum_{t \in \mathbf{T}_1} |\{w \in W \mid {}^{wF}t = t\}|,$$

the last equality since for $t \in \mathbf{T}_1$ there exists $v \in W$ such that ${}^{vF}t = t$ thus ${}^wt = t$ is equivalent to ${}^{wvF}t = t$. Thus we get

$$|(\mathbf{T}^*/W)^F| = \frac{1}{|W|} \sum_{w \in W} |\{t \in \mathbf{T}_1 \mid {}^{wF}t = t\}| = \frac{1}{|W|} \sum_{w \in W} |\mathbf{T}_w^{*F}|.$$

Thus, using 4.4.9 and using that $(-1)^{l(w)} = \det w = \det w^{-1}$ and that $\det \tau = \det \tau^{-1}$ since it is a sign, we get

$$|(\mathbf{T}^*/W)^F| = \frac{1}{|W|} \sum_{w \in W} \det(q - \tau^{-1}w \mid X(\mathbf{T}^*) \otimes \mathbb{R}).$$

If $\mathbf{T}$ is a maximal torus of $\mathbf{G}$ in duality with $\mathbf{T}^*$, we can replace $X(\mathbf{T}^*)$ by $Y(\mathbf{T})$. Further, since $Y(\mathbf{T}) = Y(Z(\mathbf{G})^0) \oplus Y(\mathbf{T} \cap \mathrm{Der}\,\mathbf{G})$ and W acts trivially on the first factor, we have

$$\det(q - \tau^{-1}w \mid Y(\mathbf{T}) \otimes \mathbb{R})$$
$$= \det(q - \tau^{-1} \mid Y(Z(\mathbf{G})^0) \otimes \mathbb{R}) \det(q - \tau^{-1}w \mid Y(\mathbf{T} \cap \operatorname{Der}\mathbf{G}) \otimes \mathbb{R})$$
$$= |Z(\mathbf{G})^{0F}| \det(q - \tau^{-1}w \mid Y(\mathbf{T} \cap \operatorname{Der}\mathbf{G}) \otimes \mathbb{R}),$$

the last equality by 4.4.9 applied to the group $Z(\mathbf{G})^0$. So we are reduced to prove $q^{\dim V} = \frac{1}{|W|} \sum_w \det(q - \tau^{-1}w \mid V)$ where $V = Y(\mathbf{T}) \otimes \mathbb{R}$, when $\mathbf{G}$ is semi-simple. Now it is well-known that

$$\det(q - \tau^{-1}w \mid V) = \sum_{i=0}^{\dim V} (-1)^i q^{\dim V - i} \operatorname{Trace}(\tau^{-1}w \mid V^{\wedge i}),$$

and since $\frac{1}{|W|} \sum_{w \in W} w$ is the projector onto the fixed points of W, we get

$$\frac{1}{|W|} \sum_W \det(q - \tau^{-1}w \mid V) = \sum_{i=0}^{\dim V} (-1)^i q^{\dim V - i} \operatorname{Trace}(\tau^{-1} \mid (V^{\wedge i})^W).$$

But it is a result of Steinberg (1968, 14.1) that if $V^W = 0$ (which is the case when $\mathbf{G}$ is semi-simple), then we also have $(V^{\wedge i})^W = 0$ for $i > 1$. Thus we only have the term for $i = 0$ in the above sum, whence the result. □

Proposition 12.4.6 *We have $\langle \chi_{(s)}, \chi_{(s)} \rangle_{\mathbf{G}^F} = |(W(s)/W^0(s))^{F^*}|$.*

Proof By definition

$$\langle \chi_{(s)}, \chi_{(s)} \rangle_{\mathbf{G}^F} = |W^0(s)|^{-2} \sum_{w, w' \in W^0(s)} \varepsilon_{\mathbf{T}_w^*} \varepsilon_{\mathbf{T}_{w'}^*} \langle R_{\mathbf{T}_w^*}^{\mathbf{G}}(s), R_{\mathbf{T}_{w'}^*}^{\mathbf{G}}(s) \rangle_{\mathbf{G}^F}$$

$$= |W^0(s)|^{-2} \sum_{w \in W^0(s)} |W(s)^{wF^*}| |W^0(s) \cap \{F^*\text{-class of } w \text{ in } W(s)\}|$$

$$= |W^0(s)|^{-2} \sum_{w \in W^0(s)} |\{v \in W(s) \mid vw^{F^*}v^{-1} \in W^0(s)\}|,$$

the second equality by 9.3.1 – since the proof of 11.1.16 shows that the stabiliser of θ in $W(\mathbf{T})^F$ is isomorphic to that of s in $W(\mathbf{T}^*)^{F^*}$ for corresponding pairs $(\mathbf{T}, \theta)$ and $(\mathbf{T}^*, s)$. But, as w is in $W^0(s)$, the F^*-conjugate by $v \in W(s)$ of w is in $W^0(s)$ if and only if the image of v in $W(s)/W^0(s)$ is F^*-stable, so we get

$$\langle \chi_{(s)}, \chi_{(s)} \rangle_{\mathbf{G}^F} = |W^0(s)|^{-2} \sum_{w \in W^0(s)} |(W(s)/W^0(s))^{F^*}| |W^0(s)|,$$

whence the result. □

Proposition 12.4.7 *For any $s \in \mathbf{G}^{*F^*}$ and $z \in H^1(F, Z(\mathbf{G}))$ we have $\langle \chi_{(s)}, \Gamma_z \rangle_{\mathbf{G}^F} = 1$.*

Proof We shall use the following property of Green functions.

Lemma 12.4.8 *The value of any Deligne–Lusztig character at a regular unipotent element is 1.*

Sketch of the proof. The proof is in Deligne and Lusztig (1976, 9.16). Its main ingredient is the use of a compactification of the variety $\mathcal{L}^{-1}(\mathbf{U})/\mathbf{T}^F$. The trace of u on $H_c^*(\mathcal{L}^{-1}(\mathbf{U})/\mathbf{T}^F)$ is then computed, using the fact that u, being regular unipotent, has only one fixed point a on that compactification, and by taking local coordinates at a. The proof in Deligne and Lusztig (1976) is given for a Frobenius endomorphism, but applies verbatim for a Frobenius root. $\square$

By definition we have

$$\langle \chi_{(s)}, \Gamma_z \rangle_{\mathbf{G}^F} = |W^0(s)|^{-1} \sum_{w \in W^0(s)} \langle \varepsilon_{\mathbf{G}} \varepsilon_{\mathbf{T}_w^*} R_{\mathbf{T}_w^*}^{\mathbf{G}}(s), \Gamma_z \rangle_{\mathbf{G}^F}.$$

But

$$\varepsilon_{\mathbf{G}} \varepsilon_{\mathbf{T}_w^*} R_{\mathbf{T}_w^*}^{\mathbf{G}}(s) = D_{\mathbf{G}}(R_{\mathbf{T}_w^*}^{\mathbf{G}}(s))$$

by 10.2.1 and the proof of 11.1.16, so the above scalar product is equal to

$$\langle D_{\mathbf{G}}(R_{\mathbf{T}_w^*}^{\mathbf{G}}(s)), \Gamma_z \rangle_{\mathbf{G}^F} = \langle R_{\mathbf{T}_w^*}^{\mathbf{G}}(s), D_{\mathbf{G}}(\Gamma_z) \rangle_{\mathbf{G}^F}.$$

If we express $D_{\mathbf{G}}(\Gamma_z)$ as a linear combination of the $\gamma_{z'}$ as in the proof of 12.3.10, we get $\sum_{z'} c_{z,z'} \langle R_{\mathbf{T}_w^*}^{\mathbf{G}}(s), \gamma_{z'} \rangle_{\mathbf{G}^F}$; but – by 12.4.8 – $\langle R_{\mathbf{T}_w^*}^{\mathbf{G}}(s), \gamma_{z'} \rangle_{\mathbf{G}^F}$ is equal to 1 and by the proof of 12.3.10, $\sum_{z'} c_{z,z'}$ is equal to 1, whence the result. $\square$

We can now prove the following proposition.

Proposition 12.4.9 *We have*

$$|Z(\mathbf{G})^F|/|Z(\mathbf{G})^{0F}| \sum_{(s)} |(W(s)/W^0(s))^{F^*}|^{-1} \chi_{(s)} = \sum_{z \in H^1(F, Z(\mathbf{G}))} \Gamma_z,$$

where in the sum of the left-hand side (s) runs over semi-simple conjugacy classes of $\mathbf{G}^{*F^*}$.

Proof To prove the result we shall show that the scalar product of the two sides is equal to the scalar product of each side with itself. We first compute the scalar product of the left-hand side with itself. By 12.4.4 this is equal to

$$|Z(\mathbf{G})^F|^2/|Z(\mathbf{G})^{0F}|^2 \sum_{(s)} |(W(s)/W^0(s))^{F^*}|^{-2} \langle \chi_{(s)}, \chi_{(s)} \rangle_{\mathbf{G}^F},$$

which, by 12.4.6 equals

$$|Z(\mathbf{G})^F|^2/|Z(\mathbf{G})^{0F}|^2 \sum_{(s)} |(W(s)/W^0(s))^{F^*}|^{-2} |(W(s)/W^0(s))^{F^*}|$$

$$= |Z(\mathbf{G})^F|^2/|Z(\mathbf{G})^{0F}|^2 \sum_{(s)} |(W(s)/W^0(s))^{F^*}|^{-1}.$$

By 4.2.15(ii), 3.5.2 and 12.2.13 – which applies since $W(s)/W^0(s)$ is abelian, see 11.2.1(iii) – $|(W(s)/W^0(s))^{F^*}|$ is the number of $\mathbf{G}^{*F^*}$-classes in the geometric conjugacy class of s, the last sum is equal to

$$|Z(\mathbf{G})^F|^2/|Z(\mathbf{G})^{0F}|^2|\{\text{ geometric semi-simple classes of } \mathbf{G}^{*F^*} \}|.$$

By 12.4.5, this equals

$$|Z(\mathbf{G})^F||Z(\mathbf{G})^F/Z(\mathbf{G})^{0F}|q^l,$$

which, by 12.2.13, can be written

$$|Z(\mathbf{G})^F||H^1(F,Z(\mathbf{G}))|q^l.$$

The scalar product of the right-hand side with itself was computed in 12.3.10 and has the same value. The scalar product of one side with the other one is easily computed from 12.4.7: it is the product of $|Z(\mathbf{G})^F/Z(\mathbf{G})^{0F}|$, the number of geometric conjugacy classes of semi-simple elements of $\mathbf{G}^*$ and the number of z, which gives again $|Z(\mathbf{G})^F|q^l|H^1(F,Z(\mathbf{G}))|$.   □

Corollary 12.4.10   *Assume that the centre of* $\mathbf{G}$ *is connected. Then:*

(i) $\chi_{(s)}$ *is an irreducible character of* $\mathbf{G}^F$ *for any* (s).
(ii) *The (unique) Gelfand–Graev representation of* $\mathbf{G}^F$ *is the sum of all the* $\chi_{(s)}$ *(which are all the regular characters).*
(iii) *The dual of the Gelfand–Graev representation (which is* $\pi_u^{\mathbf{G}^F}$ *if the characteristic is good; see 12.3.12) is the sum of all the semi-simple characters of* $\mathbf{G}^F$ *up to signs.*

Proof By hypothesis, from 11.2.2(ii) and 3.5.2, for all s we have $|(W(s)/W^0(s))^{F^*}| = 1$, thus $\langle \chi_{(s)}, \chi_{(s)} \rangle_{\mathbf{G}^F} = 1$ by 12.4.6. By the disjunction of the geometric series, which are in the present case equal to the $\mathbf{G}^{*F^*}$-series, $\chi_{(s)}$ which is the projection of the unique Gelfand–Graev character on the series $\mathcal{E}(\mathbf{G}^F,(s))$ is a true character. The corollary is then straightforward (note that in the present case we do not actually have to use 12.4.4).   □

We now want to get a result similar to 12.4.10 for groups with a non-connected centre. First we shall embed $\mathbf{G}$ in a group with connected centre and the same derived group. The construction is as follows: take any torus $\mathbf{S}$ containing $Z(\mathbf{G})$ and put $\tilde{\mathbf{G}} = \mathbf{G} \times_{Z(\mathbf{G})} \mathbf{S}$, where $Z(\mathbf{G})$ acts by translation on $\mathbf{G}$ and on $\mathbf{S}$. The centre of $\tilde{\mathbf{G}}$ is isomorphic to $\mathbf{S}$ and clearly $\mathrm{Der}\,\tilde{\mathbf{G}} = \mathrm{Der}\,\mathbf{G}$. If $\mathbf{T}$ is a maximal torus of $\mathbf{G}$ then it is contained in a unique maximal torus $\tilde{\mathbf{T}} = \mathbf{T} \times_{Z(\mathbf{G})} \mathbf{S}$ (see 1.2.3) of $\tilde{\mathbf{G}}$ and $\mathbf{T} = \tilde{\mathbf{T}} \cap \mathbf{G}$. We shall choose, in addition, $\mathbf{S}$ F-stable and extend F to $\tilde{\mathbf{G}}$ in the obvious way. We then have a bijection $\mathbf{T} \mapsto \tilde{\mathbf{T}}$ from the set of F-stable maximal tori of $\mathbf{G}$ onto that of $\tilde{\mathbf{G}}$. If θ is a character of

$\tilde{\mathbf{T}}^F$ (where $\mathbf{T}$ is an F-stable maximal torus of $\mathbf{G}$), then by 11.3.10 the restriction of $R_{\tilde{\mathbf{T}}}^{\tilde{\mathbf{G}}}(\theta)$ to $\mathbf{G}^F$ is $R_{\mathbf{T}}^{\mathbf{G}}(\mathrm{Res}_{\mathbf{T}^F}^{\tilde{\mathbf{T}}^F}\theta)$.

We now consider dual groups. If $\tilde{\mathbf{T}}^*$ is a torus dual to $\tilde{\mathbf{T}}$ and $\mathbf{T}^*$ a torus dual to $\mathbf{T}$, then $X(\mathbf{T})$ is a quotient of $X(\tilde{\mathbf{T}})$, so $Y(\mathbf{T}^*)$ is a quotient of $Y(\tilde{\mathbf{T}}^*)$, so $\mathbf{T}^*$ is naturally a quotient of $\tilde{\mathbf{T}}^*$ by 1.2.10. Let $\tilde{\mathbf{G}}^*$ be a group dual to $\tilde{\mathbf{G}}$ (the torus $\tilde{\mathbf{T}}^*$ of $\tilde{\mathbf{G}}^*$ being dual of the torus $\tilde{\mathbf{T}}$ of $\tilde{\mathbf{G}}$); then the kernel of $\tilde{\mathbf{T}}^* \to \mathbf{T}^*$ is a central torus of $\tilde{\mathbf{G}}^*$ and the quotient $\mathbf{G}^*$ of $\tilde{\mathbf{G}}^*$ by this central torus is clearly dual to $\mathbf{G}$ (the torus $\mathbf{T}^*$ being dual of $\mathbf{T}$). We shall assume also that $\tilde{\mathbf{G}}^*$ has a Frobenius root F^* dual to F, so the same is true for $\mathbf{G}^*$. Note that the root systems of $\tilde{\mathbf{G}}^*$ (relative to $\tilde{\mathbf{T}}^*$) and of $\mathbf{G}^*$ (relative to $\mathbf{T}^*$) are the same.

Proposition 12.4.11 *For any $s \in \mathbf{G}^{*F^*}$ the class function $\chi_{(s)}$ on $\mathbf{G}^F$ is a proper character.*

Proof Let $\tilde{s} \in \tilde{\mathbf{G}}^{*F^*}$ be a semi-simple element whose image is $s \in \mathbf{G}^{*F^*}$ ($\tilde{s}$ exists as the kernel of $\tilde{\mathbf{G}}^* \to \mathbf{G}^*$ is connected); we shall show that $\chi_{(s)}$ is the restriction of the character $\chi_{(\tilde{s})}$ of $\tilde{\mathbf{G}}^F$, which implies the result by 12.4.10(i). Since the centre of $\tilde{\mathbf{G}}$ is connected, the centralisers of semi-simple elements in $\tilde{\mathbf{G}}^*$ are connected, hence by 3.5.2 $W(\tilde{s})$ is the group generated by the reflections of $\tilde{\mathbf{G}}^*$ with respect to the roots which are trivial on $\tilde{s}$ so is the same as $W^0(s)$. Since $\tilde{s}$ maps to s, for any maximal torus $\mathbf{T}^* \ni s$ the character $R_{\mathbf{T}^*}^{\mathbf{G}}(s)$ is the restriction to $\mathbf{G}^F$ of the character $R_{\tilde{\mathbf{T}}}^{\tilde{\mathbf{G}}}(\tilde{s})$, and, from the definitions of $\chi_{(s)}$ and $\chi_{(\tilde{s})}$ the result follows. □

Once we know that $\chi_{(s)}$ is a proper character, we can give the decomposition of the Gelfand–Graev representation in general.

Theorem 12.4.12 *For any $z \in H^1(F, Z(\mathbf{G}))$ and any semi-simple conjugacy class (s) of $\mathbf{G}^{*F^*}$, there is exactly one irreducible common constituent $\chi_{s,z}$ of $\chi_{(s)}$ and Γ_z; it has multiplicity 1 in both $\chi_{(s)}$ and Γ_z and we have*

$$\Gamma_z = \sum_{(s)} \chi_{s,z},$$

where (s) runs over the set of semi-simple classes of $\mathbf{G}^{*F^*}$.

Proof The result is straightforward by 12.4.7 and 12.4.9. □

Note that we have not given a parametrisation of the regular characters: indeed it is possible to have $\chi_{s,z} = \chi_{s,z'}$. For instance, when $s = 1$ we have $\chi_{(s)} = \mathrm{St}_{\mathbf{G}^F}$ as observed in the proof of 10.2.6. It can be shown that $\chi_{s,z} = \chi_{s,z'}$ if and only if $z^{-1}z'$ is in a certain subgroup of $H^1(F, Z(\mathbf{G}))$ which can be related to $C_{\mathbf{G}^*}(s)/C_{\mathbf{G}^*}(s)^0$ via duality, see Digne et al. (1992, Proposition 3.12(ii)).

Proof of 12.4.4 Let s be an element of $\mathbf{G}^{*F^*}$ and $\tilde{s}$ be a preimage of s in $\tilde{\mathbf{G}}^{*F^*}$. The series $\mathcal{E}(\mathbf{G}^F, (s)_{\mathbf{G}^{*F^*}})$ consists, by 11.3.10, of the irreducible constituents of the restrictions of elements in $\mathcal{E}(\tilde{\mathbf{G}}^F, (\tilde{s}))$. Assume that two series $\mathcal{E}(\mathbf{G}^F, (s)_{\mathbf{G}^{*F^*}})$ and $\mathcal{E}(\mathbf{G}^F, (s')_{\mathbf{G}^{*F^*}})$ have a non-empty intersection. Let $\tilde{s}'$ be a preimage in $\tilde{\mathbf{G}}^{*F^*}$ of s'; then there exist two irreducible characters $\chi \in \mathcal{E}(\tilde{\mathbf{G}}^F, (\tilde{s}))$ and $\chi' \in \mathcal{E}(\tilde{\mathbf{G}}^F, (\tilde{s}'))$ whose restrictions to $\mathbf{G}^F$ have a common constituent. By Clifford's theory (see 11.3.9) we see that the restrictions of two irreducible characters χ and χ' of $\tilde{\mathbf{G}}^F$ to $\mathbf{G}^F$ are disjoint or equal and that they are equal if and only if $\chi' = \chi.\zeta$, where ζ is a linear character of $\tilde{\mathbf{G}}^F/\mathbf{G}^F$. By 11.4.12 a character of $\tilde{\mathbf{G}}^F/\mathbf{G}^F$ is of the form $\hat{z}$ with $z \in Z(\tilde{\mathbf{G}}^{*F^*})$, since such a character is trivial on $\mathrm{Der}(\tilde{\mathbf{G}})^F$. Since multiplication by $\hat{z}$ maps $R_{\tilde{\mathbf{T}}}^{\tilde{\mathbf{G}}}(\tilde{s})$ to $R_{\tilde{\mathbf{T}}}^{\tilde{\mathbf{G}}}(\tilde{s}z)$, the series $\mathcal{E}(\tilde{\mathbf{G}}^F, (\tilde{s}'))$ and $\mathcal{E}(\tilde{\mathbf{G}}^F, (\tilde{s}z))$ have to be equal as they have $\chi\hat{z}$ as a common element. But $\hat{z}$ is trivial on $\mathbf{G}^F$, so we see that the series $\mathcal{E}(\tilde{\mathbf{G}}^F, (\tilde{s}))$ and $\mathcal{E}(\tilde{\mathbf{G}}^F, (\tilde{s}z))$ have the same restrictions, and thus $\mathcal{E}(\mathbf{G}^F, (s)_{\mathbf{G}^{*F^*}}) = \mathcal{E}(\mathbf{G}^F, (s')_{\mathbf{G}^{*F^*}})$.   □

Theorem 12.4.13 *Let s and s' be two semi-simple elements of $\mathbf{G}^{*F^*}$. Then two Deligne–Lusztig characters $R_{\mathbf{T}^*}^{\mathbf{G}}(s)$ and $R_{\mathbf{T}'^*}^{\mathbf{G}}(s')$ have a common constituent if and only if s and s' are $\mathbf{G}^{*F^*}$-conjugate.*

Proof The "only if" part is 12.4.4. A similar computation to that of 12.4.6 shows that for any $w \in W^0(s)$ we have $\langle \chi_{(s)}, \chi_{(s)} \rangle_{\mathbf{G}^F} = \langle \chi_{(s)}, R_{\mathbf{T}_w^*}^{\mathbf{G}}(s) \rangle_{\mathbf{G}^F}$. As $R_{\mathbf{T}_w^*}^{\mathbf{G}}(s)$ is invariant by conjugation under $\tilde{\mathbf{G}}^F$, all the irreducible constituents of $\chi_{(s)}$ have the same multiplicity in $R_{\mathbf{T}_w^*}^{\mathbf{G}}(s)$, so they must have multiplicity 1. Thus we see that, when s and s' are conjugate in $\mathbf{G}^{*F^*}$, all irreducible constituents of $\chi_{(s)}$ ($= \chi_{(s')}$) occur in both $R_{\mathbf{T}^*}^{\mathbf{G}}(s)$ and $R_{\mathbf{T}'^*}^{\mathbf{G}}(s')$.   □

12.5 The Character Table of $\mathbf{SL}_2(\mathbb{F}_q)$

The greater part of the character table of $\mathbf{SL}_2(\mathbb{F}_q)$ (Table 12.1) will be obtained by restriction from that of $\mathbf{GL}_2(\mathbb{F}_q)$ and by Clifford theory, using that $\mathbf{GL}_2(\mathbb{F}_q)/\mathbf{SL}_2(\mathbb{F}_q) \simeq \mathbb{F}_q^\times$.

As in 11.7.4, F will denote the Frobenius endomorphism raising the matrix entries to their qth power, $\mathbf{T}$ the diagonal torus, and $\{1, s\}$ the elements of $W(\mathbf{T})$. This time s can be represented by the matrix $\begin{pmatrix} 0 & -1 \\ 1 & 0 \end{pmatrix}$.

The group $\mathbf{T}^F$, formed of the matrices $\mathrm{diag}(x, x^{-1})$ where $x \in \mathbb{F}_q^\times$ is isomorphic to $\mathbb{F}_q^\times$. An element of $\mathrm{Irr}(\mathbf{T}^F)$ is given by $\alpha \in \mathrm{Irr}(\mathbb{F}_q^\times)$ and s acts by changing α into α^{-1}.

By 9.3.1, the Deligne–Lusztig characters $R_{\mathbf{T}}^{\mathbf{G}}(\alpha)$ are pairwise orthogonal for distinct subsets $\{\alpha, \alpha^{-1}\} \subset \text{Irr}(\mathbb{F}_q^{\times})$; they are irreducible when $\alpha^2 \neq \mathbf{1}$, and otherwise have two distinct irreducible constituents.

For the other $\mathbf{G}^F$-conjugacy class of maximal tori, we have

$$\mathbf{T}_s^F \simeq \mathbf{T}^{sF} = \{\text{diag}(t_1, t_1^q) \mid t_1^{q+1} = 1\} \simeq \mu_{q+1}.$$

The element s acts by $\omega \mapsto \omega^q = \omega^{-1}$ on $\text{Irr}(\mu_{q+1})$.

By 9.3.1, the Deligne–Lusztig characters $R_{\mathbf{T}_s}^{\mathbf{G}}(\omega)$ are pairwise orthogonal for distinct subsets $\{\omega, \omega^{-1}\} \subset \text{Irr}(\mu_{q+1})$; they are the negative (by 7.1.6) of irreducible characters when $\omega^2 \neq \mathbf{1}$ and when they are not irreducible they have two irreducible constituents. When they are irreducible the characters $R_{\mathbf{T}}^{\mathbf{G}}(\alpha, \beta)$ and $-R_{\mathbf{T}_s}^{\mathbf{G}}(\omega)$ are all distinct since they are in different Lusztig series.

We now decompose the non-irreducible $R_{\mathbf{T}}^{\mathbf{G}}(\theta)$. By 11.3.8, the unipotent characters of $\mathbf{SL}_2(\mathbb{F}_q)$ are the restriction of those of $\mathbf{GL}_2(\mathbb{F}_q)$, and we still have $R_{\mathbf{T}}^{\mathbf{G}}(\mathbf{1}) = \mathbf{1}_{\mathbf{G}^F} + \text{St}_{\mathbf{G}^F}$ and $R_{\mathbf{T}_s}^{\mathbf{G}}(\mathbf{1}) = \mathbf{1}_{\mathbf{G}^F} - \text{St}_{\mathbf{G}^F}$. In characteristic 2, any character of order 2 is trivial and thus there are no other non-irreducible Deligne–Lusztig characters. Otherwise, let α_0 (resp. ω_0) be the non-trivial character of order 2 of $\mathbb{F}_q^{\times}$ (resp. μ_{q+1}). By 11.3.10, for any F-stable maximal torus $\tilde{\mathbf{T}}$ of $\mathbf{GL}_2$ and any $\theta \in \text{Irr}(\tilde{\mathbf{T}}^F)$ we have

$$R_{\tilde{\mathbf{T}} \cap \mathbf{SL}_2}^{\mathbf{SL}_2}(\text{Res}_{\tilde{\mathbf{T}}^F \cap \mathbf{SL}_2}^{\tilde{\mathbf{T}}^F} \theta) = \text{Res}_{\mathbf{SL}_2^F}^{\mathbf{GL}_2^F}(R_{\tilde{\mathbf{T}}}^{\mathbf{GL}_2}(\theta)).$$

Applying this to an extension of α_0, being the restriction of an irreducible character, $R_{\mathbf{T}}^{\mathbf{G}}(\alpha_0)$ is the sum of two distinct irreducible characters $\chi_{\alpha_0}^+$ and $\chi_{\alpha_0}^-$; similarly $-R_{\mathbf{T}_s}^{\mathbf{G}}(\omega_0) = \chi_{\omega_0}^+ + \chi_{\omega_0}^-$; the characters $\chi_{\alpha_0}^{\pm}$ must be distinct from $\chi_{\omega_0}^{\pm}$, since $R_{\mathbf{T}_s}^{\mathbf{G}}(\omega_0)$ and $R_{\mathbf{T}}^{\mathbf{G}}(\alpha_0)$ are orthogonal. Actually $R_{\mathbf{T}}^{\mathbf{G}}(\alpha_0) = \chi_{(s_1)}$ and $R_{\mathbf{T}_s}^{\mathbf{G}}(\omega_0) = \chi_{(s_2)}$ where s_1 is the element of order 2 of the split torus of $\mathbf{PGL}_2$ and s_2 is the element of order 2 of the non-split torus of $\mathbf{PGL}_2$, and, since s_1 and s_2 are not rationally conjugate, they are in distinct rational series (see 12.4.2, 12.4.4).

The list of irreducible characters of $\mathbf{SL}_2(\mathbb{F}_q)$ is thus

- $R_{\mathbf{T}}^{\mathbf{G}}(\alpha)$ attached to $\{\alpha, \alpha^{-1}\} \subset \text{Irr}(\mathbb{F}_q^{\times})$ of size 2.
- $-R_{\mathbf{T}_s}^{\mathbf{G}}(\omega)$ attached to $\{\omega, \omega^{-1}\} \subset \text{Irr}(\mu_{q+1})$ of size 2.
- The characters $\mathbf{1}_{\mathbf{G}^F}$ and $\text{St}_{\mathbf{G}^F}$.
- If q is odd, the two irreducible constituents $\chi_{\alpha_0}^+$ and $\chi_{\alpha_0}^-$ of $R_{\mathbf{T}}^{\mathbf{G}}(\alpha_0)$.
- If q is odd, the two irreducible constituents $\chi_{\omega_0}^+$ and $\chi_{\omega_0}^-$ of $-R_{\mathbf{T}_s}^{\mathbf{G}}(\omega_0)$.

We now give the list of conjugacy classes of $\mathbf{SL}_2(\mathbb{F}_q)$, by describing how the classes of $\mathbf{GL}_2(\mathbb{F}_q)$ which are in $\mathbf{SL}_2(\mathbb{F}_q)$ split. The semi-simple classes do not split since they have a connected centraliser since the dual group $\mathbf{PGL}_2$

has a connected centre, see 11.2.2(ii). But the class of $\begin{pmatrix} 1 & 1 \\ 0 & 1 \end{pmatrix}$ which has been analysed in 4.2.18(ii) splits in odd characteristic.

The list of semi-simple classes we get in $\mathbf{SL}_2(\mathbb{F}_q)$ is:

- The centre, formed of $\mathrm{diag}(1,1)$ and in odd characteristic $\mathrm{diag}(-1,-1)$.
- In $\mathbf{T}^F$ the classes $\mathrm{diag}(a,a^{-1})$ for $\{a,a^{-1}\} \subset \mathbb{F}_q^\times$ of size 2.
- In $\mathbf{T}_s^F$, the classes conjugate over $\overline{\mathbb{F}}_q$ to $\mathrm{diag}(x,x^{-1})$ for $\{x,x^{-1}\} \subset \mu_{q+1}$ of size 2.

In addition, we have the non-semi-simple classes zu where z is central and $u \neq 1$ unipotent. From the above descriptions there are 4 such classes in odd characteristic, and only one in characteristic 2 – these unipotent elements are regular so we could also take the analysis from 12.2.15. We get thus in odd characteristic the four classes $\begin{pmatrix} a & b \\ 0 & a \end{pmatrix}$ with $a \in \{-1,1\}$ and $b \in (\mathbb{F}_q^\times)^2$ or $b \in \mathbb{F}_q^\times - (\mathbb{F}_q^\times)^2$.

Since the classes do not split, in characteristic 2 we get the character table of $\mathbf{SL}_2(\mathbb{F}_q)$ by restriction form that of $\mathbf{GL}_2(\mathbb{F}_q)$. Thus from now on and in Table 12.1. we assume q odd.

Most of the table of $\mathbf{SL}_2(\mathbb{F}_q)$ (q odd) is obtained by restriction from $\mathbf{GL}_2(\mathbb{F}_q)$, using 11.3.10; the only values which are not given by restriction are those of $\chi_{\alpha_0}^\pm$ and $\chi_{\omega_0}^\pm$. Since $\chi_{\alpha_0}^+ + \chi_{\alpha_0}^-$ is the restriction of an irreducible character of $\mathbf{GL}_2(\mathbb{F}_q)$, the characters $\chi_{\alpha_0}^+$ and $\chi_{\alpha_0}^-$ are conjugate under $\mathbf{GL}_2(\mathbb{F}_q)$ – because $\mathbf{SL}_2(\mathbb{F}_q)$ is a normal subgroup of $\mathbf{GL}_2(\mathbb{F}_q)$. They have thus the same dimension, and more generally the same value on all classes of $\mathbf{SL}_2(\mathbb{F}_q)$ that are invariant under the action of $\mathbf{GL}_2(\mathbb{F}_q)$; in particular they also have the same central character α_0. These remarks show that to get all their values it is enough to compute their values on the classes $\begin{pmatrix} 1 & b \\ 0 & 1 \end{pmatrix}$. The same analysis holds for $\chi_{\omega_0}^+$ and $\chi_{\omega_0}^-$. These classes are regular unipotent; if 1 and z are the two elements of $H^1(F,Z(\mathbf{SL}_2))$ and we choose for u_1 – see definition after 12.2.17 – the class of $\begin{pmatrix} 1 & 1 \\ 0 & 1 \end{pmatrix}$, then u_z is the class of $\begin{pmatrix} 1 & x \\ 0 & 1 \end{pmatrix}$ with $x \in \mathbb{F}_q^\times - (\mathbb{F}_q^\times)^2$. For any $\chi \in \mathrm{Irr}(\mathbf{SL}_2(\mathbb{F}_q))$, by 12.3.10(i) we have

$$\left. \begin{aligned} -\chi(u_1) &= \sigma_1 \langle D_{\mathbf{G}}(\chi), \Gamma_1 \rangle_{\mathbf{G}^F} + \sigma_z \langle D_{\mathbf{G}}(\chi), \Gamma_z \rangle_{\mathbf{G}^F} \\ \text{and } -\chi(u_z) &= \sigma_1 \langle D_{\mathbf{G}}(\chi), \Gamma_z \rangle_{\mathbf{G}^F} + \sigma_z \langle D_{\mathbf{G}}(\chi), \Gamma_1 \rangle_{\mathbf{G}^F}. \end{aligned} \right\} \quad (12.5.1)$$

As we have seen $\chi_{\alpha_0}^+$ and $\chi_{\alpha_0}^-$ are the two constituents of $\chi_{(s_1)}$ and $\chi_{\omega_0}^+$ and $\chi_{\omega_0}^-$ are the two constituents of $\chi_{(s_2)}$. Let us choose for $\chi_{\alpha_0}^+$ (resp. $\chi_{\omega_0}^+$) the common constituent of $\chi_{(s_1)}$ and Γ_1 (resp. of $\chi_{(s_2)}$ and Γ_1) – see 12.4.12.

Table 12.1. *Character Table of* $\mathbf{SL}_2(\mathbb{F}_q)$ *for q odd (note that* $|\mathbf{SL}_2(\mathbb{F}_q)| = q(q-1)(q+1)$*)*

Classes	$\begin{pmatrix} a & 0 \\ 0 & a \end{pmatrix}$ $a \in \{1,-1\}$	$\begin{pmatrix} a & 0 \\ 0 & a^{-1} \end{pmatrix}$ $a \in \mathbb{F}_q^\times$ $a \neq \{1,-1\}$	$\begin{pmatrix} x & 0 \\ 0 & ^F x \end{pmatrix}$ $x \cdot {}^F x = 1$ $x \neq {}^F x$	$\begin{pmatrix} a & b \\ 0 & a \end{pmatrix}$ $a \in \{1,-1\}$, $b \in \{1,x\}$ with $x \in \mathbb{F}_q^\times - (\mathbb{F}_q^\times)^2$
Number of classes of this type	2	$(q-3)/2$	$(q-1)/2$	4
Cardinal of the class	1	$q(q+1)$	$q(q-1)$	$(q^2-1)/2$
$R_{\mathbf{T}}^{\mathbf{G}}(\alpha)$ $\alpha \in \mathrm{Irr}(\mathbb{F}_q^\times)$ $\alpha^2 \neq 1$	$(q+1)\alpha(a)$	$\alpha(a) + \alpha\left(\dfrac{1}{a}\right)$	0	$\alpha(a)$
$\chi_{\alpha_0}^\varepsilon$ $\varepsilon \in \{1,-1\}$	$\dfrac{q+1}{2}\alpha_0(a)$	$\alpha_0(a)$	0	$\dfrac{\alpha_0(a)}{2}\left(1+\varepsilon\,\alpha_0(ab)\sqrt{\alpha_0(-1)q}\right)$
$-R_{\mathbf{T}_s}^{\mathbf{G}}(\omega)$ $\omega \in \mathrm{Irr}(\mu_{q+1})$ $\omega^2 \neq 1$	$(q-1)\omega(a)$	0	$-\omega(x) - \omega({}^F x)$	$-\omega(a)$
$\chi_{\omega_0}^\varepsilon$ $\varepsilon \in \{1,-1\}$	$\dfrac{q-1}{2}\omega_0(a)$	0	$-\omega_0(x)$	$\dfrac{\omega_0(a)}{2}\left(-1+\varepsilon\,\alpha_0(ab)\sqrt{\alpha_0(-1)q}\right)$
$1_{\mathbf{G}^F}$	1	1	1	1
$\mathrm{St}_{\mathbf{G}^F}$	q	1	-1	0

Since by 9.3.3, $\chi_{\omega_0}^+$ and $\chi_{\omega_0}^-$ are cuspidal, we have $D_{\mathbf{G}}(\chi_{\omega_0}^+) = -\chi_{\omega_0}^+$ and $D_{\mathbf{G}}(\chi_{\omega_0}^-) = -\chi_{\omega_0}^-$. We have $D_{\mathbf{G}}(\chi_{\alpha_0}^+) = R_{\mathbf{T}}^{\mathbf{G}}{}^* R_{\mathbf{T}}^{\mathbf{G}}\chi_{\alpha_0}^+ - \chi_{\alpha_0}^+$. Now, by Frobenius reciprocity α_0 has multiplicity 1 in $^* R_{\mathbf{T}}^{\mathbf{G}}(\chi_{\alpha_0}^+)$, hence $^* R_{\mathbf{T}}^{\mathbf{G}}\chi_{\alpha_0}^+ = \alpha_0$ since $^* R_{\mathbf{T}}^{\mathbf{G}}$ preserves Harish-Chandra series –see 5.3.9– and α_0 is the only character of $\mathbf{T}^F$ in its series. So we get $D_{\mathbf{G}}(\chi_{\alpha_0}^+) = \chi_{\alpha_0}^-$ and $D_{\mathbf{G}}(\chi_{\alpha_0}^-) = \chi_{\alpha_0}^+$. Thus (12.5.1) gives

$$\chi_{\alpha_0}^+(u_z) = \chi_{\alpha_0}^-(u_1) = -\chi_{\omega_0}^+(u_1) = -\chi_{\omega_0}^-(u_z) = -\sigma_1$$

and

$$\chi_{\alpha_0}^+(u_1) = \chi_{\alpha_0}^-(u_z) = -\chi_{\omega_0}^+(u_z) = -\chi_{\omega_0}^-(u_1) = -\sigma_z.$$

To compute σ_1 and σ_z, we notice that if $\chi_1 \in \mathrm{Irr}(\mathbb{F}_q^+)$ is a character as defined above 12.3.3, then by 12.3.11 we have

$$\sigma_1 = |Z(\mathbf{G}^F)|^{-1}\Gamma_1(u_1) = \sum_{x \in (\mathbb{F}_q^\times)^2} \chi_1(x),$$

and

$$\sigma_z = |Z(\mathbf{G}^F)|^{-1}\Gamma_z(u_1) = |Z(\mathbf{G}^F)|^{-1}\Gamma_1(u_z) = \sum_{x \in \mathbb{F}_q \backslash (\mathbb{F}_q)^2} \chi_1(x),$$

the second equality since $\Gamma_z = {}^t\Gamma_1$ and $u_z = {}^t u_1$, where t is as in the remark after 12.3.3. To compute them notice that $\sigma_1 + \sigma_z = -1$ and that $\sigma_1 - \sigma_z$ is the Gauss sum

$$\sum_{x \in \mathbb{F}_q} \alpha_0(x)\chi_1(x),$$

so $(\sigma_1 - \sigma_z)^2 = \alpha_0(-1)q$ – see for example Digne et al. (1997, above 2.8).

In Table 12.1 we denote by $\sqrt{\alpha_0(-1)q}$ the square root of $\alpha_0(-1)q$ which is equal to $\sigma_1 - \sigma_z$.

Notes

Many results about regular elements can be found in Steinberg (1974). The Gelfand–Graev representation was first introduced by Gelfand and Graev (1962), who proved that it is multiplicity free in the case of $\mathbf{SL}_n$; this result was proved in general by Steinberg. All results about semi-simple characters and the decomposition of the Gelfand–Graev representation in the case of a group with connected centre are due to Deligne and Lusztig (1976). The Gelfand–Graev representations for groups with non-connected centre were studied in Digne et al. (1992), using earlier work of Lehrer (1978).

For an interesting account of the theory of representations of finite reductive groups in the setting of $\mathbf{SL}_2$, see Bonnafé (2010).

13

Green Functions

In this chapter we explain the Lusztig–Shoji algorithm for computing Green functions. First, we must relate the values $R_{\tilde{\chi}}(1)$, see 11.6.1 and 11.6.4, to the invariant theory of W.

13.1 Invariants

Let V be a vector space of dimension n over a field k of characteristic 0, and let $W \subset \mathbf{GL}(V)$ be a finite group. Then the action of W extends naturally to the **symmetric algebra** SV which, choosing a basis $x_1,\ldots,x_n$ of V, is isomorphic to the polynomial algebra $k[x_1,\ldots,x_n]$. Thus SV is graded by polynomial degree with $(SV)_1 = V$ and $(SV)_0 = k$.

A **pseudo-reflection** is an element $s \in \mathbf{GL}(V)$ such that $\mathrm{Ker}(s-1)$ is a hyperplane; if $k \subset \mathbb{R}$, a pseudo-reflection of finite order is a reflection. If $W \subset \mathbf{GL}(V)$ is a finite group, we will denote by $\mathrm{Ref}(W)$ the set of pseudo-reflections in W.

Theorem 13.1.1 *Assume $W \subset \mathbf{GL}(V)$ is a finite subgroup generated by $\mathrm{Ref}(W)$. Then the subalgebra $SV^W \subset SV$ of W-invariants is a polynomial algebra $k[f_1,\ldots,f_n]$ where $f_1,\ldots,f_n \in SV$ are algebraically independent homogeneous polynomials.*

References This theorem has been proved case-by-case by Shephard and Todd (1954), by classifying all finite groups generated by pseudo-reflections, and then in a case-free way by Chevalley (1955). □

In the situation of the theorem the f_i are called **fundamental invariants** for W. They are not unique but their degrees are – a consequence of Lemma 13.1.4 below; these are called the **reflection degrees** of W.

The converse of the above theorem is easier and we will prove it here.

Theorem 13.1.2   *Assume $W \subset \mathbf{GL}(V)$ is a finite subgroup and $SV^W \simeq k[f_1, \ldots, f_n]$ where $f_1, \ldots, f_n \in SV$ are algebraically independent homogeneous polynomials of degree $d_1, \ldots, d_n$. Then:*

(i) *W is generated by $\mathrm{Ref}(W)$.*

(ii) *$|W| = d_1 \ldots d_n$.*

(iii) *$|\mathrm{Ref}(W)| = \sum_{i=1}^{n}(d_i - 1)$.*

Proof The tool for proving this theorem is to use Hilbert–Poincaré series. If $X = \bigoplus_{i \geq 0} X_i$ is a graded k-vector space such that all X_i are finite dimensional, and $g \in \mathrm{End}(X)$ is a graded endomorphism (that is, it preserves the X_i), we define $P_X(g) := \sum_{i \geq 0} \mathrm{Trace}(g \mid X_i)t^i \in k[[t]]$, and $P_X := P_X(\mathrm{I}_X) = \sum_{i \geq 0} \dim(X_i)t^i$.

Lemma 13.1.3   (Molien)

(i) *If $w \in \mathrm{End}(V)$ is semi-simple, then*
$$P_{SV}(w) = \det(\mathrm{I}_V - wt \mid V \otimes k[t])^{-1}.$$

(ii) *For $W \subset \mathbf{GL}(V)$ a finite group, we have*
$$P_{SV^W} = |W|^{-1} \sum_{w \in W} \det(\mathrm{I}_V - wt \mid V \otimes k[t])^{-1}.$$

Proof For (i) we first notice that if $V = V_1 \oplus V_2$, then $SV = SV_1 \otimes SV_2$, from which it follows that if w respects the decomposition $V_1 \oplus V_2$ we have $P_{SV}(w) = P_{SV_1}(w)P_{SV_2}(w)$. Since both sides of formula (i) do not change if we extend k, we may assume k algebraically closed and diagonalise w; let $x_1, \ldots, x_n$ be a basis of V formed of eigenvectors of w for the eigenvalues $\lambda_1, \ldots, \lambda_n$. Then $P_{SV}(w) = P_{k[x_1]}(\lambda_1) \ldots P_{k[x_n]}(\lambda_n)$ and it is enough to observe that $P_{k[x]}(\lambda) = \sum_{i \geq 0} \lambda^i t^i = \frac{1}{1-\lambda t}$.

(ii) is an immediate consequence of (i) since $|W|^{-1} \sum_{w \in W} w$ is a projector on the W-invariants for any kW-module. Thus the graded dimension of SV^W is given by the graded trace of this projector, which is formula (ii). □

Using 13.1.3(ii) we see that for any finite group W, the formal series in t given by P_{SV^W} has a Laurent expansion at 1 whose first terms are $\frac{1}{|W|}$
$\left(\frac{1}{(1-t)^n} + \frac{|\mathrm{Ref}(W)|}{2(1-t)^{n-1}} + \cdots \right)$. Here the first term corresponds to $1 \in W$ and the second to $\mathrm{Ref}(W)$, taking in account that each $s \in \mathrm{Ref}(W)$ of non-trivial eigenvalue λ contributes a term $\frac{1}{(1-\lambda)(1-t)^{n-1}}$, and that if $s \neq s^{-1}$ we have terms $\frac{1}{1-\lambda} + \frac{1}{1-\lambda^{-1}} = 1$ and if $s = s^{-1}$ then $\lambda = -1$ thus $\frac{1}{1-\lambda} = \frac{1}{2}$ whence the factor $\frac{1}{2}$.

We now compare with the Laurent expansion at 1 of the series $P_{k[f_1,\ldots,f_n]} = P_{k[f_1]}\cdots P_{k[f_n]} = \dfrac{1}{1-t^{d_1}}\cdots\dfrac{1}{1-t^{d_n}}$. The first terms are $\dfrac{1}{(1-t)^n d_1\ldots d_n} + \dfrac{\sum_{i=1}^{n}(d_i - 1)}{(1-t)^{n-1}2d_1\ldots d_n} + \cdots$. Here we obtain the first term by using that $(1-t^d) = (1-t)(1+t+\cdots+t^{d-1})$, and to get the second one we multiply the series by $(t-1)^n$, derive with respect to t and set $t = 1$.

Comparing both developments, it follows that if for a finite group W we have $SW \simeq k[f_1,\ldots,f_n]$ then $|W| = d_1\ldots d_n$ and $|\operatorname{Ref}(W)| = \sum_{i=1}^{n}(d_i - 1)$. We have proved (ii) and (iii) of 13.1.2. We now prove (i).

Lemma 13.1.4 *Let* $k[f_1,\ldots,f_n] \subset k[f_1',\ldots,f_n']$ *be two polynomial subalgebras of SV where the f_i and f_i' are homogeneous, the degrees of the f_i are $d_1 \leq \cdots \leq d_n$, and the degrees of the f_i' are $d_1' \leq \cdots \leq d_n'$. Then for all i we have $d_i \geq d_i'$.*

Proof There are polynomials $p_1,\ldots,p_n$ such that $f_i = p_i(f_1',\ldots,f_n')$. It follows that $d_i \geq d_i'$ otherwise for an i such that $d_i < d_i'$ then $p_1,\ldots,p_i$ can only involve $f_1',\ldots,f_{i-1}'$ which is impossible since $f_1,\ldots,f_i$ are algebraically independent. □

Let W' be the subgroup of W generated by $\operatorname{Ref}(W)$. By 13.1.1, we have $SV^{W'} = k[f_1',\ldots,f_n']$ for some homogeneous f_i' of degrees $d_1',\ldots,d_n'$ and by the above lemma $d_i \geq d_i'$ if we order them in ascending order. Now we have $|\operatorname{Ref}(W)| = \sum_i(d_i - 1) \geq \sum_i(d_i' - 1) = |\operatorname{Ref}(W')|$ and as the extreme terms are equal we must have $d_i = d_i'$ thus $|W| = d_1\ldots d_n = |W'|$ and $W = W'$ is generated by $\operatorname{Ref}(W)$. □

Exercise 13.1.5 Show that (i) of Molien's Lemma 13.1.3 remains true without assuming w semi-simple.

We will need the following definition.

Definition 13.1.6 Let I be the ideal of SV generated by the W-invariants of positive degree. Then $SV_W := SV/I$ is called the **coinvariant algebra** of W.

Since I is a graded ideal, the coinvariant algebra inherits a grading from SV. This leads to the following theorem.

Theorem 13.1.7 *As a kW-module, SV_W is isomorphic to the regular representation of W, and there is a graded W-equivariant isomorphism of SV^W-modules $SV_W \otimes SV^W \simeq SV$.*

References See Chevalley (1955) or Hiller (1982, II §3). □

Corollary 13.1.8 $P_{SV_W} = \prod_{i=1}^{n}\dfrac{t^{d_i}-1}{t-1}$ *and for $w \in W$ we have*

$$P_{SV_W}(w) = \frac{\prod_{i=1}^{n}(1-t^{d_i})}{\det(1 - wt \mid V \otimes k[t])}.$$

Proof By the previous theorem we have $P_{SV}(w) = P_{SV_W}(w)P_{SV^W}(w)$ and $P_{SV^W}(w) = P_{SV^W} = \frac{1}{\prod_{i=1}^{n}(1-t^{d_i})}$, see the proof of 13.1.2.   □

Definition 13.1.9 For $\chi \in \mathrm{Irr}(W)$, we define the **fake degree** Feg χ of χ to be the graded multiplicity of χ in SV_W.

Corollary 13.1.10 *For* $\chi \in \mathrm{Irr}(W)$ *we have*

$$\mathrm{Feg}\,\chi = \langle \chi, P_{SV_W} \rangle_W = \frac{\prod_{i=1}^{n}(1-t^{d_i})}{|W|} \sum_{w \in W} \frac{\chi(w)}{\det(1 - w^{-1}t \mid V \otimes k[t])}.$$

Proof By definition, $\chi(1)$ Feg χ is the graded dimension of the χ-isotypic part of SV_W, that is $\sum_i \langle \chi, (SV_W)_i \rangle_W t^i = \langle \chi, P_{SV_W} \rangle_W$ and we obtain the result using the formula for $P_{SV_W}(w)$ in 13.1.8.   □

In order to apply this theory to a reductive group with a Frobenius root, we need to generalise it to reflection cosets.

Proposition 13.1.11 *Let* $\tau \in N_{\mathbf{GL}(V)}(W)$ *be an element of finite order. Assume k large enough so that it contains roots of unity of the same order as τ. Then it is possible to find fundamental invariants for W which are τ-eigenvectors. The corresponding pairs of degrees and eigenvalues* $(d_1, \varepsilon_1), \ldots, (d_n, \varepsilon_n)$ *depend only on the coset $W\tau$ and are called the* **generalised reflection degrees** *of $W\tau$.*

Proof We show the proposition by induction on the degree. Assume we have chosen invariants which are eigenvectors of τ for degrees less than d. Since τ normalises W it preserves the space $(SV^W)_d$ of homogeneous invariants of degree d. The subspace V_1 of $(SV^W)_d$ spanned by monomials in the invariants of degree less than d is τ-stable by induction. Since τ is of finite order V_1 has a τ-stable complement V_2 by Maschke's theorem. A basis of V_2 formed of τ-eigenvectors will give fundamental invariants of degree d.   □

We assume now that k is a subfield of $\mathbb{C}$. Let $\tilde{\chi}$ be as in 11.6.4 an extension to $W \rtimes \langle \tau \rangle$ of $\chi \in \mathrm{Irr}(W)^{\tau}$. We then generalise the definition of the fake degree to functions like $w \mapsto \tilde{\chi}(w\tau)$ by using the scalar product 11.6.2 on $C(W\tau)$.

Definition 13.1.12 If we denote by $P_{SV_W}(.\,\tau)$ the function $w \mapsto P_{SV_W}(w\tau)$, we define Feg $\tilde{\chi} := \langle \tilde{\chi}, P_{SV_W}(.\,\tau) \rangle_{W\tau}$.

We note that by 11.6.3 Feg $\tilde{\chi}$ is a polynomial which coefficients algebraic integers – it is a polynomial since SV_W is finite dimensional, see 13.1.7.

Proposition 13.1.13 *If* $(d_1, \varepsilon_1), \ldots, (d_n, \varepsilon_n)$ *are the generalised reflection degrees of $W\tau$, we have*

$$\mathrm{Feg}\,\tilde{\chi} = \frac{\prod_{i=1}^{n}(1 - \varepsilon_i t^{d_i})}{|W|} \sum_{w \in W} \frac{\tilde{\chi}(w\tau)}{\det(1 - (w\tau)^{-1}t \mid V \otimes k[t])}.$$

Proof The proof is the same as for 13.1.10, except that instead of P_{SVW} we use $P_{SV^w}(\tau) = (\prod_{i=1}^n (1 - \varepsilon_i t^{d_i}))^{-1}$. $\square$

In view of the above formula and the remark before 13.1.13, if $\tilde{\chi}$ takes integral values and τ is real, Feg $\tilde{\chi}$ is a polynomial with integer coefficients.

Formula (i) below was given in Steinberg (1968, 11.16), with a different proof.

Theorem 13.1.14 *Let* **G** *be a connected reductive group over* $\overline{\mathbb{F}}_q$, *and let* F *be a Frobenius root acting as* $q\tau$ *on* $X(\mathbf{T})$ *where* $\mathbf{T}$ *is a quasi-split torus of* **G**. *Let* $(d_1, \varepsilon_1), \ldots, (d_n, \varepsilon_n)$ *be the generalised reflection degrees of* $W\tau$ *on* $X(\mathbf{T}) \otimes \mathbb{C}$. *Then*

(i) $|\mathbf{G}^F| = q^{\sum_{i=1}^n (d_i-1)} \prod_{i=1}^n (q^{d_i} - \varepsilon_i)$.

(ii) $R_{\tilde{\chi}}(1) = \text{Feg } \tilde{\chi}(q)$.

Proof We first compute $R_{\tilde{\chi}}(1)$. We have

$$R_{\tilde{\chi}}(1) = |W|^{-1} \sum_{w \in W} \tilde{\chi}(w\tau) R_{\mathbf{T}_w}^{\mathbf{G}}(1)(1)$$

$$= |W|^{-1} \sum_{w \in W} \tilde{\chi}(w\tau) \varepsilon_{\mathbf{G}} \varepsilon_{\mathbf{T}_w} \frac{|\mathbf{G}^F|_{p'}}{|\mathbf{T}_w^F|}$$

$$= \frac{|\mathbf{G}^F|_{p'}}{|W|} \sum_{w \in W} \frac{\tilde{\chi}(w\tau)}{\det \tau \det(wq\tau - 1 \mid X(\mathbf{T}))}$$

$$= \frac{|\mathbf{G}^F|_{p'}}{\det \tau \prod_{i=1}^n (\varepsilon_i q^{d_i} - 1)} \text{Feg } \tilde{\chi}(q)$$

where the first equality is the definition 11.6.1, the second is by 10.2.2, the third by 7.1.6 and 4.4.9, and the last is by 13.1.13 taking into account that $w\tau$ and $(w\tau)^{-1}$ being real and of finite order have the same set of eigenvalues, thus $\det(wq\tau - 1) = (-1)^n \det(1 - (w\tau)^{-1}q)$.

We now apply the above computation to the function $\tilde{\chi}$ defined by $\tilde{\chi}(w\tau) = 1$ for all $w \in W$, an extension of the trivial character of W. Then by 10.2.5 we have $R_{\tilde{\chi}}(1) = 1$ and by definition, using that the only component of SV_W where W acts trivially is the component of degree 0 where τ acts trivially too, we have Feg $\tilde{\chi} = 1$. We thus get (ii) of the theorem and $|\mathbf{G}^F|_{p'} = \det \tau \prod_{i=1}^n (\varepsilon_i q^{d_i} - 1)$. This last equality is equivalent to (i) of the theorem: indeed we have $|\mathbf{G}^F|_p = q^{l(w_0)} = q^{|\text{Ref}(W)|}$ by 4.4.1(iii) and, by the proof of 13.1.11, the ε_i attached to a given degree are globally invariant by complex conjugacy, thus by inverse, thus $\det \tau \prod_{i=1}^n (\varepsilon_i q^{d_i} - 1) = (\det \tau \prod_i \varepsilon_i) \prod_{i=1}^n (q^{d_i} - \varepsilon_i)$, so that to get the positive integer $|\mathbf{G}^F|$ we must have $\det \tau = \prod_{i=1}^n \varepsilon_i^{-1}$. One could also deduce this last formula from the fact that τ has a regular vector for the eigenvalue 1 – see the proof of 7.1.7 and Springer (1974, 6.4 (v)). □

Remark 13.1.15   The reflection degrees of W on the orthogonal of $\Phi^\vee$ are all equal to 1. One often gives a formula for 13.1.14(i) which only involves the non-trivial degrees $d_{r+1}, \ldots d_n$ where $r = \mathrm{rank}(Z(\mathbf{G})^0)$:

$$|\mathbf{G}^F| = |Z(\mathbf{G})^{0F}| q^{\sum_{i=r+1}^{n} (d_i - 1)} \prod_{i=r+1}^{n} (q^{d_i} - \varepsilon_i).$$

This is obtained by using formula 13.1.14(i) for the group $\mathbf{G}/Z(\mathbf{G})^0$.

Remark 13.1.16   A proof for $|\mathbf{G}^F|_{p'}$ in the spirit of this book would be to use the Lefschetz formula for computing $|(\mathbf{G}/\mathbf{B})^F|$, and to use that $H_c^*(\mathbf{G}/\mathbf{B})$ is isomorphic to the coinvariant algebra. However this last fact is beyond the scope of this book.

Examples 13.1.17   For applying Theorem 13.1.14, one needs a list of generalised degrees. Here they are for quasi-simple groups (we give just the degrees when the group is split):

- A_n: $2, \ldots, n+1$.
- B_n, C_n: $2, 4, \ldots, 2n$.
- D_n: $2, 4, \ldots, 2n-2$ and n.
- ^2A_n: $(2,1), (3,-1), \ldots, (n+1, (-1)^{n+1})$.
- ^2D_n: $(2,1), (4,1), \ldots, (2n-2, 1)$ and $(n, -1)$.
- $^3D_4$: $(2,1), (4, \zeta_3), (6,1), (4, \zeta_3^2)$.
- E_6: $2, 5, 6, 8, 9, 12$.
- $^2E_6$: $(2,1), (5,-1), (6,1), (8,1), (9,-1), (12,1)$.
- E_7: $2, 6, 8, 10, 12, 14, 18$.
- E_8: $2, 8, 12, 14, 18, 20, 24, 30$.
- F_4: $2, 6, 8, 12$.
- G_2: $2, 6$.
- $^2B_2$: $(2,1), (4,-1)$.
- $^2F_4$: $(2,1), (6,-1), (8,1), (12,-1)$.
- $^2G_2$: $(2,1), (6,-1)$.

Exercise 13.1.18   This exercise recovers four lines of the above list.

(i) Show that the group $G(de, e, n)$ of monomial $n \times n$ complex matrices whose entries are in μ_{de} and whose product of entries is in μ_d is generated by pseudo-reflections in its action on $\mathbb{C}^n$.

(ii) Show that the elementary symmetric functions $f_1, \ldots, f_n$ where f_i is the coefficient of t^{n-i} in the polynomial $(t-x_1) \ldots (t-x_n)$ is a set of fundamental invariants of the symmetric group $\mathfrak{S}_n \simeq G(1, 1, n)$.

(iii) Show that the fundamental invariants of $G(de,e,n)$ are $\{f_j(x_1^{de},\ldots,x_n^{de}) \mid j = 1,\ldots,n-1\}$ and $f_n(x_1^{de},\ldots,x_n^{de})$. Hint: compute $|G(de,e,n)|$ by writing $G(de,e,n)$ as the semi-direct product of a group of diagonal matrices by $\mathfrak{S}_n$.

(iv) Show that $W(A_{n-1}) \simeq G(1,1,n)$, $W(B_n) \simeq G(2,1,n)$, $W(D_n) \simeq G(2,2,n)$, $W(G_2) = G(6,6,2)$.

13.2 Green Functions and the Springer Correspondence

A consequence of the results of this chapter is that one can compute in principle the character values of $R_{\mathbf{T}}^{\mathbf{G}}(\theta)$ on all elements of $\mathbf{G}^F$, at least in good characteristic and when F is a Frobenius morphism. By the character formula 10.1.2 the problem is in principle reduced to the computation of the Green functions $Q_{\mathbf{T}}^{\mathbf{G}}(u,1)$. By the character formula, applied to $g = u$ (thus $s = 1$) we have $R_{\mathbf{T}}^{\mathbf{G}}(\theta)(u) = Q_{\mathbf{T}}^{\mathbf{G}}(u,1)$ for any $\theta \in \mathrm{Irr}(\mathbf{T}^F)$; it follows that these Green functions take their values in $\mathbb{Z}$. Our goal now is to explain the Lusztig–Shoji algorithm to compute these Green functions. We extend $Q_{\mathbf{T}}^{\mathbf{G}}$ by 0 on non-unipotent elements by the following definition:

Definition 13.2.1 Let $\mathbf{G}$ be a connected reductive group with a Frobenius root F. If $\mathbf{T}$ is an F-stable maximal torus of $\mathbf{G}$ we define the (one-variable) Green function $Q_{\mathbf{T}}^{\mathbf{G}}$ as $R_{\mathbf{T}}^{\mathbf{G}}(1) \cdot \chi_p$ where χ_p is as in 7.3.2(ii) the characteristic function of the set $\mathbf{G}_u^F$ of unipotent elements.

Proposition 13.2.2 *For F-stable maximal tori $\mathbf{T}, \mathbf{T}'$ of $\mathbf{G}$ we have*

$$\langle Q_{\mathbf{T}}^{\mathbf{G}}, Q_{\mathbf{T}'}^{\mathbf{G}} \rangle_{\mathbf{G}^F} = \begin{cases} \frac{|W(\mathbf{T})^F|}{|\mathbf{T}^F|} & \text{if } \mathbf{T} \text{ and } \mathbf{T}' \text{ are } \mathbf{G}^F\text{-conjugate,} \\ 0 & \text{otherwise.} \end{cases}$$

Proof Since $\mathrm{reg}_{\mathbf{T}^F} = \sum_{\theta \in \mathrm{Irr}(\mathbf{T}^F)} \theta$, we have for u unipotent $Q_{\mathbf{T}}^{\mathbf{G}}(u) = |\mathbf{T}^F|^{-1} R_{\mathbf{T}}^{\mathbf{G}}(\mathrm{reg}_{\mathbf{T}^F})(u)$. This equality holds everywhere since $R_{\mathbf{T}}^{\mathbf{G}}(\mathrm{reg}_{\mathbf{T}^F}) = R_{\mathbf{T}}^{\mathbf{G}}(\chi_p^{\mathbf{T}} \cdot \mathrm{reg}_{\mathbf{T}^F}) = R_{\mathbf{T}}^{\mathbf{G}}(\mathrm{reg}_{\mathbf{T}^F}) \cdot \chi_p^{\mathbf{G}}$ where the second equality is by 10.1.6 which holds since $\chi_p \in C(\mathbf{G}^F)_{p'}$. Now $\langle R_{\mathbf{T}}^{\mathbf{G}}(\mathrm{reg}_{\mathbf{T}^F}), R_{\mathbf{T}'}^{\mathbf{G}}(\mathrm{reg}_{\mathbf{T}'^F}) \rangle_{\mathbf{G}^F} = \langle \mathrm{reg}_{\mathbf{T}^F}, {}^*R_{\mathbf{T}}^{\mathbf{G}} \circ R_{\mathbf{T}'}^{\mathbf{G}}(\mathrm{reg}_{\mathbf{T}'^F}) \rangle_{\mathbf{T}^F}$ and by the Mackey formula 9.2.6 we have, using that $\mathrm{reg}_{\mathbf{T}^F}$ is invariant by conjugation by $W(\mathbf{T})^F$,

$$ {}^*R_{\mathbf{T}}^{\mathbf{G}} \circ R_{\mathbf{T}'}^{\mathbf{G}}(\mathrm{reg}_{\mathbf{T}'^F}) = \begin{cases} |W(\mathbf{T})^F| \, \mathrm{reg}_{\mathbf{T}^F} & \text{if } \mathbf{T} \text{ and } \mathbf{T}' \text{ are } \mathbf{G}^F\text{-conjugate} \\ 0 & \text{otherwise,} \end{cases}$$

which gives the result. $\qquad\square$

The computation of Green functions relies on the above scalar product formula and an approach due to Springer relating Green functions to the cohomology of Springer fibres (see 12.2.1).

Theorem 13.2.3

(i) *For $u \in \mathbf{G}_u$ there is a natural action of W on $H_c^*(\mathcal{B}_u)$.*

(ii) *Assume that F is a Frobenius endomorphism; then for $u \in \mathbf{G}_u^F$ we have*
$$Q_{\mathbf{T}_w}^{\mathbf{G}}(u) = \mathrm{Trace}(wF \mid H_c^*(\mathcal{B}_u)).$$

Note that the right-hand side of the above formula is not a number of fixed points, since W does not act on $\mathcal{B}_u$ itself. Note also that the variety $\mathcal{B}_u$ depends up to isomorphism only on the $\mathbf{G}$-conjugacy class of u, but the action of F on it, for $u \in \mathbf{G}_u^F$, depends on the $\mathbf{G}^F$-class of u. Finally note that we use compact support cohomology H_c (to fit with Chapter 8) while the cited literature uses ordinary cohomology; they coincide since the varieties $\mathcal{B}_u$ are projective.

Proof

(i) was proved in Springer (1976),

(ii) of this difficult theorem was conjectured in Springer (1976), proved for p and q large enough in Kazhdan (1977), and finally proved in Lusztig (1986c) and Shoji (1995, II, Theorem 5.5). See Shoji (1987) for the equivalence of the statements in these papers and the statement 13.2.3(ii).

$\square$

The Springer Correspondence

Until the end of this chapter we assume that F is a Frobenius endomorphism so that 13.2.3(ii) is valid.

The group $C_{\mathbf{G}}(u)$ acts on $\mathcal{B}_u$. By 8.1.14, the group $C_{\mathbf{G}}(u)^0$ acts trivially on $H_c^*(\mathcal{B}_u)$, thus the action of $C_{\mathbf{G}}(u)$ factors through $A(u) := C_{\mathbf{G}}(u)/C_{\mathbf{G}}(u)^0$. Springer (1976) has shown that the actions of $A(u)$ and of W on $H_c^*(\mathcal{B}_u)$ commute.

Proposition 13.2.4 *For $u \in \mathbf{G}_u$, all the irreducible components of $\mathcal{B}_u$ have the same dimension $\beta_u := \frac{1}{2}(\dim C_{\mathbf{G}}(u) - \mathrm{rank}\,\mathbf{G})$.*

Reference See Spaltenstein (1977). $\square$

We will denote by $I(\mathcal{B}_u)$ the set of irreducible components of $\mathcal{B}_u$. It results from 13.2.4 and from 8.1.12(ii) that $H_c^{2\beta_u}(\mathcal{B}_u)$ has a basis indexed by $I(\mathcal{B}_u)$, on which $A(u)$ acts by permutation.

Theorem 13.2.5 *For $\phi \in \mathrm{Irr}(A(u))$, the representation $\chi_{u,\phi}$ of W afforded by the ϕ-isotypic component $H_c^{2\beta_u}(\mathcal{B}_u)_\phi$ is irreducible or 0. If $\mathrm{Irr}(A(u))^0$ denotes $\{\phi \in \mathrm{Irr}(A(u)) \mid \chi_{u,\phi} \neq 0\}$, there is a bijection – called the **Springer correspondence** – between $\mathrm{Irr}(W)$ and the set of $\chi_{u,\phi}$ when u runs over representatives of the $\mathbf{G}$-conjugacy classes in $\mathbf{G}_u$, and ϕ runs over $\mathrm{Irr}(A(u))^0$.*

References   See Springer (1976) and Lusztig (1984b).   □

Since (by definition) $C_G(u)$ acts by inner automorphisms on $A(u)$, the representation $\chi_{u,\phi}$ depends only on the G-conjugacy class C of u. We will denote it by $\chi_{C,\phi}$ from now on.

Note that $\mathbf{1} \in \mathrm{Irr}(A(u)^0)$ since a permutation representation always involves the trivial character – with multiplicity the number of orbits of $A(u)$ on $I(\mathcal{B}_u)$, which is thus equal to $\chi_{u,1}(1)$.

We will now assume that $Z(\mathbf{G}) = \{1\}$; this is not a restriction for computing the Green functions since by 11.3.8 they factor through $\mathbf{G}^F/Z(\mathbf{G})^F$. The advantage is that the groups $A(u)$ are smaller in this case, and it happens more often that $\mathrm{Irr}(A(u))^0 = \mathrm{Irr}(A(u))$.

Since $Z(\mathbf{G}) = \{1\}$, by 2.3.9 $(\mathbf{G}, F)$ is a direct product of "**descent of scalars**", that is groups of the form $\mathbf{G}_1 \times \cdots \times \mathbf{G}_1$ with n factors permuted cyclically by F. Since $\mathbf{G}^F \simeq \mathbf{G}_1^{F^n}$ it is enough to study the characters of $(\mathbf{G}_1, F^n)$, so we may and will assume that $\mathbf{G}$ is quasi-simple.

The Springer correspondence has been computed in good characteristic by Springer (1976, 7.16) for G_2 and A_n, by Shoji (1979) for the other classical groups and Shoji (1982) for F_4, and by Alvis et al. (1982) for E_6, E_7 and E_8; a summary can be found in the tables of Carter (1985, 13.3). It has been computed in bad characteristic – where the description of unipotent classes is different – by Lusztig and Spaltenstein (1985) and Spaltenstein (1985). It gives global information on $H_c^*(\mathcal{B}_u)$ through the following theorem from Borho and MacPherson (1981).

Theorem 13.2.6

(i) $\chi_{C,\phi}$ *does not occur in the $\overline{\mathbb{Q}}_\ell W$-module $H_c^i(\mathcal{B}_u)$ for any $u \in C, i < 2\beta_u$.*
(ii) *Assume that $\chi_{C,\phi}$ occurs in $H_c^i(\mathcal{B}_v)$ for some i and $v \in \mathbf{G}_u$. Then $v \in \overline{C}$, the Zariski closure of C.*

To make the Lusztig–Shoji algorithm work we now just need information on the action of F on $A(u)$ and $I(\mathcal{B}_u)$ for F-stable u; all eigenvalues of F on $H_c^{2\beta_u}(\mathcal{B}_u)$ are of absolute value q^{β_u} and F acts on $H_c^{2\beta_u}(\mathcal{B}_u)$ as the scalar q^{β_u} times a permutation of the basis corresponding to the permutation induced by F on $I(\mathcal{B}_u)$, see 8.1.12(iii).

Proposition 13.2.7 *Let $C \subset \mathbf{G}_u$ be an F-stable conjugacy class. There exists always $u \in C^F$ such that F acts trivially on $A(u)$.*

Proof For classical adjoint groups, $A(u)$ is always of the form $(\mathbb{Z}/2\mathbb{Z})^i$ for some i; the result is proved in Shoji (1983, 3.1). For exceptional adjoint groups in good characteristic we always have $A(u) \simeq \mathfrak{S}_i$ for $1 \leq i \leq 5$. Since these groups have only inner automorphisms, there exists an u in each geometric class such that F acts trivially on $A(u)$. Indeed if F acts on $A(u)$ by ad a then it acts trivially on $A(^g u)$ where $g^{-1}{}^F g \in C_G(u)$ is a representative of a^{-1}. For exceptional adjoint groups in bad characteristic see Liebeck and Seitz (2012, Lemma 20.16). □

Proposition 13.2.8 *Assume $(\mathbf{G}, F)$ split and the characteristic good for $\mathbf{G}$. Let C be an F-stable $\mathbf{G}$-conjugacy class in $\mathbf{G}_u$. Then there is a $\mathbf{G}^F$-class $C_0 \subset C^F$ such that F acts trivially on $I(\mathcal{B}_u)$ for $u \in C_0$ (and this class is unique), unless $\mathbf{G}$ is of type E_8 and C is the class called $D_8(a_3)$ by Mizuno (1980) and called $E_8(b_6)$ in Carter (1985, §13.1.) We will call such a $\mathbf{G}^F$-class **split**. For u split, the action of F on $A(u)$ is trivial.*

References This is proved in Springer (1976) for G_2, in Shoji (1982) for F_4, in Shoji (1983) for classical groups and in Beynon and Spaltenstein (1984) for E_6, E_7 and E_8. □

By definition and 8.1.12, if the $\mathbf{G}^F$-class of u is split then F acts by the scalar q^{β_u} on $H_c^{2\beta_u}(\mathcal{B}_u)$. For $u \in C$, where C is the class $E_8(b_6)$, we have $A(u) \simeq \mathfrak{S}_3$ and, by Beynon and Spaltenstein (1984), there exists $u \in C^F$ such that F acts by q^{β_u} on $H_c^{2\beta_u}(\mathcal{B}_u)$ except on the sgn-isotypic component of $A(u)$ when $q \equiv -1$ (mod 3), in which case F acts by $-q^{\beta_u}$ on that component.

Assume now that $(\mathbf{G}, F)$ is not split; then if W is the Weyl group of a quasi-split torus $\mathbf{T}$, there exists τ such that $F = q\tau$ on $X(\mathbf{T})$. Let C be an F-stable conjugacy class in $\mathbf{G}_u$ and let $u \in C^F$ be such that F acts trivially on $A(u)$ and for $\phi \in \mathrm{Irr}(A(u))^0$ consider the ϕ-isotypic component of $H_c^{2\beta_u}(\mathcal{B}_u)$. It is an irreducible $\overline{\mathbb{Q}}_\ell(W \times A(u))$-module of character $\chi_{C,\phi} \otimes \phi$; since F commutes with $A(u)$, the action of F on this component is of the form $q^{\beta_u} \tau_1 \otimes I$ where τ_1 defines a certain extension $\tilde{\chi}_{C,\phi}$ of $\chi_{C,\phi}$ to $W \rtimes \langle \tau \rangle$. Beynon and Spaltenstein (1984) and Shoji (1983) describe a particular choice of u giving rise to a particular extension; we still say that this u is **split** for F and we will call the corresponding $\tilde{\chi}_{C,\phi}$ the **almost preferred** extension of $\chi_{C,\phi}$. The reason for this terminology is that Lusztig (1986a, 17.2) defined a "preferred extension" which coincides with our almost preferred extension except in $^2E_6$ for the 3 characters of $W(E_6)$ such that $a_{\chi_{C,\phi}} \not\equiv \beta_u$ (mod 2) – see above 14.1.1 for the definition of a_χ. We need to choose the almost preferred extension for 13.2.10(iii) to hold.

Note that since the Green functions take rational values (since by definition $Q_{\mathbf{T}}^{\mathbf{G}}(u,1)$ is a Lefschetz number divided by $|\mathbf{T}^F|$ so is a rational number, see 8.1.6), as well as the characters of $A(u)$ (see the proof of 13.2.7) and of W, the almost preferred extension takes rational values.

To have uniform notations for the split and non-split cases, when $(\mathbf{G},F)$ is split we will define $\tau = \text{Id}$ and let $\tilde{\chi}_{C,\phi} = \chi_{C,\phi}$.

If u is a split element of C^F, where C is an F-stable $\mathbf{G}$-conjugacy class of unipotent elements, by 4.2.15(ii) the $\mathbf{G}^F$-orbits in C^F are indexed by the conjugacy classes of $A(u)$, where we can index u itself by 1. For the class $E_8(b_6)$ we index the $\mathbf{G}^F$-orbits by the F-conjugacy classes of $A(u)$, taking as a reference unipotent the element u described above.

Notation 13.2.9 *For $\iota = (C,\phi)$ where $C \subset \mathbf{G}_u$ is an F-stable $\mathbf{G}$-conjugacy class and $\phi \in \text{Irr}(A(u))$ we write*

- C_ι *for* C,
- β_ι *for* $\beta_u, u \in C$,
- $\mathcal{Y}_\iota$ *for the function on $\mathbf{G}^F$ which vanishes outside C^F and is equal to $\phi(a)$ on the $\mathbf{G}^F$-class indexed by $a \in H^1(F,A(u))$, where $u \in C^F$ is a unipotent element chosen as above (split except for $C = E_8(b_6)$)*,
- $\tilde{\chi}_\iota$ *for* $\tilde{\chi}_{C,\phi}$ *with* $\phi \in \text{Irr}(A(u))^0, u \in C$.

We will denote by $\mathcal{I}$ the set of all $\iota = (C,\phi)$ where C runs over the F-stable unipotent classes and ϕ runs over $\text{Irr}(A(u))$ for $u \in C$, and by $\mathcal{I}_0$ the subset where $\phi \in \text{Irr}(A(u))^0$ – thus $\mathcal{I}_0$ identifies with the image of the Springer correspondence. Note that the Springer correspondence is F-equivariant in that we have $F(\chi_{C,1}) = \chi_{F(C),1}$; since F induces an inner automorphism of W except for types $^3D_4$ and $^2D_{2n}$, it follows that except for these cases all classes in $\mathbf{G}_u$ are F-stable; see, for example Liebeck and Seitz (2012, Theorem 7.1(i), Theorem 7.3(i) and the tables).

Proposition 13.2.10

(i) *The set $\{\mathcal{Y}_\iota\}_{\iota \in \mathcal{I}}$ is a basis of the $\mathbf{G}^F$-class functions on $\mathbf{G}_u^F$.*

(ii) *For $\iota \in \mathcal{I}_0$, the restriction of $R_{\tilde{\chi}_\iota}$ to $C_{\iota'}^F$ is 0 unless $C_{\iota'}$ is in the Zariski closure of C_ι, where $R_{\tilde{\chi}_\iota}$ is the function defined in 11.6.1 (note that since $\mathbf{G}$ is semi-simple, the action of τ on W has same order as τ).*

(iii) *For $\iota \in \mathcal{I}_0$, the restriction of $R_{\tilde{\chi}_\iota}$ to C_ι^F is equal to $\varepsilon_\iota q^{\beta_\iota} \mathcal{Y}_\iota$ where ε_ι is a sign.*

   In good characteristic, $\varepsilon_\iota = 1$ except for $\iota = (E_8(b_6),\text{sgn})$ in which case $\varepsilon_\iota \equiv q \pmod 3$. In bad characteristic, $\varepsilon_\iota = 1$ except possibly for $\iota = (E_8(b_6),\text{sgn})$ where the value of ε_ι is still unknown.

Proof For (i), by the second orthogonality relation for characters of $A(u)$, where u is a reference unipotent in C^F, the function $\sum_{\phi \in \mathrm{Irr}(A(u))} \overline{\phi(a)} \mathcal{Y}_{C,\phi}$ is a multiple of the characteristic function of the $\mathbf{G}^F$-class in C parametrised by $a \in A(u)$.

We prove (ii). For a $\overline{\mathbb{Q}}_\ell W \rtimes \langle F \rangle$-module M we denote by $\mathrm{Trace}(.\,F \mid M)$ the function $w \mapsto \mathrm{Trace}(wF \mid M)$. For $u \in C_{\iota'}^F$ we have $R_{\tilde{\chi}_\iota}(u) = \langle \mathrm{Trace}(.\,F \mid H_c^*(\mathcal{B}_u)), \tilde{\chi}_\iota \rangle_{W\tau}$ by 13.2.3(ii), using that $\tilde{\chi}_\iota$ takes rational values. By 13.2.6(ii) this scalar product is 0 unless $C_{\iota'}$ is in the Zariski closure of C_ι.

For (iii), if u_a in the $\mathbf{G}^F$-class in C_ι is parametrised by $a \in H^1(F, A(u))$ we have

$$R_{\tilde{\chi}_\iota}(u_a) = \langle \mathrm{Trace}(.\,F \mid H_c^*(\mathcal{B}_{u_a})), \tilde{\chi}_\iota \rangle_{W\tau} = \langle \mathrm{Trace}(.\,F \mid H_c^{2\beta_u}(\mathcal{B}_{u_a})), \tilde{\chi}_\iota \rangle_{W\tau}$$

where the second equality is by 13.2.6(i).

Assume now that the characteristic is good. If C_ι is not the class $E_8(b_6)$, it results from the fact that u is split that F acts as $q^{\beta_u} \tau_1 \otimes I$ on the ϕ-isotypic component $H_c^{2\beta_u}(\mathcal{B}_u)_\phi$ – see the definition of the almost preferred extension. Thus it results from the definitions that F acts as $q^{\beta_u} \tau_1 \otimes \phi(a)$ on $H^{2\beta_u}(\mathcal{B}_{u_a})_\phi$. It follows that $R_{\tilde{\chi}_\iota}(u_a)$ is equal to $q^{\beta_u} \phi(a)$. A variation of the same computation gives the result for the class $E_8(b_6)$.

In bad characteristic a similar argument shows that $R_{\tilde{\chi}_\iota}(u_a)$ differs by a root of unity from $q^{\beta_u} \phi(a)$ and then the fact that $R_{\tilde{\chi}_\iota}(u_a)$ is a rational number shows that this root of unity is a sign. This sign has been shown to be 1 by Shoji (2007) in classical groups; it has been determined in most of the remaining cases by Geck (2019); it is still unknown for the class mentioned in type E_8 in bad characteristic.   □

13.3 The Lusztig–Shoji Algorithm

We define a preorder on $\mathcal{I}$ by the rule $\iota \le \iota'$ if $C_\iota \subset \overline{C_{\iota'}}$ (this is a partial order on unipotent classes). We write $\iota \sim \iota'$ for equivalent elements for this preorder, which happens exactly when $C_\iota = C_{\iota'}$.

We define a matrix Ω by $\Omega_{\iota,\iota'} := \langle R_{\tilde{\chi}_\iota} \chi_p, R_{\tilde{\chi}_{\iota'}} \chi_p \rangle_{\mathbf{G}^F}$. The following lemma shows that Ω can be computed just from information about the characters of W.

Lemma 13.3.1 $\Omega_{\iota,\iota'} = |\mathbf{G}^F|_{p'}^{-1} \mathrm{Feg}(\mathrm{sgn} \otimes \tilde{\chi}_\iota \otimes \tilde{\chi}_{\iota'})(q)$.

Proof By the definition $R_{\tilde{\chi}_\iota} \chi_p = |W|^{-1} \sum_{w \in W} \tilde{\chi}(w\tau) Q_{\mathbf{T}_w}^{\mathbf{G}}$, 13.2.2 and the fact that the almost preferred extensions take rational values, we get

$$\Omega_{\iota,\iota'} = |W|^{-1} \sum_{w \in W} \frac{\tilde{\chi}_\iota(w\tau) \tilde{\chi}_{\iota'}(w\tau)}{|\mathbf{T}_w^F|},$$

using also that $W(T_w)^F \simeq C_W(wF)$. This gives the statement by the second formula for Feg $\tilde{\chi}(q) = R_{\tilde{\chi}}(1)$ in the proof of 13.1.14. □

Note that the above formula shows that Ω has rational entries.

Finally we define entries $P_{\iota',\iota}$ by the condition that the decomposition of $R_{\tilde{\chi}_\iota} \cdot \chi_p$ in the basis $\mathcal{Y}_{\iota'}$ is $\sum_{\iota' \in I} P_{\iota',\iota} \mathcal{Y}_{\iota'}$; thus computing Green functions is equivalent to knowing the $P_{\iota',\iota}$.

Theorem 13.3.2 (Lusztig–Shoji algorithm)

(i) $P_{\iota',\iota} = 0$ *unless* $\iota' \in I_0$.
(ii) *We have the matrix equation* ${}^t P \Lambda P = \Omega$ *where* $P = \{P_{\iota,\iota'}\}_{\iota,\iota' \in I_0}$ *and where for* $\iota, \iota' \in I_0$ *we set* $\Lambda_{\iota,\iota'} := \langle \mathcal{Y}_\iota, \mathcal{Y}_{\iota'} \rangle_{\mathbf{G}^F}$. *This equation, given* Ω, *determines uniquely the matrices* Λ *and* P *subject to the conditions:*

- $P_{\iota,\iota'} = 0$ *unless* $\iota < \iota'$ *or* $\iota = \iota'$; *in this last case* $P_{\iota,\iota} = \varepsilon_\iota q^{\beta_\iota}$ *where* ε_ι *is as in 13.2.10(iii)*.
- $\Lambda_{\iota,\iota'} = 0$ *unless* $\iota \sim \iota'$; Λ *is invertible*.

(iii) *The matrices* P *and* Λ *have rational entries*.

Condition (i) is by definition and condition (ii) on P reflects 13.2.10(ii) and (iii) and the conditions on Λ hold by definition – it is invertible as the matrix of scalar products between linearly independent functions.

Proof Let us write Q_ι for $R_{\tilde{\chi}_\iota} \cdot \chi_p$. By Lusztig (1986b, 24.2.9) one can extend P to a matrix $\tilde{P}$, indexed by the whole of I by identifying (up to a scalar) the functions $\{Q_\iota\}_{\iota \in I_0}$ with "characteristic functions of sheaves" $IC(\overline{C}, \phi)$ which are indexed by pairs $(C, \phi) \in I_0$ and using the fact that the sheaves $IC(\overline{C}, \phi)$ make sense for any $\iota \in I$ to define functions Q_ι for any $\iota \in I$. We define $\tilde{P}_{\iota,\iota'}$ by saying that the decomposition of Q_ι in the basis $\mathcal{Y}_{\iota'}$ is $\sum_{\iota' \in I} \tilde{P}_{\iota',\iota} \mathcal{Y}_{\iota'}$.

If we extend Λ and Ω to the whole of I by $(\tilde{\Lambda}_{\iota,\iota'})_{\iota,\iota' \in I} = \langle \mathcal{Y}_\iota, \mathcal{Y}_{\iota'} \rangle_{\mathbf{G}^F}$ and $(\tilde{\Omega}_{\iota,\iota'})_{\iota,\iota' \in I} = \langle Q_\iota, Q_{\iota'} \rangle_{\mathbf{G}^F}$, then by the definitions we have the matrix equation ${}^t \tilde{P} \tilde{\Lambda} \tilde{P} = \tilde{\Omega}$.

Let $\chi_{I_0} : I \to \{0, 1\}$ be the characteristic function of I_0. The facts we need from Lusztig (1986b, §24) are:

(a) $\tilde{\Omega}_{\iota,\iota'} = 0$ if $\chi_{I_0}(\iota) \neq \chi_{I_0}(\iota')$.
(b) The function Q_ι is supported by $\overline{C_\iota}$ and the restriction of Q_ι to C_ι is equal up to a scalar to $\mathcal{Y}_\iota$.

(a) is from Lusztig (1986b, 24.3.6); (b), which extends 13.2.10(ii) and (iii), results from the definitions.

We prove now by induction on $<$ that P and Λ can be determined from the matrix equation and that

(c) $\tilde{P}_{\iota,\iota'} = 0$ if $\chi_{I_0}(\iota) \neq \chi_{I_0}(\iota')$.

(d) $\tilde{\Lambda}_{\iota,\iota'} = 0$ if $\chi_{\mathcal{I}_0}(\iota) \neq \chi_{\mathcal{I}_0}(\iota')$.

(e) P and Λ are rational.

Let us write the matrix equation by blocks indexed by classes, grouping all ι with a given C_ι to make blocks $P_{C,C'}, \Lambda_{C,C}$ and $\Omega_{C,C'}$ of the matrices $\tilde{P}, \tilde{\Lambda}$ and $\tilde{\Omega}$; note that by definition $\tilde{\Lambda}$ is block-diagonal and that $\tilde{P}$ is upper block-triangular if we order the blocks in the same way as the conjugacy classes; we get

$$\sum_{C \leq C', C \leq C''} {}^t P_{C,C'} \Lambda_{C,C} \overline{P}_{C,C''} = \Omega_{C',C''}. \tag{$*$}$$

We will do a double induction, the first over C'. Assume that we can determine $P_{C,C_0'}$ and $\Lambda_{C,C}$ for $C \leq C_0' < C'$ and that they satisfy (c),(d),(e). We show by a second induction, over C that we can determine $P_{C,C'}$ and $\Lambda_{C',C'}$ and they satisfy (c),(d),(e). Here "determine" is in the following sense: the matrices $P_{C,C'}$ and $\Lambda_{C',C'}$ are themselves by blocks according to the value of $\chi_{\mathcal{I}_0}$: we determine completely the value of the block where both indices are in $\mathcal{I}_0$ and show it satisfies (e), and we also show (c) and (d); but we do not determine completely (and don't care about) the block where none of the two indices is in $\mathcal{I}_0$. Assume that we have determined all $P_{C_0,C'}$ for $C_0 < C \leq C'$; we show that we can determine $P_{C,C'}$. Equation $(*)$ applied with $C'' = C$ can be written

$${}^t P_{C,C'} \Lambda_{C,C} \overline{P}_{C,C} = \Omega_{C',C} - \sum_{C_0 < C} {}^t P_{C_0,C'} \Lambda_{C_0,C_0} \overline{P}_{C_0,C}.$$

By induction we know all terms on the right-hand side, and they satisfy (c), (d),(e). If $C < C'$ we also know $\Lambda_{C,C}$ and $P_{C,C}$ which satisfy (d) and (e). This determines $P_{C,C'}$ and proves it satisfies (c) and (e); for (e) we use that Ω is rational which comes from 13.3.1. If $C = C'$ the left-hand side is ${}^t P_{C',C'} \Lambda_{C',C'} \overline{P}_{C',C'}$; this time we use that by (b) $P_{C',C'}$ is diagonal, with known scalars up to sign for the indices in $\mathcal{I}_0$; we thus can determine $\Lambda_{C',C'}$ (up to some unknown scalars for the indices not in $\mathcal{I}_0$) and the equality shows it satisfies (d) and (e). $\qquad\square$

Remark 13.3.3 The knowledge of Λ determines the numbers $|C_{\mathbf{G}}(u_a)^F|$ where u_a is in the $\mathbf{G}^F$-class parametrised by $a \in H^1(F, A(u))$. Indeed the definition gives

$$\Lambda_{(C,\phi),(C,\phi')} = \sum_{a \in H^1(F,A(u))} |C_{\mathbf{G}^F}(u_a)|^{-1} \phi(a)\phi'(a).$$

It follows that $\sum_{\phi \in \mathrm{Irr}(A(u))} \phi(a)\Lambda_{(C,\phi),(C,1)} = \frac{|C_{A(u)}(a)|}{|C_{\mathbf{G}^F}(u_a)|}$, where Λ is the bigger matrix used in the proof of 13.2.2, but we can restrict the sum to $(C,\phi) \in \mathcal{I}_0$ since $(C,1) \in \mathcal{I}_0$.

Remark 13.3.4   The functions Q_ι are uniform since – see the proof of 13.2.2 – $Q_{\mathbf{T}}^{\mathbf{G}} = |\mathbf{T}^F|^{-1} R_{\mathbf{T}}^{\mathbf{G}}(\text{reg}_{\mathbf{T}^F})$. The fact that the matrix P is unitriangular thus shows that the unipotently supported uniform functions are spanned by the $\mathcal{Y}_\iota$ for $\iota \in \mathcal{I}_0$. In particular the characteristic function of the geometric class C^F where $C \subset \mathbf{G}_u$ is an F-stable $\mathbf{G}$-conjugacy class, being equal to $\mathcal{Y}_{(C,1)}$, is uniform. Corollary 13.3.5 follows in turn from 10.3.4.

Corollary 13.3.5   *The characteristic function of any geometric conjugacy class of* $\mathbf{G}^F$ *is uniform.*

Remark 13.3.6   The order $\le$ on unipotent classes is easy to describe for classical adjoint groups in good characteristic. In these groups unipotent classes are determined by their Jordan form in the natural representation, except for some classes in types D_{2n}. The partition defined by Jordan blocks can be any partition of n for $\mathbf{PGL}_n$, any partition of n where odd parts have an even multiplicity for $\mathbf{PSp}_{2n}$, and any partition of n where even parts have an even multiplicity for $\mathbf{PSO}_n$; for $\mathbf{PSO}_{2n}$ the partitions with all parts even give rise to two classes. In any case, the order $\le$ is given by the **dominance order** $\vartriangleleft$ on partitions: for partitions $\lambda_1 \ge \lambda_2 \ge \ldots$ and $\mu_1 \ge \mu_2 \ge \ldots$ we have $\lambda \vartriangleleft \mu$ if and only if $\lambda_1 + \cdots + \lambda_i \le \mu_1 + \cdots + \mu_i$ for all i. For more details and proofs, see Spaltenstein (1982, I, §2 and II §8).

The order on unipotent classes of exceptional groups is harder to describe, but it can be recovered from the Lusztig–Shoji algorithm:

- It can be shown that $C \le C'$ if and only if $P_{(C,1),(C',1)} \ne 0$; see Beynon and Spaltenstein (1984, §5(E)).
- The Lusztig–Shoji algorithm does not need the order $\le$ to work; any preorder $<'$ which has the property that $C < C'$ implies $C <' C'$ and $C \sim C'$ for $<'$ implies $C = C'$ or that C and C' are not comparable for $\le$, will give the same result. We can choose for instance $C <' C'$ to be dim $C \le$ dim C'.

Example 13.3.7   In this example $\mathbf{G}$ is a group of type G_2; we assume the characteristic good for $\mathbf{G}$; that is, $p \ne 2, 3$. We denote by s and t the Coxeter generators of the Weyl group, where s corresponds to a long root and t to a short root. The list of unipotent classes, the value of β_u and the Springer correspondence can be found in Spaltenstein (1985). There are 5 unipotent classes which are ordered by $<$ as:

$$1 < A_1 < \tilde{A}_1 < G_2(a_1) < G_2,$$

the corresponding β_u are $6, 3, 2, 1, 0$ (the order can be recovered from the algorithm using remark 13.3.6).

The name of the classes comes from the Bala–Carter classification, see Carter (1985, 5.9). The class G_2 is the regular class, the class $G_2(a_1)$ is the class containing the regular class of the reductive subgroup $\mathbf{SL}_3$ (corresponding to the long roots) and the class $\tilde{A}_1$ (resp. A_1) is the class containing the regular class of a Levi with Weyl group $\langle t \rangle$ (resp. $\langle s \rangle$).

We use for $\mathrm{Irr}(W)$ the same names as in 6.4.1, that is $1, \mathrm{sgn}, \sigma, \tau, \rho, \rho'$ where τ (resp. σ) is the linear character defined by $s \mapsto -1, t \mapsto 1$ (resp. $s \mapsto 1$, $t \mapsto -1$), where ρ is the reflection representation and $\rho' = \rho \otimes \tau = \rho \otimes \sigma$.

The only conjugacy class such that $A(u) \neq 1$ is the class $G_2(a_1)$, for which $A(u) \simeq \mathfrak{S}_3$. If we denote by $1, \varepsilon, r$ the characters in $\mathrm{Irr}(\mathfrak{S}_3)$ where ε is the sign character and r the reflection representation, the Springer correspondence is given by

$$
\begin{array}{ccccccc}
\chi: & \mathrm{sgn} & \tau & \rho' & \sigma & \rho & \mathbf{1} \\
(C, \phi): & (1, \mathbf{1}) & (A_1, \mathbf{1}) & (\tilde{A}_1, \mathbf{1}) & (G_2(a_1), r) & (G_2(a_1), \mathbf{1}) & (G_2, \mathbf{1}).
\end{array}
$$

We see that $(G_2(a_1), \varepsilon) \in \mathcal{I} - \mathcal{I}_0$.

The fake degrees can be computed by formula 13.1.10 and are

$$
\begin{array}{ccccccc}
\chi: & \mathrm{sgn} & \tau & \rho' & \sigma & \rho & \mathbf{1} \\
\mathrm{Feg}\,\chi(q): & q^6 & q^3 & q^2(q^2+1) & q^3 & q(q^4+1) & 1
\end{array}
$$

This allows to compute Ω. Here is $|\mathbf{G}^F|_{p'}\Omega$:

$$
\begin{pmatrix}
q^6 & q^3 & q^4 + q^2 & q^3 & q^5 + q & 1 \\
q^3 & q^6 & q^5 + q & 1 & q^4 + q^2 & q^3 \\
q^4 + q^2 & q^5 + q & (q^2+1)(q^4+1) & q^5 + q & q(q^2+1)^2 & q^4 + q^2 \\
q^3 & 1 & q^5 + q & q^6 & q^4 + q^2 & q^3 \\
q^5 + q & q^4 + q^2 & q(q^2+1)^2 & q^4 + q^2 & (q^2+1)(q^4+1) & q^5 + q \\
1 & q^3 & q^4 + q^2 & q^3 & q^5 + q & q^6
\end{pmatrix}.
$$

This in turn allows to compute P, giving the decomposition of R_χ in the basis $\mathcal{Y}_t$:

	$\mathcal{Y}_{(1,1)}$	$\mathcal{Y}_{(A_1,1)}$	$\mathcal{Y}_{(\tilde{A}_1,1)}$	$\mathcal{Y}_{(G_2(a_1),r)}$	$\mathcal{Y}_{(G_2(a_1),1)}$	$\mathcal{Y}_{(G_2,1)}$
St	q^6	.	.	.	.	.
R_τ	q^3	q^3	.	.	.	.
$R_{\rho'}$	$q^4 + q^2$	q^2	q^2	.	.	.
R_σ	q^3	.	q	q	.	.
R_ρ	$q^5 + q$	q	q	.	q	.
$\mathbf{1}$	1	1	1	.	1	1

and Λ which turns out to be diagonal, with diagonal terms

$$
\begin{array}{cccccc}
(1, \mathbf{1}) & (A_1, \mathbf{1}) & (\tilde{A}_1, \mathbf{1}) & (G_2(a_1), r) & (G_2(a_1), \mathbf{1}) & (G_2, \mathbf{1}) \\
\hline
\frac{1}{|\mathbf{G}^F|} & \frac{1}{q^6(q^2-1)} & \frac{1}{q^4(q^2-1)} & \frac{1}{q^4} & \frac{1}{q^4} & \frac{1}{q^2}
\end{array}
$$

from which one can conclude, for example, that for $u \in G_2(a_1)$, the group $C_{\mathbf{G}}(u)^0$ is unipotent of dimension 4.

Example 13.3.8   We give now the same example of a group of type G_2, but this time in characteristic 3. The unipotent variety $\mathbf{G}_u$ is quite different from that in good characteristic; there is a sixth unipotent class $(\tilde{A}_1)_3$ which cannot be distinguished from the class A_1 for the partial order $\leq$, and has same $\beta_u = 3$. This is the notation for classes in Spaltenstein (1985); in Liebeck and Seitz (2012) the names $\tilde{A}_1$ and $(\tilde{A}_1)_3$ are interchanged.

This time for the class $G_2(a_1)$ we have $A(u) \simeq \mathbb{Z}/2\mathbb{Z}$, but there is another class with $A(u) \neq 1$, the regular class G_2 with $A(u) \simeq \mathbb{Z}/3\mathbb{Z}$.

The Springer correspondence is given by

$\chi:$	sgn	τ	σ	ρ'	ρ	$\mathbf{1}$
$(C,\phi):$	$(1,\mathbf{1})$	$(A_1,\mathbf{1})$	$((\tilde{A}_1)_3,\mathbf{1})$	$(\tilde{A}_1,\mathbf{1})$	$(G_2(a_1),\mathbf{1})$	$(G_2,\mathbf{1})$

thus only the pairs with $\phi = \mathbf{1}$ are in $\mathcal{I}_0$.

The decomposition of the R_χ in the basis $\mathcal{Y}_\iota$ is given by:

	$\mathcal{Y}_{(1,1)}$	$\mathcal{Y}_{(A_1,1)}$	$\mathcal{Y}_{((\tilde{A}_1)_3,1)}$	$\mathcal{Y}_{(\tilde{A}_1,1)}$	$\mathcal{Y}_{(G_2(a_1),1)}$	$\mathcal{Y}_{(G_2,1)}$
St	q^6	.	.	.	.	.
R_τ	q^3	q^3	.	.	.	.
R_σ	q^3	.	q^3	.	.	.
$R_{\rho'}$	$q^4 + q^2$	q^2	q^2	q^2	.	.
R_ρ	$q^5 + q$	q	q	q	q	.
$\mathbf{1}$	1	1	1	1	1	1

And the matrix Λ is diagonal with diagonal terms:

$(1,\mathbf{1})$	$(A_1,\mathbf{1})$	$((\tilde{A}_1)_3,\mathbf{1})$	$(\tilde{A}_1,\mathbf{1})$	$(G_2(a_1),\mathbf{1})$	$(G_2,\mathbf{1})$
$\frac{1}{|\mathbf{G}^F|}$	$\frac{1}{q^6(q^2-1)}$	$\frac{1}{q^6(q^2-1)}$	$\frac{1}{q^6}$	$\frac{1}{q^4}$	$\frac{1}{q^2}$

note that the centraliser of an element of the class $\tilde{A}_1$ turns out to be unipotent in this case.

Notes

The Lusztig–Shoji algorithm bears this name since it is an improvement by Lusztig (1986b) of an algorithm first formulated in Shoji (1982).

14

The Decomposition of Deligne–Lusztig Characters

In this chapter we describe the results of Lusztig giving an explicit description of the decomposition of Deligne–Lusztig characters into irreducible characters. Using the results of Chapter 13, this gives the scalar product of any character with the functions $\mathcal{Y}_\iota, \iota \in \mathcal{I}_0$, and in particular gives the "average value" of any irreducible character on any geometric conjugacy class.

Theorem 11.5.1 reduces the decomposition problem to unipotent characters. We give the classification of unipotent characters and the decomposition of unipotent Deligne–Lusztig characters; we did this for the general linear and unitary groups in Theorem 11.7.2, but the general case is more complicated: the matrix expressing the decomposition of $R_{\tilde{x}}$ into irreducible unipotent characters has blocks which group both characters of W and unipotent characters in "Lusztig families", and these blocks are attached to "special" unipotent classes.

14.1 Lusztig Families and Special Unipotent Classes

We now explain these two concepts. Let $\mathbf{G}$ be a connected reductive group with Weyl group W. Let $\chi = \chi_\iota \in \mathrm{Irr}(W)$ correspond to $\iota = (C, \phi)$ by the Springer correspondence. We attach to χ three numerical invariants: β_u, the dimension of $\mathcal{B}_u$ for $u \in C$, b_χ, the valuation at t of $\mathrm{Feg}(\chi)$, which can also be defined as the smallest degree i such that χ occurs in $(SV)_i$, and a_χ, the valuation at q of the generic degree of χ when all the parameters q_s are equal to q and all q'_s equal to -1 – the generic degree is also the degree of γ_χ if $(\mathbf{G}, F)$ is split, see 6.3.14.

Proposition 14.1.1 *Assume the characteristic good for* $\mathbf{G}$. *We have* $a_\chi \le \beta_u \le b_\chi$ *where the right inequality is an equality if and only if* $\phi = \mathbf{1}$.

References For the right inequality see Borho and MacPherson (1981, Corollaire 4); the left inequality results from the particular case where C
is special which is Lusztig (1984a, 13.1.2) and from Geck and Malle (1999,
Proposition 2.2). □

Definition 14.1.2 $\chi \in \mathrm{Irr}(W)$ is **special** if $a_\chi = b_\chi$. A class $C \subset \mathbf{G}_u$ is **special**
if $\chi_{C,1}$ is special.

Assume now that $\mathbf{G}$ is over $\overline{\mathbb{F}}_q$, with a Frobenius F.

Theorem 14.1.3 *Assume the characteristic good for* $\mathbf{G}$. *Let* $\gamma \in \mathcal{E}(\mathbf{G}^F, 1)$ *be
a unipotent character. Then there is a unique unipotent class* $C_\gamma \subset \mathbf{G}_u$ *called
the* **unipotent support** *of* γ, *such that*

- *There is* $u \in C_\gamma^F$ *such that* $\gamma(u) \neq 0$.
- *For any unipotent class* C' *and* $u \in C'^F$ *we have* $\gamma(u) = 0$ *unless* $C' = C_\gamma$ *or*
  $\dim C' < \dim C_\gamma$.

The class C_γ *is special.*

References This has been proved in Lusztig (1992) with some restrictions on
q, then just assuming p good by Geck (1996). □

The above theorem can be extended to bad characteristic if one replaces the
condition $\gamma(u) = 0$ by the condition "the weighted average $\sum_{a \in A(u)} \gamma(u_a)$ of γ
on C_γ is 0" see Geck and Malle (1999, 5.1 and 5.2).

The set of unipotent characters with a given unipotent support is called a
Lusztig family.

A Lusztig family of unipotent characters is also the set of indices of columns
for a block of the matrix of coefficients of the $R_{\tilde{\chi}}$ on unipotent characters. The
set of $\chi \in \mathrm{Irr}(W)$ indexing the lines of the block is called a **Lusztig family** of
$\mathrm{Irr}(W)$.

To each family, Lusztig (1984a, 13.1.2) associates a quotient of $A(u)$ using
the following definition.

Definition 14.1.4 Let C be a special unipotent class, and $u \in C$. We define the
Lusztig quotient $\overline{A(u)}$ of $A(u)$ as its quotient by the normal subgroup which is
the intersection of the kernels of all $\phi \in \mathrm{Irr}(A(u))^0$ such that $a_{\chi_{C,\phi}} = \beta_u$.

Assume the characteristic is good for $\mathbf{G}$, and let Γ be the Lusztig quotient
associated with a family. Then the unipotent characters in the family, and the
scalar products between these unipotent characters and the R_χ for χ in the
family, can all be described in terms of the "Drinfeld double" of Γ.

Definition 14.1.5 For Γ a finite group, the **Drinfeld double** $D\Gamma$ of Γ over a
field k is the algebra $\mathrm{Hom}_k(k\Gamma, k) \rtimes k\Gamma$.

$\mathrm{Hom}_k(k\Gamma,k)$ identifies with the algebra of k-valued functions on Γ with pointwise multiplication – the algebra of Γ as an algebraic group, see 1.1.1(iv). It has a basis $\{\delta_g\}_{g\in\Gamma}$ formed of the Dirac functions $\delta_g(g') := \delta_{g,g'}$. The action of $g \in \Gamma$ which defines the semi-direct product is the right action $\delta_{g'} \mapsto \delta_{g^{-1}g'g}$.

In the following we consider only the Drinfeld double over $k = \mathbb{C}$. It is then a semi-simple algebra.

14.2 Split Groups

We first describe the situation when F is a split Frobenius.

Theorem 14.2.1 *Assume F is a split Frobenius endomorphism on* **G***. Then the unipotent characters in a family associated with the Lusztig quotient* Γ *are parametrised by the set* $\mathcal{M}(\Gamma)$ *of simple* $D\Gamma$*-modules.*

We should note that though we have defined the groups Γ parametrising the families only in good characteristic – as the Lusztig quotients attached to special unipotent classes – the description of the families and the decomposition of the Deligne–Lusztig characters is independent of the characteristic, thus the families are parametrised in any characteristic by the same groups Γ.

For detailed proofs of the following properties of $D\Gamma$, see for instance Broué (2017, Chapter 8).

Proposition 14.2.2 *The set* $\mathcal{M}(\Gamma)$ *is parametrised by representatives up to* Γ*-conjugacy of pairs* (x,χ) *where* $x \in \Gamma$ *and* $\chi \in \mathrm{Irr}(C_\Gamma(x))$.

The map from a $C_\Gamma(x)$-module X with character χ to a $D\Gamma$-module is obtained by making δ_g act on $Y = \mathrm{Ind}^\Gamma_{C_\Gamma(x)}(X) \simeq \oplus_{r\in[\Gamma/C_\Gamma(x)]} r \otimes X$ by the projection onto $\oplus_{gr=rx} r \otimes X$, which makes sense since the condition does not depend on the representative chosen in $rC_\Gamma(x)$. We will write (x,χ) for the $D\Gamma$-module Y.

$D\Gamma$ is a Hopf algebra for the comultiplication: $g \mapsto g \otimes g$ for $g \in \Gamma$ and $\delta_g \mapsto \sum_{g=g'g''} \delta_{g'} \otimes \delta_{g''}$. It follows that the category of $D\Gamma$-modules is a tensor category: if X,Y are $D\Gamma$-modules then $D\Gamma$ acts on $X \otimes Y$ via the comultiplication. Further this category is braided, which means there are natural isomorphisms $X \otimes Y \xrightarrow{\sim} Y \otimes X$ given by $x \otimes y \mapsto \sum_{g\in\Gamma} gy \otimes \delta_g x$. This implies that the Grothendieck ring of the category of $D\Gamma$-modules is a commutative algebra. The **Lusztig Fourier transform** attached to Γ is the (normalised) character table of this Grothendieck group defined as follows: the characters of the Grothendieck group are themselves parametrised by the simple modules, and if we denote by $\Phi_{(x,\chi)}$ the character parametrised by (x,χ) we have

$$\Phi_{(x,\chi)}((y,\psi)) = \frac{|\Gamma|}{\dim\left((x,\chi)\right)} \frac{1}{|C_\Gamma(x)||C_\Gamma(y)|} \sum_{\{g\in\Gamma\,|\,[^g x, y]=1\}} \chi(g^{-1}yg)\psi(^g x).$$

The normalised character table considered by Lusztig is the matrix

$$\{(x,\chi),(y,\psi)\} := \frac{\dim\left((x,\chi)\right)}{|\Gamma|}\Phi_{(x,\overline{\chi})}((y,\psi))$$

which is unitary and Hermitian.

Lusztig defines a bijection between the characters of $\mathrm{Irr}(W)$ in the family attached to Γ and a subset $\mathcal{M}_0 \subset \mathcal{M}(\Gamma)$, such that

Theorem 14.2.3 *If $\phi \in \mathrm{Irr}(W)$ is parametrised by (x,χ) and $\gamma_{(y,\psi)} \in \mathcal{E}(\mathbf{G}^F,1)$ is parametrised by (y,ψ), then $\langle R_\phi, \gamma_{(y,\psi)}\rangle_{\mathbf{G}^F} = \{(x,\chi),(y,\psi)\}$.*

Since $(\mathbf{G},F)$ is split, that $(y,\psi) \in \mathcal{M}_0$ is equivalent to $\gamma_{(y,\psi)}$ being in the principal series.

The list of unipotent characters and their parameters in $\mathcal{M}(\Gamma)$ was determined in Lusztig (1984a); see pages 110 and 372 of loc. cit. for the following example.

Example 14.2.4   We describe the decomposition of R_χ in a simple group $\mathbf{G}$ of type G_2. We always have $R_1 = \mathbf{1}$ and $R_{\mathrm{sgn}} = \mathrm{St}_{\mathbf{G}^F}$, see 10.2.5. These two characters are thus alone in their family. The other four characters in $\mathrm{Irr}(W)$ form a family, attached to the special class $G_2(a_1)$, see 13.3.7. For this class we have $\overline{A(u)} = A(u) = \mathfrak{S}_3$. The set $\mathcal{M}(\mathfrak{S}_3)$ is

$$\mathcal{M}(\mathfrak{S}_3) = \{(1,\mathbf{1}),(1,\varepsilon),(1,r),(g_2,\mathbf{1}),(g_2,\varepsilon),(g_3,\mathbf{1}),(g_3,\zeta_3),(g_3,\zeta_3^2)\},$$

where $\mathrm{Irr}(\mathfrak{S}_3) = \{\mathbf{1},\varepsilon,r\}$, where g_2 (resp. g_3) is an element of order 2 (resp. of order 3) of $\mathfrak{S}_3$ and $\mathrm{Irr}(C_{\mathfrak{S}_3}(g_2)) = \{\mathbf{1},\varepsilon\}$ (resp. $\mathrm{Irr}(C_{\mathfrak{S}_3}(g_3)) = \{\mathbf{1},\zeta_3,\zeta_3^2\}$).

The bijection from the family in $\mathrm{Irr}(W)$ to the subset $\mathcal{M}_0$ is defined by

$$\rho \mapsto (1,\mathbf{1}), \quad \rho' \mapsto (g_2,\mathbf{1}), \quad \tau \mapsto (g_3,\mathbf{1}), \quad \sigma \mapsto (1,r).$$

Table 14.1 is the pairing $\{,\}$. The rows of Table 14.1 correspond to the unipotent characters. The first four rows are the principal series characters in the family. The next four rows are the cuspidal unipotent characters which are labelled by subscripts $[\zeta]$ where ζ is a root of unity which is attached to a unipotent character by the following theorem Lusztig (1978, 3.9).

Theorem 14.2.5 *Let F be a Frobenius root on $\mathbf{G}$ attached to the number q and let F^δ be the smallest power which is split. Let $\rho \in \mathcal{E}(\mathbf{G}^F,1)$. Then for any F-stable torus $\mathbf{T}$ and any (non-necessarily F-stable) Borel subgroup $\mathbf{B}$ containing $\mathbf{T}$, any eigenvalue of F^δ on the ρ-isotypic component of $H_c^i(\mathbf{X}_\mathbf{B})$ is of the form $\zeta_\rho q^{\delta j/2}$ for some integer j and some root of unity ζ_ρ, where ζ_ρ and j (mod 2) depend only on ρ and not on $\mathbf{T},\mathbf{B}$ or i.*

Table 14.1. *Decomposition of the R_χ into unipotent characters*

	dim	R_ρ	$R_{\rho'}$	R_τ	R_σ	$R_{(1,\varepsilon)}$	$R_{(g_2,\varepsilon)}$	$R_{(g_3,\zeta_3)}$	$R_{(g_3,\zeta_3^2)}$
dim		$q\phi_8$	$q^2\phi_4$	q^3	q^3	0	0	0	0
γ_ρ	$\frac{q}{6}\phi_2^2\phi_3$	$\frac16$	$\frac12$	$\frac13$	$\frac13$	$\frac16$	$\frac12$	$\frac13$	$\frac13$
$\gamma_{\rho'}$	$\frac{q}{2}\phi_2^2\phi_6$	$\frac12$	$\frac12$	·	·	$-\frac12$	$-\frac12$	·	·
γ_τ	$\frac{q}{3}\phi_3\phi_6$	$\frac13$	·	$\frac23$	$-\frac13$	$\frac13$	·	$-\frac13$	$-\frac13$
γ_σ	$\frac{q}{3}\phi_3\phi_6$	$\frac13$	·	$-\frac13$	$\frac23$	$\frac13$	·	$-\frac13$	$-\frac13$
$\gamma_{[-1]}$	$\frac{q}{6}\phi_1^2\phi_3$	$\frac16$	$-\frac12$	$\frac13$	$\frac13$	$\frac16$	$-\frac12$	$\frac13$	$\frac13$
$\gamma_{[1]}$	$\frac{q}{6}\phi_1^2\phi_6$	$\frac12$	$-\frac12$	·	·	$-\frac12$	$\frac12$	·	·
$\gamma_{[\zeta_3]}$	$\frac{q}{3}\phi_1^2\phi_2^2$	$\frac13$	·	$-\frac13$	$-\frac13$	$\frac13$	·	$\frac23$	$-\frac13$
$\gamma_{[\zeta_3^2]}$	$\frac{q}{3}\phi_1^2\phi_2^2$	$\frac13$	·	$-\frac13$	$-\frac13$	$\frac13$	·	$-\frac13$	$\frac23$

The second column of Table 14.1 indicates the dimensions of the unipotent characters; we have expressed them using cyclotomic polynomials ϕ_i to save space – see the character table of the Hecke algebra (Table 6.2 in Chapter 6).

The next four columns within Table 14.1 represent the decompositions of the R_χ into unipotent characters. The last four data columns represent functions which complete the R_χ to make an orthonormal basis of the unipotent class functions – by which we mean the functions spanned by $\mathcal{E}(\mathbf{G}^F,1)$. These functions are orthogonal to all uniform functions, in particular as indicated, their value at 1 is 0. Let us describe these functions when the characteristic is good for $\mathbf{G}$. We have $R_{(1,\varepsilon)} = q\mathcal{Y}_{(G_2(a_1),\varepsilon)}$ (see 13.2.9). The functions $R_{(g_3,\zeta_3)}$ and $R_{(g_3,\zeta_3^2)}$ are supported on the class of an element with Jordan decomposition su where $C_\mathbf{G}(s) \simeq \mathbf{SL}_3$ and u is a regular unipotent element of $\mathbf{SL}_3$; they correspond to the non-trivial characters of $A(u) \simeq \mathbb{Z}/3\mathbb{Z}$. Finally $R_{(g_2,\varepsilon)}$ is supported on the class of an element su where $C_\mathbf{G}(s) \simeq \mathbf{SL}_2 \times_{Z(\mathbf{SL}_2)} \mathbf{SL}_2$ and u is a regular unipotent element of $C_\mathbf{G}(s)$; it corresponds to the non-trivial character of $A(u) \simeq \mathbb{Z}/2\mathbb{Z}$.

Functions like the ones above have been constructed in general by Lusztig as the characteristic functions of some perverse sheaves, called **character sheaves**, see Lusztig (1986b).

We have now gathered enough information to give a part of the character table of $\mathbf{G}^F$ in good characteristic, the values of unipotent characters on unipotent elements. In Table 14.2 we have denoted by $G_2(a_1)$ the $\mathbf{G}^F$-class

Table 14.2. *Values of unipotent characters on unipotent classes*

	1	A_1	$\tilde{A}_1$	$G_2(a_1)_3$	$G_2(a_1)_2$	$G_2(a_1)$	G_2
1	1	1	1	1	1	1	1
St	q^6	.	.	.	.	.	.
σ	$\frac{q}{3}\phi_3\phi_6$	$-\frac{q}{2}\phi_1\phi_2$	q	$2q$	.	.	.
τ	$\frac{q}{3}\phi_3\phi_6$	$\frac{q}{3}(2q^2+1)$	.	.	.	q	.
ρ	$\frac{q}{6}\phi_2^2\phi_3$	$\frac{q}{6}(2q+1)\phi_2$	$\frac{q}{2}\phi_2$	q	.	.	.
ρ'	$\frac{q}{2}\phi_2^2\phi_6$	$\frac{q}{2}\phi_2$	$\frac{q}{2}\phi_2$	.	q	.	.
$\gamma_{[-1]}$	$\frac{q}{2}\phi_1^2\phi_3$	$-\frac{q}{2}\phi_1$	$-\frac{q}{2}\phi_1$	.	q	.	.
$\gamma_{[1]}$	$\frac{q}{6}\phi_1^2\phi_6$	$\frac{q}{6}(2q-1)\phi_1$	$-\frac{q}{2}\phi_1$	q	.	.	.
$\gamma_{[\zeta_3]}$	$\frac{q}{3}\phi_1^2\phi_2^2$	$-\frac{q}{3}\phi_1\phi_2$	.	.	.	q	.
$\gamma_{[\zeta_3^2]}$	$\frac{q}{3}\phi_1^2\phi_2^2$	$-\frac{q}{3}\phi_1\phi_2$	.	.	.	q	.

of a split element u in the geometric unipotent class $G_2(a_1)$, and by $G_2(a_1)_2$ (resp. $G_2(a_1)_3$) the $\mathbf{G}^F$-class parametrised by an element of order 2 (resp. 3) of $C_{\mathbf{G}}(u)/C_{\mathbf{G}}(u)^0 \simeq \mathfrak{S}_3$.

14.3 Twisted Groups

We now assume that F is a Frobenius root and $(\mathbf{G}, F)$ is not split. If $C \subset \mathbf{G}_u$ is an F-stable special class, then F induces an automorphism of $\overline{A(u)}$ for $u \in C^F$, thus induces an automorphism of $D\Gamma$.

Since one can reduce the description of irreducible characters for a descent of scalars to the case of a quasi-simple group, we will focus on the case of quasi-simple groups. Then, if F is a Frobenius morphism, one can always find some $u \in C^F$ such that the action of F on $A(u)$ is trivial, as observed in 13.2.7. In this case F will also act trivially on $\Gamma = \overline{A(u)}$ thus on $D\Gamma$ and the description of families is the same as in the untwisted case; in particular the sets $\mathcal{M}(\Gamma)$ and $\mathcal{M}_0(\Gamma)$ are defined in the same way.

For the "very twisted" cases $^2B_2, ^2G_2$ and $^2F_4$ the characters still have a unipotent support, which determines families, but the families are not described by the corresponding group $\overline{A(u)}$. The decomposition of Deligne–Lusztig characters has been described in Lusztig (1984a, Appendix). The families of unipotent characters of $\mathbf{G}^F$ are parametrised by the F-stable families of W; for each F-stable family of W, F induces an involution on $\mathcal{M}(\Gamma)$, which is the involution induced on $\mathrm{Irr}(W)$ for $\mathcal{M}_0$, and leaves the other elements of $\mathcal{M}(\Gamma)$ invariant excepted for the family in F_4 attached to $\Gamma = \mathfrak{S}_4$, where two more elements are exchanged. The unipotent characters of $\mathbf{G}^F$ are parametrised by the elements of $\mathcal{M}(\Gamma)$ fixed by this involution. An interpretation of the "Fourier matrix" expressing the decomposition of the $R_{\tilde{\chi}}$ into unipotent characters has been given in Geck and Malle (2003).

Notes

The description of families given in this chapter does not correspond to the historical development. The decomposition of R_χ and the families were described in Lusztig (1984a) using Kazhdan–Lusztig theory and intersection cohomology of Deligne–Lusztig varieties. The geometric explanation we give was only shown later.

References

Adams, J., and He, X. 2017. Lifting of elements of Weyl groups. *J. Algebra*, **485**, 142–165.

Alvis, D. 1980. Duality in the character ring of a finite Chevalley group. In *The Santa Cruz Conference on Finite Groups*. Proc. Symp. in Pure Math., vol. 37, 353–357, University of California, Santa Cruz, CA.

Alvis, D., Lusztig, G., and Spaltenstein, N. 1982. On Springer's correspondence for simple groups of type $E_n(n = 6,7,8)$. *Math. Proc. Camb. Phil. Soc.*, **92**, 65–78.

Benson, C. T., and Curtis, C. W. 1972. On the degrees and rationality of certain characters of finite Chevalley groups. *Trans. Amer. Math. Soc.*, **165**, 251–273.

Beynon, W. M., and Spaltenstein, N. 1984. Green functions of finite Chevalley groups of type $E_n(n = 6,7,8)$. *J. Algebra*, **88**, 584–614.

Bonnafé, C. 2005a. Actions of relative Weyl groups II. *J. Group Theory*, **8**, 351–387.

Bonnafé, C. 2005b. Quasi-isolated elements in reductive groups. *Comm. Algebra*, **33**, 2315–2337.

Bonnafé, C. 2006. Sur les caractères des groupes réductifs finis à centre non connexe: applications aux groupes spéciaux linéaires et unitaires. *Astérisque*, **306**, 1–174.

Bonnafé, C. 2010. *Representations of* $\mathbf{SL}_2(\mathbb{F}_q)$. London, Springer Verlag.

Bonnafé, C., and Michel, J. 2011. A computational proof of the Mackey formula for $q > 2$. *J. Algebra*, **327**, 506–526.

Bonnafé, C., and Rouquier, R. 2006. On the irreducibility of Deligne–Lusztig varieties. *C. R. Math. Acad. Sci. Paris*, **343**, 37–39.

Borel, A. 1991. *Linear Algebraic Groups*. Second enlarged ed. Graduate Texts in Mathematics, no. 126. New York, Springer Verlag.

Borel, A., and Tits, J. 1965. Groupes réductifs. *Publ. Math. Inst. Hautes Études Sci.*, **27**, 55–160.

Borel, A., Carter, R., Curtis, C. W., Iwahori, N., Springer, T. A., and Steinberg, R. 1970. *Seminar on Algebraic Groups and Related Topics*. Lecture Notes in Mathematics, no. 131. New York, Springer Verlag.

Borho, W., and MacPherson, R. 1981. Représentations des groupes de Weyl et homologie d'intersection pour les variétés nilpotentes. *C. R. Math. Acad. Sci. Paris*, **292**, 707–710.

Bourbaki, N. 1968. *Groupes et algèbres de Lie,* chapters IV, V, VI. Hermann, Paris.

Bourbaki, N. 1971. *Algèbre,* chapters 1–3. Hermann/Addison-Wesley, Paris.

Bourbaki, N. 1975. *Algèbre commutative,* chapters 5–7. Springer Verlag, Berlin.

Bourbaki, N. 1981. *Algèbre,* chapter 8. Springer Verlag, Berlin.

Broué, M. 2017. *On Characters of Finite Groups.* Mathematical Lectures from Peking University. Springer Verlag.

Broué, M., Malle, G., and Michel, J. 1999. Towards Spetses I. *Transformation Groups,* **4**, 157–218.

Carter, R. W. 1985. *Finite Groups of Lie Type: Conjugacy Classes and Complex Characters.* Chichester, Wiley-Interscience.

Chevalley, C. 1955. Invariants of finite reflection groups. *Amer. J. Math.,* **77**, 778–782.

Chevalley, C. 1994. Sur les décompositions cellulaires des espaces G/B. In *Algebraic Groups and Their Generalizations: Classical Methods (University Park, PA, 1991).* Proc. Symp. Pure Math., vol. 56, 1–23. American Mathematical Society, Providence, RI.

Chevalley, C. 2005. *Classification des groupes algébriques semi-simples.* Collected works, vol. 3. Springer Verlag, Berlin.

Curtis, C. W. 1980. Truncation and duality in the character ring of a finite group of Lie type. *J. Algebra,* **62**, 320–332.

Curtis, C. W., and Reiner, I. 1981. *Methods of Representation Theory I.* Wiley-Interscience, New York.

Deligne, P. 1977. *Cohomologie étale.* Lecture Notes in Mathematics, no. 569. Springer Verlag. Séminaire de Géométrie Algébrique du Bois-Marie SGA $4\frac{1}{2}$, Avec la collaboration de J. F. Boutot, A. Grothendieck, L. Illusie et J. L. Verdier.

Deligne, P. 1980. La conjecture de Weil II. *Publ. Math. Inst. Hautes Études Sci.,* **43**, 137–252.

Deligne, P., and Lusztig, G. 1976. Representations of reductive groups over finite fields. *Ann. of Math.,* **103**, 103–161.

Deligne, P., and Lusztig, G. 1982. Duality for representations of a reductive group over a finite field. *J. Algebra,* **74**, 284–291.

Deligne, P., and Lusztig, G. 1983. Duality for representations of a reductive group over a finite field II. *J. Algebra,* **81**, 540–545.

Deodhar, V. 1989. A note on subgroups generated by reflections in Coxeter groups. *Arch. Math.,* **53**, 543–546.

Deriziotis, D. I. 1984. Conjugacy classes and centralizers of semi-simple elements in finite groups of Lie type. *Vorlesungen aus dem Fachbereich Mathematik der Universität Essen,* **11**.

Digne, F., and Michel, J. 1982. Remarques sur la dualité de Curtis. *J. Algebra,* **79**, 151–160.

Digne, F., and Michel, J. 1983. Foncteur de Lusztig et fonctions de Green généralisées. *C. R. Math. Acad. Sci. Paris,* **297**, 89–92.

Digne, F., and Michel, J. 1987. Foncteurs de Lusztig et caractères des groupes linéaires et unitaires sur un corps fini. *J. Algebra,* **107**, 217–255.

Digne, F., and Michel, J. 1990. On Lusztig's parametrization of characters of finite groups of Lie type. *Asterisque,* **181–182**, 113–156.

Digne, F., and Michel, J. 1994. Groupes réductifs non connexes. *Ann. ENS,* **27**, 345–406.

Digne, F., Lehrer, G. I., and Michel, J. 1992. The characters of the group of rational points of a reductive group with non-connected centre. *J. Reine Angew. Math.,* **425**, 155–192.

Digne, F., Lehrer, G. I., and Michel, J. 1997. On Gelfand–Graev characters of reductive groups with disconnected centre. *J. Reine Angew. Math.*, **491**, 131–147.

Digne, F., Michel, J., and Rouquier, R. 2007. Cohomologie des variétés de Deligne–Lusztig. *Adv. Math.*, **209**, 749–822.

Dipper, R., and Du, J. 1993. Harish-Chandra vertices. *J. Reine Angew. Math.*, **437**, 101–130.

Dyer, M. 1990. Reflection subgroups of Coxeter systems. *J. Algebra*, **135**, 57–73.

Geck, M. 1993. A note on Harish-Chandra induction. *Manuscripta Math.*, **80**, 393–401.

Geck, M. 1996. On the average value of the irreducible characters of finite groups of Lie type on geometric unipotent classes. *Doc. Math. J. DMV*, **1**, 293–317.

Geck, M. 2003. *An Introduction to Algebraic Geometry and Algebraic Groups*. Oxford Graduate Texts in Mathematics, Vol. 10. Oxford University Press, Oxford.

Geck, M. 2019. *Computing Green Functions in Small Characteristic*. Arxiv: 1904.06970 [math.RT].

Geck, M., and Jacon, N. 2011. *Representations of Hecke Algebras at Roots of Unity*. Algebra and Applications, No. 15. Springer Verlag, London.

Geck, M., and Malle, G. 1999. On the existence of a unipotent support for the irreducible characters of a finite group of Lie type. *Trans. Amer. Math. Soc.*, **352**, 429–456.

Geck, M., and Malle, G. 2003. Fourier transforms and Frobenius eigenvalues for finite Coxeter groups. *J. Algebra*, **260**, 162–193.

Geck, M., and Pfeiffer, G. 2000. *Characters of Finite Coxeter Groups and Iwahori–Hecke Algebras*. London Mathematical Society Monographs, Vol. 21. Oxford University Press, New York.

Geck, M., Hiss, G., Lübeck, F., Malle, G., and Pfeiffer, G. 1996a. CHEVIE A system for computing and processing generic character tables for finite groups of Lie type. *Appl. Algebra Eng. Comm. Comput.*, **7**, 175–210.

Geck, M., Hiss, G., and Malle, G. 1996b. Towards a classification of the irreducible representations in non-describing characteristic of a finite group of Lie type. *Math. Z.*, **221**(3), 353–386.

Gelfand, I. M., and Graev, M. I. 1962. Construction of irreducible representations of simple algebraic groups over a finite field. *Doklady Akad. Nauk SSSR*, **147**, 529–532.

Gorenstein, D. 1980. *Finite Groups*. Chelsea Publishing, New York.

Green, J. 1955. The characters of the finite general linear groups. *Trans. Amer. Math. Soc.*, **80**, 402–447.

Grothendieck, A. 1967. Éléments de géométrie algébrique IV. Étude locale des schemas et des morphismes de schémas, quatrième partie. *Publ. Math. Inst. Hautes Études Sci.*, **32**, 5–361.

Grothendieck, A., et al. 1972–1973. *Théorie des topos et cohomologie étale des schéma*. Lecture Notes in Mathematics, Nos. 269, 270, 305. Springer Verlag, Berlin. Séminaire de Géométrie Algébrique du Bois-Marie 1963–1964 (SGA 4), Dirigé par M. Artin, A. Grothendieck et J. L. Verdier. Avec la collaboration de P. Deligne et B. Saint-Donat.

Grothendieck, A., et al. 1977. *SGA 5. Cohomologie l-adique et fonctions L*. Lecture Notes in Mathematics, No. 589. Springer Verlag, Berlin. Editor L. Illusie.

Harish-Chandra. 1970. Eisenstein series over finite fields. Pages 76–88 of: *Functional Analysis and Related Fields*. Springer Verlag, Berlin.

Hartshorne, R. 1977. *Algebraic Geometry.* Graduate Texts in Mathematics, No. 52. Springer Verlag, New York.

Hiller, H. 1982. *Geometry of Coxeter groups.* Pitman, Boston.

Howlett, R. B. 1974. On the degrees of Steinberg characters of Chevalley groups. *Math. Z.*, **135**, 125–135.

Howlett, R. B., and Lehrer, G. I. 1980. Induced cuspidal representations and generalized Hecke rings. *Invent. Math.*, **58**, 37–64.

Howlett, R. B., and Lehrer, G. I. 1994. On Harish-Chandra induction and restriction for modules of Levi subgroups. *J. Algebra*, **165**, 172–183.

Humphreys, J. E. 1975. *Linear Algebraic Groups.* Graduate Texts in Mathematics, No. 21. Springer Verlag, Berlin.

Kawanaka, N. 1982. Fourier transforms of nilpotently supported invariant functions on a simple Lie algebra over a finite field. *Invent. Math.*, **69**, 411–435.

Kazhdan, D. 1977. Proof of Springer's hypothesis. *Israel J. Math.*, **28**, 272–286.

Lang, S. 2002. *Algebra.* Revised 3rd ed. Graduate Texts in Mathematics, No. 211. Springer Verlag, New York.

Lehrer, G. I. 1978. On the characters of semisimple groups over finite fields. *Osaka J. Math.*, **15**, 77–99.

Liebeck, M., and Seitz, G. 2012. *Unipotent and Nilpotent Classes in Simple Algebraic Groups and Lie Algebras.* Mathematical Surveys and Monographs, Vol. 180. American Mathematical Society, Providence, RI.

Lou, B. 1968. The centralizer of a regular unipotent element in a semi-simple algebraic group. *Bull. Amer. Math. Soc.*, **74**, 1144–1146.

Lusztig, G. 1976a. Coxeter orbits and eigenspaces of Frobenius. *Invent. Math.*, **28**, 101–159.

Lusztig, G. 1976b. On the finiteness of the number of unipotent classes. *Invent. Math.*, **34**, 201–213.

Lusztig, G. 1977. Representations of finite classical groups. *Invent. Math.*, **43**, 125–175.

Lusztig, G. 1978. *Representations of Finite Chevalley Groups.* CBMS Regional Conference Series in Mathematics, Vol. 39. American Mathematical Society, Providence, RI. Expository lectures from the CBMS Regional Conference held at Madison, WI, August 8–12, 1977.

Lusztig, G. 1984a. *Characters of Reductive Groups Over a Finite Field.* Annals of Mathematics Studies, No. 107. Princeton University Press, Princeton, NJ.

Lusztig, G. 1984b. Intersection cohomology complexes on a reductive group. *Invent. Math.*, **75**, 205–272.

Lusztig, G. 1986a. Character sheaves IV. *Advances in Math.*, **59**, 1–63.

Lusztig, G. 1986b. Character sheaves V. *Advances in Math.*, **61**, 103–165.

Lusztig, G. 1986c. On the character values of finite Chevalley groups at unipotent elements. *J. Algebra*, **104**, 146–194.

Lusztig, G. 1988. On the representations of reductive groups with disconnected center. *Astérisque*, **168**, 157–166.

Lusztig, G. 1990. Green functions and character sheaves. *Ann. Math.*, **131**, 355–408.

Lusztig, G. 1992. A unipotent support for irreducible representations. *Adv. Math.*, **94**, 139–179.

Lusztig, G., and Spaltenstein, N. 1985. On the generalized Springer correspondence for classical groups. Pages 289–316 of: *Algebraic Groups and Related Topics (Kyoto/Nagoya, 1983)*. Adv. Stud. Pure Math., Vol. 6. North-Holland, Amsterdam.

Lusztig, G., and Srinivasan, B. 1977. The characters of the finite unitary groups. *J. Algebra*, **49**, 167–171.

Malle, G. 1990. Die unipotenten Charaktere von $^2F_4(q^2)$. *Comm. Algebra*, **18**, 2361–2381.

Malle, G. 1999. On the rationality and fake degrees of characters of cyclotomic algebras. *J. Math. Sci. Univ. Tokyo*, **6**, 647–677.

Malle, G., and Testerman, D. 2011. *Linear Algebraic Groups and Finite Groups of Lie Type*. Cambridge Studies in Advanced Mathematics, Vol. 133. Cambridge University Press, Cambridge.

Milne, J. S. 1980. *Étale cohomology*. Princeton Mathematical Series, No. 33. Princeton University Press, Princeton, NJ.

Milne, J. S. 2017. *Algebraic Groups*. Cambridge Studies in Advanced Mathematics, No. 170. Cambridge University Press, Cambridge.

Mizuno, K. 1980. The conjugate classes of unipotent elements of the Chevalley groups E_7 and E_8. *Tokyo J. Math.*, **3**, 391–459.

Serre, J.-P. 1994. *Cohomologie Galoisienne*, 5th ed. Lecture Notes in Mathematics, No. 5. Springer Verlag, Berlin.

Shephard, G. C., and Todd, J. A. 1954. Finite unitary reflection groups. *Canad. J. Math.*, **6**, 274–304.

Shoji, T. 1979. On the Springer representations of the Weyl groups of classical algebraic groups. *Comm. Algebra*, **7**, 1713–1745, 2027–2033.

Shoji, T. 1982. On the Green polynomials of a Chevalley group of type F_4. *Comm. Algebra*, **10**, 505–543.

Shoji, T. 1983. On the Green polynomials of classical groups. *Invent. Math.*, **74**, 239–267.

Shoji, T. 1987. Green functions of reductive groups over a finite field. Pages 289–301 of: *The Arcata Conference on Representations of Finite Groups (Arcata, Calif., 1986)*. Proc. Sympos. Pure Math., vol. 47. American Mathematical Society, Providence, RI.

Shoji, T. 1995. Character sheaves and almost characters of reductive groups I, II. *Adv. Math.*, **111**, 244–354.

Shoji, T. 2007. Generalized Green functions and unipotent classes for finite reductive groups, II. *Nagoya Math. J.*, **188**, 133–170.

Slodowy, P. 1980. *Simple Singularities and Simple Algebraic Groups*. Lecture Notes in Mathematics, No. 815. Springer Verlag, Berlin.

Spaltenstein, N. 1977. On the fixed point set of a unipotent element on the variety of Borel subgroups. *Topology*, **16**, 203–204.

Spaltenstein, N. 1982. *Classes unipotentes et sous-groupes de Borel*. Lecture Notes in Mathematics, No. 946. Springer Verlag, Berlin.

Spaltenstein, N. 1985. On the generalized Springer correspondence for exceptional groups. Pages 317–338 of: *Algebraic Groups and Related Topics (Kyoto/Nagoya, 1983)*. Adv. Stud. Pure Math., vol. 6. North-Holland, Amsterdam.

Springer, T. A. 1966a. A note on centralizers in semi-simple groups. *Indag. Math.*, **28**, 75–77.

Springer, T. A. 1966b. Some arithmetic results on semi-simple Lie algebras. *Publ. Math. Inst. Hautes Études Sci.*, **30**, 115–141.

Springer, T. A. 1974. Regular elements of finite reflection groups. *Invent. Math.*, **25**, 159–198.

Springer, T. A. 1976. Trigonometric sums, Green functions of finite groups and representations of Weyl groups. *Invent. Math.*, **36**, 173–207.

Springer, T. A. 1998. *Linear Algebraic Groups,* 2nd ed. Progress in Mathematics, No. 9. Birkhäuser, Boston.

Srinivasan, B. 1979. *Representations of Finite Chevalley Groups.* Lecture Notes in Mathematics, No. 764. Springer Verlag, Berlin.

Steinberg, R. 1956. Prime power representations of finite linear groups I. *Canad. J. Math.*, **8**, 580–591.

Steinberg, R. 1957. Prime power representations of finite linear groups II. *Canad. J. Math.*, **9**, 347–351.

Steinberg, R. 1968. Endomorphisms of linear algebraic groups. *Mem. Am. Math. Soc.*, **80**.

Steinberg, R. 1974. *Conjugacy Classes in Algebraic Groups.* Lecture Notes in Mathematics, No. 366. Springer Verlag, Berlin.

Steinberg, R. 2016. *Lectures on Chevalley Groups.* University Lecture Series, No. 66. American Mathematical Society, Providence, RI. Revised and corrected edition of the 1968 original.

Taylor, J. 2018. On the Mackey formula for connected centre groups. *J. Group Theory*, **21**, 439–448.

Tits, J. 1964. Algebraic simple groups and abstract groups. *Ann. Math.*, **80**, 313–329.

Tits, J. 1966. Normalisateurs de tores. I. Groupes de Coxeter étendus. *J. Algebra*, **4**, 96–116.

Zhelevinski, A. 1981. *Representations of Finite Classical Groups.* Lecture Notes in Mathematics, No. 869. Springer Verlag, Berlin.

Index